# Where Science and Magic Meet

by
Serena Roney-Dougal

GREEN
MAGIC

Where Science and Magic Meet © 2010 by Serena Roney-Dougal. This is the third edition revised and updated. All rights reserved No part of this book may be used or reproduced in any form without written permission of the author, except in the case of quotations in articles and reviews.

First published 1991 by Element Books
Second edition 1993 by Element Books
Republished by Vega 2002

www.greenmagicpublishing.com
info@greenmagicpublishing.com

Index by Guy Gotto

Typeset by Green Man Books, Dorchester
greenmangallery@lineone.net

ISBN 978-0-9561886-1-8

GREEN MAGIC

For all those who love and take care of this planet,
our Mother, the Earth

# CONTENTS

| | |
|---|---|
| Acknowledgements | vi |
| Foreword | viii |
| Introduction | 1 |
| 1. The Subliminal Mind | 8 |
| 2. Parapsychology: The Scientific Study of the Tools of Magic | 32 |
| 3. The Holographic Universe: A New/Old World-view | 63 |
| 4. The Pineal Gland: Third Eye and Psychic Chakra | 92 |
| 5. Earth Magic: Sensitivity to the Earth's Magnetic Field | 128 |
| 6. The Fairy Faith | 170 |
| 7. The Goddess Reawakening | 200 |
| 8. The Emerging Relationship between Science and Mysticism | 232 |
| Epilogue | 252 |
| Notes | 264 |
| References | 278 |
| Index | 294 |

# ACKNOWLEDGEMENTS

These are too numerous to mention everyone by name, so I would like to thank and acknowledge the inspiration of *all* those who have helped me and guided me during the past forty-three years since I first started waking up. I know who they are and love and thank them with all my heart.

In particular I would like to acknowledge my parents, Dolores Ashcroft-Nowicki, John Barrett, John Beloff, Sue Blackmore, Ben Brooksbank, John Cox, Paul Devereux, Norman Dixon, Jo Freedman, Leon Gold, Tom Graves, Marian Green, Julian Isaacs, Liz Jellinek, Kathy Jones, Stella Joy, Stanley Krippner, Terence Lee, Rollo Maughfling, John Michell, Bob Morris, Keith Mylchreest, Eleanor O'Keefe, John Palmer, Gerry Paster, Ora Paster-Bax, Dave Pate, Glen Rein, Bill Roll, Colva, Tara and Richard Roney-Dougal, Carl Sargent, Michael Shallis, Monica Sjoo, Anna Smith, Ellis Snitcher, John Steele, Rex Stanford, The Street Librarians, Charles Tart, Les Taylor, Val Thomas, Tom Troscianko, Gunther Vogl, Rhea White, Elizabeth Whitehouse, Paul Williams, Michael Winkelman and Shan Woodburn. For this third edition, I thank John Martineau for encouraging me to bring it up-to-date.

# Copyright Acknowledgements

I am most grateful to the following for permission to use excerpts from their work:

The Journal of the American Society for Psychical Research for excerpts from Batcheldor, K.J. 'Contributions to the theory of PK induction from sitter-group work,' *Journal of the American Society for Psychical Research,* 78 (2), 105-22.

Mrs. S. Bohm for permission to quote from David Bohm's book *Wholeness and the Implicate Order,* Routledge & Kegan Paul, 1982.

Excerpts from *Witchcraft: The Sixth Sense,* by Justine Glass, 1971, republished by C.W. Daniel. Used by permission of Random House Group Ltd.

Paul Devereux for extracts from *Earthlights,* Turnstone Press, 1982.

Michael Shallis for extracts from *The Electric Shock Book,* Souvenir Press, 1988.

Colin Smythe for permission to quote from *The Fairy Faith in Celtic Countries* by W.Y. Evans Wentz, 1977.

Parapsychology Foundation Inc. for excerpts from 'Anthropology of Magic and Psi Research' in *Parapsychology Review,* Vol. 14, No. l, 1983.

Danah Zohar for extracts from her book *Through the Time Barrier: A Study of Precognition and Modern Physics,* Heinemann, 1982.

Illustration from *Quantum Questions,* by Ken Wilber, ©1984, 2001. Reprinted by arrangement with Shambhala Publications Inc., Boston, MA. www.shambhala.com.

Illustration courtesy of Kevin Redpath, Glastonbury.

Illustrations courtesy of Lesley Delamont, Glastonbury.

Illustrations courtesy of Paul Devereux and Paul McCartney from *Earthlights,* Turnstone Press, 1982.

Illustration courtesy of Paul Devereux from *The Ley Hunter,* issue 75.

Illustration courtesy of Paul Devereux and Ian Thompson from *The Ley Hunter.*

Illustration courtesy of Mary Evan Picture Library.

Illustration courtesy of Danish Natural History Museum, Copenhagen.

Illustration courtesy of Brian Froud, Devon.

*Where Science and Magic Meet*

# FOREWORD

This book was originally written 20 years ago. Revising it for this edition I have been pleasantly surprised at how well it has aged. There really is very little to do, other than note that most of what I said then is still true and relevant today. There has obviously been a considerable body of research in the intervening decades, but none of it counters what I said then. Therefore, in revising the work for this edition I have added in notes giving more up-to-date references, in particular, books or review articles that give the more recent research in an easily accessible form. I have also referred to more recent articles that I have written rather than incorporating them wholesale into the book, which would have made the book far too long.

The area which has needed most revising is that concerning the pineal gland – again, in general, all of what I said previously has been corroborated, but in the intervening decades neuroscience research has further clarified and explicated the very speculative ideas I originally presented. I have not done a full review of the literally hundreds of thousands of experiments done in this area – just given a general summary of present knowledge.

The same goes for the chapter entitled "Earth Magic," because so much more research has been done on the effect of the earth's magnetic field on us and on the pineal gland. Like the neuroscience of the pineal gland this is an area which has made considerable strides in the past twenty years. And interestingly enough after a break of a few years when I have been involved in other areas of research, I am once again fully involved in doing research in this area, and am writing this whilst living at Samye Ling, a Tibetan Buddhist monastery in Scotland, engaged in looking at the effect of the earth's magnetic field on meditation and our conscious awareness of clairvoyant impressions.

The Faery Faith chapter has in the intervening decades become expanded into a book in its own right. If you are interested in the details of the psychic connections with our mythic lore, may I suggest that you take at look at it.[1]

The last two chapters have been re-written a bit, but less than I expected, because they are primarily my ideas and opinions and whilst these have changed since the original book was written, the basic ideas still stand. For my latest thoughts on the links between science and spirituality, I have written a new Epilogue for this edition.

I hope you thoroughly enjoy reading this work of mine. May this book be for the benefit of all!

*Where Science and Magic Meet*

# INTRODUCTION

Let me introduce myself. This is a speculative book, written on a foundation of scientific knowledge with leaps of intuition, opinion and feeling. These leaps are ones that are current in the popular imagination. I feel that ideas that resonate with the popular imagination are worth exploring; that they lead us on in our social, psychic and spiritual evolution. I expect that many academics, scholars and scientists will shudder at some of the ideas put forward in this book. I myself feel that these speculations are still at a very early stage in development; there is a lot I still don't fully understand or see clearly. This is all stuff that I thoroughly enjoy, that excites my imagination. It is in this spirit that I wish you to enjoy reading this book.

The meeting between science and magic is part of the evolution of our times, part of the evolution of consciousness that is happening whether we like it or not, whether we are consciously working for it or not.

I have always had a logical, analytical mind, so I studied science at school and started to train in anatomy and biochemistry in the hope of becoming a psychiatrist, when I was arrested for smoking hashish. So it was time to leave and I went to London and studied psychology, the mind being my first love and psychiatry now being barred to me.

All the time I studied at university I was leading two lives. I went to my lectures, I did my psychology projects; and I was part of the drop-out counter-culture of the 1970s, living in a squat in

*Introduction*

Chalk Farm where the first Film Co-operative started. There was a burgeoning of street theatres such as Action Space; the first community wholefood supplies shop and wholesaler; drum-makers, inflatable makers, musicians, poets, anarchic printing presses, and all the other manifestations of that incredibly creative time. Music, books, drugs, people — they all contributed to my education. I read the Don Juan books[1] as they were released, *The White Goddess*[2] from cover to cover in 1971. I didn't understand a word but it all went in and over the years has proved to have been fully assimilated. And so psychology at university was dead as a dodo — who wants to maim, torture and murder rats when one's soul is in the process of transformation? It was in my second year that a postgraduate gave two classes on parapsychology. I breathed a sigh of relief. At last, here was someone talking about the higher reaches of a human's oh-so-wonderful mind. At last, here was something that I found fascinating. Fascinate — this is a word to conjure with, connected with mesmerise, enchant and other magical words.

I was very fortunate; there was an excellent teacher at the university who was a pioneer of research into the subliminal mind — Professor Norman Dixon — and he was willing to let me study parapsychology under his guidance, comparing it with subliminal perception. This is the perception of everything around us of which we are not consciously aware — for example, the ticking of the clock goes on all the time but we notice it only when it stops. The way in which we are affected by subliminal perceptions seems to be the same way we are affected by psychic impressions. So I was lucky to have started my research with the expert on subliminal perception.

I keep on being lucky. Good luck and bad luck play a very strong part in my life — and that's fate or destiny. I was possibly the only person ever to study parapsychology as an undergraduate at London University at that time, which is a very strange sort of luck that makes me shiver a little inside, because parapsychology was totally unacceptable to the Establishment then, and there was I right in the middle of the stronghold of academia doing that which was anathema to them — studying the mind in all its glory and power. I escaped the rats.

Three years later I was a postgraduate studying parapsychology in the Psychology Department at City University, London. Someone suggested that I submit a paper for a conference. This annual conference is for those people who have contributed significant research to the field of parapsychology. As a first-year student I hadn't contributed anything but I had my ideas that I was working on. These were accepted. My air fare was paid and I was flown to America where I met all the leading researchers in the field, was accepted into the Association, and came back flying high. I couldn't believe my luck. I was so welcomed and accepted and I hadn't even done my very first experiment. I didn't really know what I was doing; I had no one to teach me or show me, but I was learning.

The second year saw me doing my major project — the one I had proposed at the conference. Friends of mine from the street of squats I lived in, in Kennington Park, helped me out. Some of them worked with me for the full three years and we went very deeply into it together. One of them said to me after one session that what I was doing was essentially a ritual. I have never forgotten that. I was astounded; I knew nothing of magic and had never thought of science in that way. And yet the more I pondered her words the more true they were. Except that in a ritual the 'leader' is normally more knowledgeable than the rest, and here was I, the coordinator, and so in a sense the leader of the experiments — the ritual — and yet I was a total ignoramus.

At about the same time, someone whom I had met who was also interested in parapsychology, Julian Isaacs, had been asked to give a talk at an occult conference and he decided that he didn't want to do it and asked me if I would take his place. I was delighted to do so, and so found myself standing up on a stage giving a talk about parapsychology to a room full of occultists. And that was the beginning of my fusion between science and magic. That talk led to another and then another, and then I wrote some pieces for the journals put out by the groups and I did a few questionnaires at the different conferences I went to and started to explore the world of magic. Not seriously studying it, but reading books, and talking with the people who were involved in that world, and slowly assimilating the essence of magic.

During the third year at university a shock occurred. I had been getting clear evidence that some of the people working with me were picking up on the information in the experiments psychically, and the process by which they were doing so seemed to be the same process by which they were picking up on subliminal information. I found this fascinating, but my supervisor and the head of department did not like it, because to them there was no such thing as psychic phenomena and so the whole of their belief structures were threatened. My supervisor said that she did not want to have me as her student, and the head of department told me to

*Introduction*

find someone else. But no one else in the department would take me on, not with the head of department wanting me out — so I had to look elsewhere. I found a supervisor at another university, which meant that according to their regulations I would have to start all over again — which actually gave me an extra three years to do my research. At the same time I left London and moved to Glastonbury, because I did not want my daughter going to a city school, and settled down to analyse the results from the experiments I had done in London. My new supervisor was very understanding and just left me to get on with it. In Glastonbury my magical education was carried into two completely different areas — Earth mysteries and psychosis.

I had first been introduced to ley lines and our magical environment by John Michell's book *A View over Atlantis*.[3] I had met some of the people involved in the Dragon Project which was investigating various aspects of Earth energies at the Rollright Stones in Oxfordshire, and now here I was living in Glastonbury, a centre of these mystical, magical energies. So I became aware of what it meant to live in a power spot. What I have noticed about Glastonbury is that, if you have a flaw in your being, it becomes accentuated living here. The traumas which I suffered during my first three years here have taught me more about the dark side of the soul and black magic and possession than any number of psychology or magical books ever could.

I also discovered what is happening magically in our culture — the whole growth of the pagan movement. I use the word pagan (I know of none that better describes what is happening) to describe people who are becoming very aware of the Earth, of the old solar and lunar festivals, and who are celebrating these festivals with song and dance and a certain amount of ritual, as well as changing their lifestyles so that they are more careful not to harm the planet on which we live. People are no longer following the old orders, the ritual methods and arcane science of the magical orders as they used to be popularly conceived in terms of Aleister Crowley, the Golden Dawn, the Rosicrucians and Dion Fortune. Instead, we are becoming aware of the movement of the moon, the changing of the seasons, herbs, sacred times in the year as represented by the old Celtic Calendar and we are creating our own rituals, our own way of being on this earth, our own spiritual growth outside of set religions or occult systems. And it is these I shall be talking about when I talk about magic. For me this is the magical way.

## Where Science and Magic Meet

The Earth is evolving and we are living this evolution in our new spiritual awareness which has more in common with the Perennial Philosophy of the mystics of all ages and all peoples throughout the history of humanity, than it has with any one particular system. The emerging women's movement is rewriting history into sistory (or at least that was a rather lovely slip of the typing fingers). We are now aware of the old matriarchal cultures, of the old pagan religions of these lands, and of the neolithic peoples who built the stone circles, the barrows and tumuli, and of their way of being. The higher the branches of our new culture grow, the deeper grow our roots. The Celtic revival has served its function and now we are looking even further back, before the Iron Age to those magical peoples who have dotted our landscape with stone megaliths and massive earthworks. The emerging magical way is linked with the magic of the Earth and its sacred sites. It has to be because we are becoming very aware of the need to be in tune with the Mother in order to prevent her destruction. The new magical way is the spiritual side of the Green movement.

I am aware that in this introduction of myself I have used all sorts of words and phrases without defining exactly what I mean by them, words such as psychic, subliminal, magic, pagan. I have used them loosely because I was introducing myself first and foremost. If you have not understood what I meant, don't worry because I shall explain.

In this book I shall first of all discuss certain aspects of parapsychology which uses the scientific method to investigate the tools of magic. In parapsychology we have the clearest bonding between science and magic. I shall show how the essential metaphysics behind modern quantum physics is akin to that of the pagan world-view that seems to be emerging at the moment. I shall present some research of mine into the pineal gland which fuses the ancient teachings of the East with regard to the chakras and the third eye or second sight of the West, with modern neurochemistry, and I shall link this with modern earth mysteries, the fairy faith and the popular modern view of paganism and witchcraft. I love doing jigsaw puzzles and I have been busy with this jigsaw for many years now. At times it feels like I am weaving, bringing together many threads or strands into one piece. The picture that is emerging is fun, fascinating and very exciting. I hope you enjoy reading about my travels as much as I have enjoyed travelling this path.

# 1. THE SUBLIMINAL MIND

Parapsychology is the study of psychic phenomena such as:

1. telepathy — mind to mind communication;

2. clairvoyance — obtaining information about, for example, lost objects, which no one else knows about;

3. precognition — knowing about something that hasn't yet happened; and

4. psychokinesis — affecting things outside of oneself just by thought alone.

These four phenomena are the basic core of a host of strange events that happen spontaneously to lots of people far more frequently than is commonly expected. Recent surveys[1] reveal that over 50 per cent of the population have experienced what seem to be a psychic, or at least unexplained, event at least once in their life, and often several times - though not many people talk about their experiences. Examples of such spontaneous events are: seeing a ghost; having a dream that actually happens later in the day or week; going to phone a friend, or talking about someone not seen in ages, and having them phone you or arrive just at that moment; what are often called hunches, or intuitions. All of these events are called psychic phenomena, and if I refer to 'psychic' it is to phenomena such as these. Sometimes I use the word psi to refer to psychic.

This is used a lot by parapsychologists these days. I like it because it has many layers of meaning and usage. First, psi is the 23rd letter of the Greek alphabet and the first letter in words such as psyche (mind, soul), psychic, and psychology (study of the mind). Second, it is used in modern quantum physics to denote what they call 'the hidden variables involved in the collapse of the wave function'. I'm not going to explain what that means just here, except to say that no one knows what these hidden variables are, though some people suspect that they might be will and consciousness, and they use the Greek symbol psi to denote them. Third, the symbol that denotes the Greek psi is found in Norse rune alphabets where it symbolises Psychic Protection. How appropriate! Fourth, Neptune's and Boadiccea's tridents are the same shape as the Greek letter psi.

## A Short History of Parapsychology

Psychic phenomena have played a part in human history for as far back as records go — and probably before. The Babylonian, Chinese, Egyptian, Roman, Greek and Celtic civilisations all accepted the psychic and spiritual as part and parcel of everyday life, and there were prophets, oracles, seers, diviners and astrologers who not only advised and helped in the marketplace, but were employed by the kings, emperors and governors of the land to advise in the rule of the realm.[2] In fact, the recent scepticism about psychic phenomena displayed by certain academics and city people (very few country people show dogmatic scepticism of psychic phenomena) is a curious anomaly in the history of humanity. It is only in the past two hundred years in the European/North American cultures that the reality of the psychic aspect of humanity and the world about us has even been questioned, let alone denied, and I explore some possible reasons for this strange aberration of our culture in some detail later. This questioning goes hand in hand with the increasing scientific and technologically based education and culture that we grow up in, and also with the increasing atheism of so many people. In denying the spiritual aspect of life so the psychic, which has always been considered a manifestation of the spirit, has been denied too.

*Victorian times*

The increasing materialism of the eighteenth and nineteenth centuries led some Cambridge academics to found the Society for Psychical Research, as parapsychology was called then. These founders conducted a massive survey of thousands of spontaneous cases of apparitions, hauntings and poltergeists, and also started a scientific study of the many physical and mental mediums who were the craze of a Christian sect called the Spiritualists at the time and who were to be found in many a Victorian parlour. They hoped, by proving the survival of personality after death through studying mediums, to counteract the increasing atheistic and materialistic spirit of the time.

This research fell into disrepute because of the number of charlatans and frauds who were encountered, but this is not to deny the vast quantity of evidence that was gathered and the deep insights concerning the working of the mind that were gleaned during this time. One such insight concerns the existence of the subconscious, or 'subliminal mind' as it is called. Most people have heard of Freud and Jung and their research into, and understanding of, the various levels of the mind gained mostly by psychoanalysis, but not many realise that their pioneering work was part of a greater whole, and that both Freud and Jung were aware of, inspired by, and even did some research into, psychical phenomena. The insights into the workings of the subconscious gained by this method have borne fruit in many self-development techniques used by so many people today.

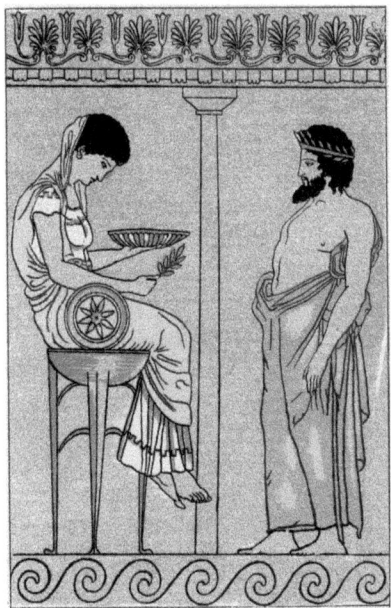

*The Delphic Oracle (Illustration supplied by the Mary Evans Picture Library)*

## 1930-50

By the 1930s the study of mediumship had declined and more and more researchers were actually setting up experiments to study telepathy, clairvoyance and precognition in their own right. Most of these were of the type where one person did a simple line drawing and another person tried to copy it while in a separate room. However, the major change that occurred at this time was the setting up of a laboratory in America within the psychology department of a university by J.B. Rhine. It was he who changed psychical research into parapsychology, and who coined the term extrasensory perception (ESP); he also brought statistical techniques to the study of psychic phenomena.[3]

With the help of a fellow psychologist he designed a pack of cards called Zener cards which have five symbols recurring five times in a pack of 25. With these he ran thousands and thousands of 'guessing' trials where different people had to guess which card was being looked at, or which card was top of the pile, or they had to guess the order of the cards in an untouched pile, or the order of the cards in the pile that would be shuffled the next day or the next week.

These carefully planned and controlled experiments were then analysed statistically and showed that people had definitely been choosing the target more often than would be expected by chance. Some of these experiments were so outstandingly successful that the only charge sceptics could bring against Rhine was that of fraud. And, as his work has been repeated by many others, this charge has to be altered to a grand conspiracy of fraud. As no one can ever prove such a charge this is not scientific and is definitely 'not cricket'. To attack a person's ethics and morality just because you do not like the findings of their research is in itself immoral and unethical.

## 1950-today

Since that time there has been an ever-increasing number of researchers and variety of research. It is now possible to study parapsychology at postgraduate level at university, and national surveys reveal that most people accept the evidence that psychic phenomena occur. Researchers since the 1950s have been concerned not so much to 'prove' the existence of psychic phenomena (which I shall call 'psi' for convenience), but rather to *understand* the whys and wherefores of the process by which we become aware of psychic information and events. It is the findings from this vast body of research that I shall be discussing. For simplicity, I shall separate it into two main sections:

1. receptive psi: telepathy, clairvoyance and precognition;
2. active psi: mind over matter, e.g. healing, psychokinesis.

## Receptive Psi

Most people when thinking about perception think of their eyes, their ears, their nose, their sense of taste and touch. All of these senses have their psychic counterpart. What is unusual about psi is that it is outside space and time as we know it and so the knowledge comes as a vision or hallucination, as a feeling or intuition, as a dream or even merely a hunch. This knowledge is not physically present and can even be about something in the future. However, the psychological processes by which one becomes aware of the psi information seem to be the same as those processes by which you become aware of normal physical sensations that are subliminal. Now, most people have heard of subliminal advertising and so know that we can perceive and be affected by stimuli and information of which we are not *consciously* aware, but what most people don't realise is that we are all, at all times, perceiving sights, sounds, smells, etc., of which we are not conscious. Our conscious minds can only cope with a minute fraction of the information that the senses are perceiving all the time. If we were aware of everything for only a fraction of a second we would be completely overwhelmed, so the mind has many different filters and barriers to ensure that only the most important information receives conscious attention.[4] We can, however, learn to become more aware of subliminal information, the most common way being through our dreams. Thus, the way in which we perceive psychic phenomena is a perfectly normal mental process; the only odd thing is how the information gets into the brain circuits in the first place.

### Personality and psi awareness

One of the first discoveries concerning the process by which one becomes aware of psi phenomena was that the more open one's *attitude*, the more likely one is to have a psi experience. Those people who are prepared to believe in the possibility of, for example, telepathy, are more likely to pick up information telepathically, while those who strongly deny the reality of psi will be far less likely to have a psychic experience. This denial also happens at the subliminal level and has been called 'perceptual defence'. It has been found that some people are so defensive that they will consistently 'miss' the target in an experiment.[5]

Let me explain this further. Imagine you are tossing a coin and guessing whether it will be heads or tails. By chance you will correctly guess 50 per cent of the time. If you guess correctly 75 per cent of the time then you are 'significantly hitting' the target because you are guessing correctly 25 per cent more often than expected by chance. If, however, you only call it correctly 25 per cent of the time then you are significantly missing the target because you should, just by chance, call it correctly 50 per cent of the time. Thus, at some level, you are stopping yourself from calling the coin correctly. Perceptual defence is a perfectly normal reaction which allows us to avoid being affected by unpleasant situations when we don't wish to deal with them. And the same process seems to happen with psychic perception.

Connected with attitude are various personality traits: artistic, musical, open-minded, relaxed, 'childlike' people who are susceptible to hypnosis are far more likely to have psychic experiences than are those who have analytical, logical minds, or are anxious or tense.

## States of consciousness

The second major discovery concerning the process by which one becomes aware of psi experiences is that they tend to occur when one is asleep, or in a very relaxed, daydreaming state of consciousness.[6] Extreme examples of this are mediums who go into trance, or Indian shamans who use certain plants to induce a trance state of consciousness for psychic purposes. It has been found that besides sleep, hypnosis facilitates awareness of psi information, as does a technique called the 'Ganzfeld', which I describe in detail later. In this state one goes into a sort of sleep state while staying alert, and the imagery and thoughts that emerge are dreamlike and are often concerned with psychic information. In general, though, just to be relaxed and quiet and meditative is sufficient to allow one to become aware of the subliminal imagery that is normally shut out of consciousness, and which is the carrier for psi, as has been found in the remote-viewing experiments by Targ and Puthoff at SRI in California.[7]

## Active Psi: Psychokinesis

Active' psi is normally called psychokinesis or PK: mind over matter. Instead of affecting things through a physical intermediary, things are affected by the mind alone. This aspect of psi has always been more controversial, an example being the metal bending by Uri Geller. Before him there were the so-called 'physical mediums' in whose presence tables floated in the air; strange knocking noises were heard; hands or even whole bodies would materialise; musical instruments would play, and so on. There are voluminous records of carefully controlled sittings with such mediums, and in certain cases one can only disbelieve the reported phenomena by doubting the veracity of the observers.

Things seen in the seance room occur spontaneously during poltergeist episodes. Researchers have noticed that poltergeists tend to occur around a focus person, who is quite frequently an adolescent, and have surmised that emotional disturbance can manifest in these strange noises, movements and the other phenomena associated with poltergeists. An example of this is in the life of Matthew Manning. He was the focus of a poltergeist outbreak at his boarding school; in time he managed to channel the energy into automatic drawing and writing, and, after Uri Geller appeared, found that he too could bend metal by stroking it gently. Then came a period when he was tested by various different researchers in America, Britain and Europe, and he found that he could affect all sorts of different materials, including enzyme reactions and tissue culture cancer cells. He has become a healer and teacher, so it means he is using his talents in the most positive way possible.

Laboratory research into PK started with J. B. Rhine's work with dice, in which people attempted to influence which dice face would land uppermost, or where the dice would fall on a grid. After thousands of trials he amassed data of such extreme statistical significance that he considered he had 'proved' the existence of something affecting the movement of the dice. Since that time researchers have become much more sophisticated and use complex machines called random-event generators some of which use natural radioactive decay as the random source, this being connected to, for instance, a circle of lights which will circle at random either clockwise or anti-clockwise. The task is to influence the lights to go in a prespecified direction.

Interesting research with sensitive strain gauges that detect the slightest movement of metals is also being developed in the wake of the 'metal benders'.[8] These strain gauges utilise what is known as the piezo-electric effect to which crystals are particularly sensitive. In this effect mechanical pressure on the material, say crystal, is converted into electricity. This electrical current is connected to a chart recorder and so you can see if there is pressure on the material, such as you would get if a spoon is being bent without being touched. Thus this is an incredibly sensitive method of assessing PK effects.

As with 'receptive psi' researchers are now beginning to understand some of the psychological processes at work, of which the following are the most important:

1. The point of concentration must be on the end result, not on the process by which that result is achieved. Thus, it has been found that the actual complexity of the process is entirely irrelevant; one merely has to be aware of the effect one wishes to achieve.

2. One must not be too serious, too ego-involved in the task. A light-hearted fun approach works wonders.

3. PK seems to happen after one has finished trying! This is known as the 'release-of-effort' effect and has a certain mischievous quality to it.

4. It helps to work in a group, so that one does not feel threatened by what is happening and the thought that it is oneself who is lifting the table, or whatever the task is. This does not seem to be important with very small-scale effects such as influencing the growth of seeds, since these effects tend to occur over a period of time after the, for example, healing session.

You will probably have noticed by now that, for all the advances being made in understanding the psychology of psi, the central question of *how* psi works is still unanswered. How does the information, that, for example, a friend I haven't seen in years is in town, come to me in my dream that morning? We have seen that the dreaming state of consciousness is admirably suited for enabling us to become aware of such intuitions, but we haven't seen how that information came to be within my subconscious mind in the first place.

In order to gain a glimpse of a possible answer some researchers are looking to the implications of the so-called 'New Physics' that is being so widely discussed at the moment by so many different people. This Alice in the Looking-Glass World posits a realm of matter in which our ordinary everyday reality is turned upside down. In this world information can pass from one particle to another instantaneously as if there were no space between the particles, so that it seems as if one of the pair 'knows' what is happening to the other even though there is no known medium of communication. This is known as the EPR paradox, and out of findings like these a new/old world-view is being forged which is called The Holographic Paradigm, and that's another chapter entirely.

## Primary Process Thinking — the Fascinating World of the Subliminal Mind

As I have just mentioned, by subconscious and subliminal I mean all those mental and bodily processes which are outside conscious awareness and control. Subliminal thought processes are best expressed as dreamlike since dreaming is our natural everyday path through to this aspect of our mind, and is the most accessible region of the subliminal. This dream type thought is called primary process by psychologists, because it is the first type of thought we experience as babies. The subliminal mind appears to be almost a second mind within. *All* subliminal stimuli, all psi stimuli, all bodily processes, everything is incorporated into this mind so as to create an awareness and understanding of our environment that is of central importance in affecting the way in which we behave. The classical Freudian unconscious is the repressed desires, childhood problems, unfulfilled wishes, and other taboo and socialisation problems that one is not allowed to realise consciously.[9]

The subliminal mind can best be developed by resting and day-dreaming. It is the source of all our creativity, art, poetry, etc. Self-transcendence is considered to be the mental attribute that distinguishes humans from animals, not that we are superior to animals, but that aspect of nature which we embody, just as tiger embodies tiger aspect of nature, crow crow aspect, so humans embody the aspect of nature that is logical, mental, self-transcendent, introspective, creative, artistic. It is an elusive, capricious human ability - the moment of insight, the flash from the unconscious that becomes conscious and gives us our mystical peak experiences, that moment in which we realise something new or understand the purpose behind life, the universe and everything, and the universal 'Why?' is answered once and for all.

I think it is important to stress that there is evidence that the subliminal mind retains every experience in its entirety. There are billions of nerve connections in the brain and so physically we can process billions of items of information at any one time — digestion, blood flow, body temperature, all sounds physically present in the environment, all sights, all taste, all touch, everything that is physically present is registered at a subliminal level. And all of this is retained. At a subliminal level we see, hear, feel and remember everything, every little detail. Then we have to add to this all of the psychic awareness which we are potentially capable of possessing. If you accept such abilities as telepathy, clairvoyance and precognition, then you must realise that we are literally potentially omniscient. This is an attribute normally given to divinity and it is true for our subliminal mind. Thus at a subliminal level we are potentially divine, potentially omniscient, potentially able to know everything that ever has been, is and will be in the whole Universe. There are at present no known limits to psychic awareness.

Let me repeat: everything, all subliminal stimuli, all psi stimuli, all body processes, everything is incorporated into this mind. Potentially at the subconscious level we are aware of everything that ever was, is, or will be — we are omniscient — we are divine.

### Why opening up to the subliminal mind is so important

We are at a time of great change. There is an explosion of interest in spiritual, magical, psychical, psychotropic and ancient knowledge that is occurring all over the Western world. Psi research is but a small part of this cultural change. Everywhere barriers between the different disciplines are being broken down as more and more people realise that they were saying the same thing - only the language is different. Thus we get books appearing with titles like *The Tao of Physics*[10] which discusses points of similarity between Eastern religious philosophies and modern quantum physics. Works by writers such as Arthur Koestler, Lawrence Le Shan, Lyall Watson and Colin Wilson[11] bridge biology, physics, psi and religion. There was a conference called 'Bridges and Boundaries' organised by an occult group, discussing the points of contact between them and humanistic psychology. The central theme to all this cultural change in which we are embroiled is the growth of consciousness, and the first step is to open the door to the subconscious *so that we are no longer at the mercy of our hang-ups, our subliminal attitudes bred into us during childhood by our parents and our society.*

There are some writers of magical books who feel the same way about this; for example, Richard Gardner in a book entitled *Evolution through the Tarot* says:

> The evolutionary process will be seen to be the constant integration of the conscious mind with that of the subconscious. A really significant fusion of our dual consciousness will undoubtedly carry with it magical powers and a tremendous increase in understanding.[12]

And Lyall Watson, who has written several very popular books about the wonderful in our world, says:

> We have a filter which structures experience and keeps reality within bearable bonds ... We may have reached a stage of development where the barrier can become semi-permeable ... (conscious) tele-pathy is an omen of a new and greater awareness. We need both conscious and unconscious mechanisms, but we perhaps have an even greater need right now of an interconnection between the two ... The process involves a new dimension of voluntariness and a deliberate adoption of trance logic; both as it happens characteristic of meditation and many of the popular new techniques of transcendence. I believe they have become growth industries in direct response to an evolutionary pressure in just this direction.[13]

We are all evolving, all the time, as individuals, as groups, as societies, as cultures, as a species. We *must* grow, change, evolve in order to be fully alive. That which is static is dying.

## Perceptual defence, psi-missing and defence mechanisms

It has been shown that our conscious mind can only cope with approximately seven things at once — and from my experience I know this to be true — and so the subliminal mind has a complex series of barriers. In psychology this is called filter theory, and the filters that protect our conscious minds from being overwhelmed are called defence mechanisms. These are essentially protective in nature, but if too rigidly enforced they can become blocks. At present we consider our defence mechanisms of repression, denial, projection, etc. to be negative blocks, and so they are, because our society has caused us to grow up stunted using only one aspect of our mind - the logical, analytical left-hemisphere part. However, in their essence these defence mechanisms are filters that protect our conscious minds and as we grow personally and spiritually so we remove the negative blocks contained within our defence mechanisms so that they can function healthily.

A really good example of how we prevent unwanted information from entering conscious experience is provided by Norman Dixon in his spot of light experiment.[14] In this experiment a person sat looking into a machine so that their right eye looked into one side and their left eye into another, a bit like a pair of binoculars. They were able to see a spot of light against a neutral background and they were asked to turn a knob until the spot of light faded into the background, so that they could no longer distinguish the spot of light from the background. The amount of brightness of the spot was measured and then, unknown to them, a word was presented subliminally so that it was physically visible, but not consciously so. Once again they were asked to turn the knob so that the spot faded into the background. Once again the brightness was noted. This was done lots of times with different words being used and it was found that people's physical thresholds at which they could perceive the spot of light varied according to the emotional meaning of the subliminal word. People actually raised their perceptual thresholds if the words, such as whore or cancer, were likely to upset them. Thus, even though they had no knowledge that words were being presented, their subliminal mind raised the perceptual threshold against taboo and unpleasant words. This is called perceptual defence, and this experiment and others like it show the enormous power of the subliminal mind to determine what we do — and do not — see, hear, think. Similarly some people lowered their perceptual threshold to the spot of light when the emotional words were presented – this is called perceptual vigilance and people who react in this way tend to also be people who are open and sensitive to psi.

There is a whole series of research in parapsychology which uses what is called the Defence Mechanism Test (DMT) in order to predict who will be most aware psychically.[15] The DMT uses a threatening picture presented subliminally, but not so subliminally that you don't know it is there, just low enough so that you know something is there but cannot quite make out what it is. You are then asked to draw what you think you can see. Then the same picture is presented again so that you can see it a little bit better and again you draw what you can see. This carries on until you can describe the picture accurately. From your descriptions of the picture while it was subliminal and from your final 'recognition threshold' it is possible to assess how defensive a person you are and the sort of defence mechanisms you use to prevent threatening information from reaching consciousness. This test is so accurate at assessing this that it was used in Scandinavia for trainee pilots. Those who were defensive and raised their thresholds when frightened or threatened and so likely to cause accidents, i.e. accident-prone people, were discovered by this method and not accepted for training.

Also, the mind strives for constancy, to perceive what it knows. If you wear coloured or inverting spectacles, after a few days you will see the world normally again; the brain 'ignores' the crazy frequency or perspective sent by the retina. In other words, subliminal perception alters our conscious perception. And the choice of what we perceive is there in the subliminal mind. That is why sometimes when you talk to someone they quite literally cannot hear what you are saying even though they aren't physically deaf!

This sort of defence goes on at the psychic level too and is called 'psi-missing', already mentioned briefly. In psi-missing someone will guess the target incorrectly more often than you expect by chance. In the research I have done, which included people who psi-missed, I found several reasons for this. The most important was attitude. One person I worked with when doing my Phd did not believe in psychic abilities and so he consistently missed the target to show there was nothing in it. Amusingly, one day he met someone who told him how significantly he was scoring, and as soon as he understood what she was saying, he then started to score precisely at chance. He was *very* psychic in order to be able to do that so well. His reason for so behaving was to reduce what is called in psychology 'cognitive dissonance'.

In cognitive dissonance if something happens that is contrary to your belief system you have to do something to reduce the discomfort felt at having your beliefs shattered. People who are open and flexible will alter their beliefs, but others are so firmly stuck in their beliefs that they have to somehow deny their perceptions. Calling a psychic a fraud or charlatan is the most common method used. There is a group of people who call themselves The Committee for the Scientific Investigation for Claims of the Paranormal (CSICOP — or Psi cops!). Reading their magazine *The Sceptical Inquirer* is a fascinating study of such attitudes.

Remember, we are normally at the mercy of our unconscious and its fears, anxieties, etc. This is the filter level, the various barriers that are so necessary for our mental survival in its bodily form. Those who do not have good filters become psychotic. Psychics are people with weak filters who can cope and who can learn to turn off the information, put up the barriers when needed in order to live their lives.

Doing psi tests teaches us a lot about our personal filters. When we psi-miss we show quite clearly the method by which we deny conscious knowledge of the information that we know at a subliminal level — some people use denial, some repression, some projection, some turn it into its opposite, etc. It is absolutely fascinating to see how we defend ourselves from unwanted information.

There are two points about the subliminal mind that are really important to note:

1. The subliminal mind utilises a totally different way of thinking from our 'normal' everyday waking thoughts, called primary process that is essentially dream thought — vivid, concrete

imagery, paralogical, symbolic.

2. Most of our everyday experience, the way we behave, our moods, our thoughts even, is totally governed by the subliminal mind — we only see what the subconscious wants us to see, we only think what the subconscious wants us to think, we only hear what the subconscious wants us to hear — the rest is filtered out of conscious awareness by our various defence mechanisms and other filters and barriers that prevent our conscious mind from being overwhelmed by the totality of the universe that is within our subliminal minds.

A practical example of this is given by an experience we have all had. Most people at some time in their life have had a route which they have travelled frequently; going to school or to work. I used to bicycle twenty miles through London every day for two years. Every day, once I had established the best back street route, I followed the same route which involved crossing main roads, negotiating one way streets, crossing parks, and so on. Busy, tricky, complicated business bicycling in London, particularly with rush-hour traffic. One day I had to go somewhere that was just off my normal route, about half to two-thirds of the way along my journey. It was only when I reached the college that I realised I had travelled right past the place where I needed to turn off. I had done the journey in a daydream in which I was not even consciously aware of where I was, let alone all the traffic and so I had completely missed the turn-off from my route. From then on I noticed how I travelled and found that I was actually safer when I was on 'automatic pilot,' because I responded to everything from a subliminal level which had faster reaction times and was more aware than my conscious mind. If you have never noticed your automatic pilot before, watch out for it now.

*Symbolism and Archetypes*

As stated earlier, the subliminal mind is the source of all our creativity, art, poetry. Self-transcendence, best expressed through the arts, is the joining of the subliminal with the conscious mind — so that our divine wisdom is expressed, made manifest, in this external see-touch reality we live in.

If, as suggested by physicists such as Eddington, Schrodinger and Wigner,[16] the universe is ultimately mental, the unconscious may represent the contact of our individual minds with the mind stuff composing the universe, and therefore, our minds are coextensive with the universe.

The language of symbolism is the language that is common to all peoples from all times and in all cultures. Psychologists call it primary process because it is a more basic way of thinking than the verbal way that we nowadays automatically think of as being thought. It is probably the first type of conscious thought process we have as a baby. In fact the verbal is only one of many ways of thinking, and its prominence in our minds is an aspect of our present society and culture. Our minds are far more intricate, and the language of symbolism, dream-type scenarios and archetypes are in fact far more true of our primary aspect of mind. As I have mentioned, this primary-process thought appears to be a common method of thought across all cultures. The Tower of Babel that separates us is language, the verbal thought-form. The symbolic, imaginal, pictorial dream thought-form joins us all into one race. Neuroscientists are now beginning to suspect that in fact the primary process mind is our default mind, and it is only when we are awake and our senses are engaged that we use secondary logical, analytical thought processes. Given half a chance and we will slip into primary process thinking, such as daydreaming![17]

The subliminal mind thinks in archetypal symbols — to me this is the world mind, the language of Mother Earth. Every culture through all time that we know of has had the same basic symbols, the same archetypes — this is the underlying common language of all people - this is the language, the thought processes of our planet, common to all thinking beings. The ancient gods, goddesses and myths reside in the unconscious — that is why modern people who venture into the subliminal mind encounter the same gods and goddesses, monsters and magical beings more or less as the ancients knew them. When we speak of spirits, daemons, we do it with the understanding that we are speaking of psychological phenomena. They are as powerful as the subliminal mind — i.e. they are potentially omnipotent.

To some extent this could be explained by our seeing what we expect to see, but in other cases there seems to be more than that. The archetypes have a definite resonance, and there are some accounts of Western people who have partaken of a South American divination potion who have experienced the same archetypes as the traditional Indian tribes people experience. The more one becomes aware of dreams, and other states of consciousness in which the subliminal mind is paramount, the more one begins to understand symbolic thought, the more strongly it resonates, the more the old myths and legends make sense at a personal practical level.

Jung, with his ideas on the collective unconscious,[18] describes this level very well indeed. The subliminal mind thinks in symbols. One symbol, like a poem, may contain an ocean of meaning.

Let me illustrate this with the word 'cup'. If I say cup to you in an ordinary sentence such as 'May I have a cup of tea please', you know exactly what I mean. The cup may be a delicate china one — or it may be a hulking great mug, but it is essentially a cup. However, if you use the word cup in poetry or in dream then it takes its symbolic form as vessel, as grail, as cauldron; my cup runneth over — love, emotion, female. A whole deep archetypal aspect of our lives is symbolised by *cup* as in the suit of cups in a pack of Tarot cards.

The subliminal mind speaks in verse — plainsong. The right hemisphere can sing poetry but not speak words. This has to do with en-chant-ment. Magic as enchantment uses this aspect of our subliminal minds. Through chant — the use of specific words in their archetypal, symbolic sense — you can induce trance: particularly at the speed and rhythm of 2 beats per second. Listen to the rhythm and music rather than the words of a 'good' preacher and you will hear what I mean. The conscious mind and its convictions have no power to withstand language aimed at the subconscious. This was utilised, for example, by Adolf Hitler and is practised by fundamentalist preachers.

Poetry uses language symbolically. This language of symbolism is not merely the language of dreams, it is also the language through which psychic awareness emerges most readily and it is the language of magic. So we must be most careful about the symbolism we use. For example: although I lived for three years with an astrologer I do not easily identify with modern astrology. I finally, after a decade of struggling with 'astro-babble', found out the reason why. We use Greek and Roman symbols and they just do not resonate so strongly with me at the deeper levels of my psyche as do the Celtic archetypes, though I am learning slowly the deeper mythological aspects of them!

Our collective archetypes live on in our myths and legends. The ancient gods were primarily personified forces or aspects of nature and personality and as such are archetypes. Through these we touch on the deeper layers of our psyche. It seems to me really important that British people rediscover their own archetypes for as children we have all been given Danish (Hans Christian Anderson) and German (Brothers Grimm) fairy tales, or myths, but how many of us know the ancient British legends other than those of Robin Hood and King Arthur. How many have read 'The Mabinogian', the Welsh myths, or about the Tuatha de Danaan, the Irish myths, or the Scottish myths such as are found in Andrew Lang's or in Campbell's collections.[19]

One aspect of our British mythology that I have been exploring is that of the archetypes of fairies, the most ancient myths of our race. I have found that there are several different strands in this lore; the psychic, spiritual; the elemental, devic which is worldwide; the ancestor aspect which is basic to all animistic religions; the divinity aspect; and trace memories of the pre-Celtic peoples of these lands. It seems as if British witchcraft is a relic of their knowledge. I shall discuss this in a later chapter.

## Brain Connections with Mind

Basically the brain is composed of three main parts, the old brain which is common to all animals, the midbrain which developed with mammals, and the new brain (cerebral cortex), which is most developed in primates, dolphins and humans and governs our everyday thought processes. All those involved in research into the subliminal mind, and most parapsychologists, emphasise the limbic system as being that area of the brain which seems to be the seat of the subconscious — the mid-brain emotional centres. It is our emotional, instinctive, gut-feeling level of mind.[20]

The subliminal mind is connected with the mid-brain (subcortical) and with an aspect of our nervous system which is called autonomic. The autonomic nervous system (ANS) is not connected with feeling heat, or pain or sending signals to my fingers to type these words or any other sensory or motor aspect of my body — that is the work of the central nervous system the main channel of which runs in the spine. The autonomic nervous system is concerned with keeping the body running smoothly at levels outside of conscious control; our hormonal system, which is linked with levels of stress, tension, relaxation; muscles such as those used in birthing of children or the beating of our heart, or digestion.

There are two aspects to the autonomic nervous system; the sympathetic and the parasympathetic. Basically the sympathetic gets us moving, and the parasympathetic slows us down. Working in harmony together we are as alert as we need to be and as relaxed as possible — the ideal state of relaxed alertness. This system can be measured using what is popularly known as a lie detector or psychogalvanometer (EDA, which stands for electro-dermal activity) which measures sweating on the palms of the hands. This sweating is linked to the autonomic nervous system and to emotional states of which we are often unconscious. In subliminal perception research and in parapsychology it has been found that people will react to a subliminal or psychic input at this level while consciously being totally unaware that they are responding to any information input at all.[21] A lot of research into psychic healing has used this technique. Basically a person was connected to the EDA machine and seated in a comfortable chair. In another room another person tried at random intervals to make the person more relaxed, or more alert, or

CONSCIOUSNESS

| THE CONSCIOUS MIND | all that we are aware of at any particular moment |
|---|---|
| THE SUBLIMINAL MIND | |
| PRECONSCIOUS | Stimuli which are above the recognition threshold and of which one can easily become aware. |
| SUBCONSCIOUS | Stimuli which are above the awareness threshold and below the recognition threshold; retrieved through dream, word association, hypnosis, etc. |
| MOMENTARY UNCONSCIOUS | All info both subliminal and psychic which is below the awareness threshold as it is present moment by moment; can be retrieved through hypnosis, dreams, etc. |
| PERSONAL UNCONSCIOUS | All info about a person's life from the moment of conception onwards, stored in memory, which creates the total of their individual personality. |
| FAMILY UNCONSCIOUS | all info specific to one's family, one's ancestral inheritance. |
| SOCIAL UNCONSIOUS | all info pertaining to the society into which one is born and grows up, e.g. inner city ghetto, country landowner. |
| TRIBAL UNCONSCIOUS | everything pertaining to one's tribe, e.g. people who speak the same dialect, Scottish clans. |
| NATIONAL UNCONSCIOUS | The nation that one perceives one belongs to, however small or large, e.g. Welsh, British. |
| CULTURAL UNCONSCIOUS | e.g. English speaking Western world which includes Australia, America, etc. |
| CROSS-CULTURAL UNCONSCIOUS | e.g. all industrialised nations, or all hunter-gatherer societies. |
| COLLECTIVE UNCONSCIOUS | common to all humanity at all times, revealed through myths and archeypes, thought by some to be the chthonic mind. |

**Consciousness Chart: The Subliminal Mind (illustration courtesy of Serena Roney Dougal)**

did nothing at all. These efforts were successful according to changes on the EDA.[22]

More recent research has shown that if the person was just looked at via CCTV their EDA responded – scientific verification of the old saying that your ears burn when someone is talking about you![23] For my PhD thesis I compared subliminal perception with telepathy and one of the things that interested me was the close similarity in the psychological processes by which the subliminal or psychic information came through into consciousness. I used an EDA machine and sometimes people responded at this physiological level to a subliminal stimulus and sometimes to a psychic one also. It all depended on how much the target meant to them at an emotional level. I feel that it is very relevant to today to understand the psychic and the magical in these physical terms and so I have followed up these ideas and have found that there is a location in the brain which is connected with our dreams, visions, fantasies, and psychic awareness. My research has led me to the pineal gland, otherwise known as the third eye, which is closely connected with the hypothalamus, part of the limbic system. I don't yet understand fully how the psychic links in with the physical but I feel that the dawning of realisation is very close and later in the book I shall explain exactly what I have found so far.

These ideas are confirmed by Andreas Mavromatis[24] who has specifically studied what is known as the hypnagogic state. This is the state many people experience just as they are falling asleep. In this state you have extremely vivid imagery, larger and brighter than life images, which are sort of dreamlike and yet you are sort of awake. He believes that the hypnagogic state implicates the subcortical structures that I have been talking about because hemispheric lateralisation (that is psychological processes that are typical to either one or the other side of the new brain) is not relevant. What he means is that the left hemisphere of our neocortex is known to be involved in language and in logical, linear, analytical type thought, and the right hemisphere is known to be involved in music, art and holistic, non-verbal type thought. The hypnagogic state with its dream-type thought (primary process) is closer to right hemisphere type cognition but could underly both of these and also be linked with the mid-brain structures[25]; the subliminal brain uses a non-verbal symbolic cognition (imagination).

## Techniques for Becoming Aware of the Subliminal Mind

1. *Biofeedback* Conscious control over subconscious bodily processes — heart, muscles, blood flow, brain waves, etc. This is one aspect of the subliminal mind that can easily be brought under conscious control, and this is done by most yogis in early stages of training. The benefits are enormous as Elmer and Alice Green[26] say:

> I began to suspect that unusual powers of control over normally involuntary physiological processes such as self-healing were accompanied by an awareness of normally unconscious psychological processes — awareness not only of oneself but of others too.

Biofeedback is essentially equivalent to yoga. In learning to control the body we are starting a whole process of self-development and self-awareness, which generalises from the specific to the normally unconscious activities such as psi. Remember our minds can control any body process.

A good example of what biofeedback can achieve is given by one of the very first experiments performed with biofeedback by Kubie. In 1943 he used biofeedback to help people get into a hypnagogic state, in order to get information from the subconscious without all the distortions that occur in dreams, by feeding back amplified breathing sounds. This is a technique that anyone with a tape recorder can do for themselves any time they wish to. It's great fun as well as being an excellent method for learning to create greater contact with the subliminal mind.

2. *Ganzfeld* This is similar to another yogic technique called Yoga Nidra, or psychic sleep. The Ganzfeld is a very dramatic and powerful method for taking one into the hypnagogic state. In the Ganzfeld technique, one listens to what is called 'white noise' through headphones. I have found that a tape of a waterfall is just as effective and much more pleasant to listen to. At the same time you cover your eyes with translucent goggles (such as halved ping-pong balls) and switch on a red light so that you can only see a homogenous red glow. Then sit back, relax, and enjoy yourself, talking out your thoughts and recording them on a tape recorder.

The quality and quantity of subconscious mentation that emerges is dependent on the personality. In the Ganzfeld first of all senses are stripped away; then the wittering mind; so that fantasies, imagery, memories, daydreams and other expressions of the subliminal mind can come into play. You will find that after a while and with practice you can distinguish psi impressions from imagination, memories, subliminal impressions and so on.

3. *Yoga Nidra* In this technique the guide leads you round your senses, and body, breath, touch, so that the concentration becomes totally absorbed in internal sensations and a space is cleared for subliminal mentation.

The hypnagogic state which all these techniques induce is called the fourth state of consciousness by the Tantrics. Mavromatis, who has made a special study of the hypnagogic state has this to say about it:

> In order to acquire continuity of consciousness unaffected by lapses into unconscious states, you must hold yourself at the junction of all the states, which constitutes the links between sleeping, dreaming and waking: the half sleep or Fourth State. (From a tenth-century Tantric text – this state is called 'turiya').[27]

It is this holding of the mind that is the tricky part. It is a bit like walking a tightrope as you drift too far into sleep and then come to with a jolt and are too much awake, and then get the exact state for a while and then drift off again, and so on. Only lots of practice works.

4. *Autogenic technique* Together with self-hypnosis this stresses the use of visualisation and the power of the imagination in developing voluntary control of unconscious behaviours. This power of imagination in linking the conscious with the subconscious is strongly emphasised in the occult traditions. In the autogenic technique too, there is a developmental process. First, one learns control over bodily functions, then over mental functions through meditation and visualisation.

5. *Meditation* This is the process par excellence by which the conscious and the subconscious join, and is the state Rhea White describes[28] which most psychics and sensitives utilise when attempting to perceive psychically. As Gardner puts it: 'The conscious mind feeding internally upon the subconscious equals meditation.'[29]

6. *Hypnosis* The same process of psychic sleep, although in this case not through self-development and discipline, but through surrender of one's will to another, and perhaps this is the reason why self-hypnosis is considered to be of greater value in the long term.

Mesmerism is the original term for hypnotism. Under hypnosis one can heal; be anaesthetised; recall memories; eliminate hunger;

speed/slow heart and other autonomic body functions, e.g. blood-sugar levels; menstruation control; conception/ abortion; lactation; etc. and even fool the senses into seeing what's not there and vice versa.

Finley Hurley, in his book, *Sorcery*,[30] notes that in hypnosis magic has returned via the back door - crept into science in disguise. The rationalists banned magic but they didn't recognise it when it returned. The subconscious 'knows' far more than our conscious mind - it is biochemist, doctor, cell biologist, etc.

7. *Trance* This is a technique used throughout the ages to harness the subliminal mind for magical purposes. There are two sorts of trance: possession trance as we know from shamanism, or voodoo for example, in which the person is possessed by a deity and has no memory afterwards of what happened; and the lighter trance used today by mediums or 'channellers,' in which the person is fully aware during the trance of what is going on.

During shamanistic trance a song often emerges — which is of particular symbolic and healing relevance to the person. Whether the trance is a going out or a going in is immaterial. Dancing, drumming, and chanting are three common techniques for entering trance.

8. *Psychotropic plants* In most tribal cultures on this planet, to the best of my knowledge, there are techniques for religious and or healing purposes whereby the tribe as a whole or a selected leader will enter an altered state of consciousness by ingesting a psycho-tropic plant, and through the altered state which this induces will heal or divine the future, or whatever other matter needs to be attended to.

Siberian shamans utilise alcohol and Amanita muscaria; European witches used belladonna, thorn apple, henbane, mandrake; North American Indians use tobacco, peyote and psilocybin, South Americans use a mixture composed of DMT and harmala alkaloids (yage, ayahuasca, daime, etc.), and so on. In the Eleusinian mysteries the bread that was passed around to the participants was probably contaminated with ergot, which is the natural precursor of LSD.

In fact in every animistic mystery religion the people experienced their heaven or hell directly, and used plants to enable themselves to become directly transformed within a religious setting. As Lyall Watson says:

> Every mystical tradition and most of the new ideas about the growth of mind refer to higher states of consciousness and the techniques for getting

there all seem to revolve around two things. The first is a method of knocking some kind of hole in the barrier between the conscious and the unconscious levels of mind and opening this door on demand and at will. The second and more difficult step is to reconcile the conflicting interests.[31]

## Magic and the Subliminal Mind: Connections between Science, Occultism and Mysticism

Magic works through the subliminal mind - join conscious and subconscious and you get superconscious. The subliminal mind is the arena in which magic acts. Opening up to this subliminal mind — changing our state of awareness to allow more sensations, perceptions, thoughts, ideas, to become conscious — enables us to control our perceptions to a greater degree and to become more aware of our environment. Most esoteric traditions and recent experimentation in parapsychology share this concept: that greater awareness of the subconscious, and an ability to control its functioning more, will eventually enable us to live at a level of mind in which we can utilise our psi abilities with some form of conscious control, e.g. siddhis of the East, magic of the West.

Let me quote one very well-known parapsychologist who used to work at the parapsychology centre (Rhine Research Centre) that was started in the 1930s by J.B. Rhine, and which has always been considered to be the most scientifically establishment-minded in its methods of all the parapsychology research centres:

> The recognition that psi usually operates at the level of the unconscious raises the possibility that an understanding of the dynamics of the unconscious may give us insights into the way psi manifests in our consciousness.[32]

But it is not only in terms of understanding how psi works, as Lyall Watson says:

> Like the unconscious dynamic factors that go into the determination of our behaviour, psi factors also may influence our behaviour whether or not we are able to discern such an influence.[33]

There are several theories that state that we are at all times being influenced by factors that are psychically perceived at all times. Two of these theories are by Rex Stanford. The most appropriate is what he calls the Psi-Mediated Instrumental Response or PMIR theory. In this he proposes that our behaviour (instrumental response) is being influenced by psychic information (psi-mediated) all the time, just as Lyall Watson states. And Stanford[34] has done experiments that show that this is so. Our behaviour is influenced all the time by subconscious psi influences, however much we might defend against this realisation by saying: 'I'm not psychic' The truth is that we are so ignorant of the subliminal mind that we must begin to open ourselves up to this conception that we are psychic all of the time — we are just not consciously aware of it — in order to help the process of becoming aware of this level of our being so that it becomes a mutual feedback system. Recent research into luck has corroborated this earlier work.[35]

This subliminal influence which is ever-present in our everyday behaviour is the basis for magic. It has been researched by psychologists and other scientists in disciplines that are apparently totally unconnected with the psychic or magic. As Finley Hurley says:

> Sorcery is alive today and thrives as well in the seemingly uncongenial ambience of scientific laboratories as it does amongst Maqui brujos or those who gather on the moors when the moon is full. And why not? Sorcery is not a rogue in the Western scheme of reality; and had no conception of it ever existed, it could now be derived from the scientific literature.[36]

> Despite an education that seduces the conscious fragment of our minds, the unconscious remembers the old ways and its laws continue to govern our lives.[37]

All the magical lores — of contagion (once touched always linked), of similarity (things that look alike share essential characteristics) and personification (everything contains a spirit aspect) - are lores of the subliminal mind: that is the way our minds work at the subliminal level. Magic works through the subliminal mind, and the process of opening up to this subliminal mind — changing our state of awareness to allow more sensations, perceptions, thoughts, ideas to become conscious — enables us to control our perceptions to a greater degree and to become more aware of our environment. Most esoteric traditions and recent experimentation in parapsychology share this concept that greater awareness of the subconscious, and an ability to control its functioning more, will eventually enable us to live at a level of mind in which we have the second sight and other psychic abilities whenever we need it — we can know whatever we need to know. But we *must not misuse it*.

Parapsychology is translating the language of the yogis and that of the magicians into a scientific language that is acceptable to modern-day Western people. This opens a door that has for too long been closed. We all need a path that we can relate to, and most people today find it easier to alter deeply ingrained patterns of belief and attitude that are held so deeply that they do not even realise that they are holding them, i.e. attitudes and belief systems held within the subliminal mind. If an experiment proves to them that something is fact - is true - is the way the world is, then they can incorporate that information more easily into their everyday lives and actually change the way in which they live in this world. Science is a method for earthing, grounding esoteric knowledge and so is a path for materialistic Westerners who are so heavily grounded into matter that they need a path using that language to help them out of the material and into the spiritual. And by materialistic Westerners I mean *us*. Parapsychology is a bridge on that path as is the recent scientific technique of biofeedback.

By earthing or grounding esoteric and mystical knowledge, I mean taking that knowledge and making sense of it in such a way that it becomes of practical use in our self-development in a very basic way. I give clear examples of this in the research I am doing, and which I present in a later chapter on the pineal gland and the chakra system. Basically the pineal gland has loads of mystical, mythical, esoteric and folk lore surrounding it - that it is the seat of the soul, the psychic centre, the ajna chakra and so on. By finding out as much as I could about the scientific knowledge concerning the pineal gland I began to understand the esoteric lore in a very practical way that makes so much sense. In fact, I had a typical Eureka experience, when suddenly everything clicked — ah! that's what they mean!

There are various aspects in magic that can be clearly linked to this understanding that the essence of magic, working with psychic abilities, is learning to work with one's subliminal mind.

> Spells often appear to work according to a least-action principle in which their goals are attained through 'coincidences' or 'accidents' by the shortest and easiest route. Every spell will nevertheless remain essentially an experiment. And as in any experiment with people, its results will not be uniform. A spell is carried by a normally weak signal, telepathy, mediated by the labyrinthine chthonic minds of both the sorcerer and his subject, and it remains vulnerable to a plethora of unknown influences. Against these the sorcerer will strive to increase the power of the signal by emotion, concentration and visualisation, and to increase the receptivity of the unconscious by using its laws and symbols.[38]

There is a question mark as to whether spells, like hypnosis, are more than suggestion. What this question mark is saying is that the power of suggestion is so strong, since it affects the subliminal mind, that spells and hypnosis are probably at least utilising that subliminal mind principle if nothing else. This is not to say that there *is* nothing else, by the way, merely that at present it is difficult to say.

The power of suggestion is shown only too well by the power of glamour and illusion — delusion and hallucinations. Some people are more suggestible than others, but all can be affected. Some times of day are more suitable than others and personal relationship makes the suggestion (spell) stronger.

Finley Hurley suggests that many spells appear to work through coincidence. I discuss coincidences in greater detail elsewhere, my basic conclusions being that every coincidence is best understood by the Jungian idea of synchronicity, and that these ideas underpin a magical philosophy that is in essence the same as the philosophy that is emerging from modern physics. Spells are basically wishes, or prayers, under another name, in another guise. The principles behind the efficacy of wishing, praying or casting spells is basically that of PK, which I discuss in greater detail next chapter. The essential point to note here is that this aspect of PK is most effective when allowed to proceed through the subliminal mind. In any ritual, spell, prayer, or wish, all generative power resides in thought fuelled by deep emotion.

Some people are best as healers, some as gardeners, some as counsellors, etc. In magic as in life we have our individual differences.

In conclusion: part of the mysterious core of magic and psychic phenomena is the controller in the unconscious. The soft, sensitive, dark, deep, underground, collective unconscious, way of thinking and being which describes the psychic way of functioning is considered by today's Western culture to be an inherently female way of thinking. Perhaps this is why male shamanic cults, magicians, yogis, have to use methods and techniques for attaining this mode of being which many women seem able to slip into easily and naturally.

As we become more open to the subliminal and in greater control of our minds, emotions, feelings and intuitions, as we become stronger and more focused in our own natures we must become very careful about what we think, feel, do and say. A word or thought or emotion can kill or cure coming from a fully developed mind. No wonder our world is in such a state when you think of all the billions of people and all their negative thoughts and emotions.

# 2. PARAPSYCHOLOGY: THE SCIENTIFIC STUDY OF THE TOOLS OF MAGIC

## Meeting Points Between Parapsychologists, Anthropologists and Occultists

> We are at a time of great change. Just look around you at the explosion of interest in spiritual, magical, psychical, psychotropic, and ancient knowledge that is occurring all over the Western world. This is our evolution as a sentient race. Psi research is but a small part of this cultural change. Everywhere barriers between the different disciplines are being broken down as more and more people realise that they are saying the same thing — only the language is different.

This is what I was saying in 1979 at a lecture to the Society for Psychical Research.[1]

The next month I was addressing a room full of occultists (the Society of Light,[2] a group descended from William Butler who had been with Dion Fortune and left her Society of Inner Light to form the SOL) and saying the following:

> I very much see the path of humanity at the moment as needing to harmoniously unite the global approach with the analytic, the right hemisphere with the left, the subconscious with the conscious, the magical with the scientific, or whatever other analogy or concept appeals to you. Psi research is part of this cultural shift away from materialism into the realms of the minds, especially of the various other

forms of consciousness into which our being can occasionally go, either spontaneously or as the result of a technique.

So, right from the beginning I was aware of the need to break down barriers, to tell academic researchers about what was happening in the world, and to tell the magical people about what was happening in science. This is what I told them:

> Research into psi processes is very much a phenomenon of our time. Since the 1960s there has been an incredible cultural swing towards 'consciousness' development. There is hypnotherapy, radionics, spiritual healing, occult techniques, yoga, astrology, leylines, EST, Silva Mind Control, Don Juan type shamanism, dowsing, primal therapy, to mention but a few of the now popular techniques for self-healing and understanding and growth.[3] (And in the past 30 years this list has multiplied enormously.)
>
> Among the occult groups there are an unprecedented number of people who believe in and have experienced psychic phenomena. This means that the reality consensus of the magical groups is totally in line with the parapsychologists who are attempting through the use of the scientific method to bring acceptance of psi phenomena to the establishment sector of Western culture. If parapsychologists are to succeed in their task, not only will psi be accepted in our culture, but the numbers who are thereby opening up to magic will be multiplied a thousandfold. This is a very serious responsibility which we must all look at *now*. Psi research is a technique by which the mechanistic Western people can allow themselves to reach into and develop themselves. In itself it is not dangerous. But a belief system that accepts psi is much closer to accepting magic and that requires high ethical and moral values. If we are to maintain the progression of spiritual growth that is occurring now, we must join hands in understanding what is happening to the culture in which we live.[4]

In the past few years some parapsychologists have done a considerable amount of work constructing the necessary bridge. A few of these have been directly concerned with magic as it is practised in the West, but mostly the work has created a new branch of parapsychology with the wonderful mouthful of a title, 'anthropological parapsychology' — or, if you prefer, 'parapsychological anthropology'. Despite the nomenclature, these people are opening up whole new spheres of knowledge, attitude and approach to things magical.

*Parapsychology: The Scientific Study of the Tools of the Mind*

## *The Gulf*

The main reason, from the parapsychological perspective, why there has been such a gulf between parapsychology and the magical disciplines has been the perceived need for parapsychology to be academically respectable - and this has been quite a struggle which is by no means over yet, although a broader acceptance of the existence of psi phenomena both among academics and the general population is now manifest. This is especially so in Britain where there are now 80 people who have PhDs in parapsychology, of whom a third are now working as lecturers in 16 Universities, mainly in Psychology departments. Parapsychologists in general have been more concerned with establishing for themselves a domain secure from the emotionally charged attacks of their critics, than in openly venturing into areas which, although they may lead to valuable data and methods, have become confused with the 'dreaded' popular superstitions. However, Rex Stanford,[5] one of the leading researchers in parapsychology in the USA, considers that investigators should endeavour personally to experience as many psi phenomena as they can, and to study the practices and beliefs of magic, religion and mysticism of diverse cultures and times as they relate to possible psi phenomena, while not 'naively' adopting the beliefs in these areas, for direct translation into hypotheses. He urges that such a study might indirectly or directly yield important clues about psi processes, at the psychological level and perhaps also at more basic levels.

The main reason from the magical perspective why the gulf exists is the prevailing fear of the establishment left over from the days of persecution, and the feeling that it is a one-way benefit for the scientists and that the magicians will learn and gain nothing from the contact — will be the losers in fact. This is well exemplified by the following: When I gave a talk at a *Quest* conference in London in 1979,[6] *Quest* being a journal published quarterly over the past forty years specifically for people interested in the occult in general, the first question I was asked after I had finished my talk on parapsychology and the links I see between science and magic was: 'The occult has a lot to give parapsychology — but what has parapsychology to give to the occult?'

At the time I was bothered by this question and totally incapable of answering it because to me it seemed so totally irrelevant. Psi research is merely another path, another way toward understanding and growth, which utilises the scientific methodology, technology and language. This scientific way is one that is readily related to by many in our twentieth-century Western world, and so it is a path to the Mysteries that is appropriate for our time — a synthesis of the old and the new — of the left and the right. Now I understand the relevance of the question and my answer is that without the earthing of science, magic becomes meaningless. We are here on this planet in material bodies. We must manifest our spirituality in practical ways in our daily lives. Aleister Crowley was an incredible magician, and a complete bastard. Science is the grounding of our knowledge. We can see through its method exactly *what is*. We can be clear what is material fact, what is imagination, what is psyche, what is soma. And so we become clearer at which level we are working. And for those who do not readily understand a magical or mystical language, for whom such a language does not make much sense, then science can be a true path to greater spirituality — paradoxical as that might sound.

## *The bridge: parapsychological anthropology*

Jeffrey Mishlove, a Californian parapsychologist, in *Psi Development Systems*[7] considers that, while a desire for distance is understandable from a political and social perspective, this has had a stifling impact on the study of occultism from a historical, sociological or other academic perspective, and the research into occult techniques for developing psi to see if they really do work and if we can learn from them more about how psi itself works.

In *Psi Development Systems,* Mishlove compares approximately 22 systems purporting to develop and train one's psychic faculties. The list includes Spiritualism, Theosophy, Anthroposophy, Rosicrucians, Franz Bardon's system of magic, Scientology, TM, Philippine healing, Silva mind control, ceremonial magic, and so on. He compares all these methods from a wide variety of aspects, such as the value placed on psi development; the use of secrecy; the emphasis or otherwise on quick results; the emphasis on ethics; the religious framework; the use of altered states of consciousness (ASCs); the logical framework within which an understanding of psi is incorporated; physical exercises; the emphasis on personality change and development; the use of tools, props or instruments; the utilisation of group dynamics; the emphasis on verifiable psi results; selection procedures for initiates or candidates; personal instruction, and other parameters.

He noticed that invariably every group which does some form of psi training has also taught a way of understanding spiritual, metaphysical and mythological figures, forces and energies. This teaching quite simply gives people a way of seeing archetypal aspects of our being, and a language with which to describe them. Thus Egyptian ritual magic uses the archetypal figures of Isis, Horus, Osiris, Ra, Anubis, Thoth and so on; the Celtic uses the Triple Goddess and the Horned God, both of whom have many names and aspects and are connected with the whole fairy faith and elemental devic worlds. These archetypal aspects seem to be totally bound up with our awareness at the psychic level of being.

Mishlove speculates that the teaching of archetypal symbolism as a way of perceiving a magical reality may be related to the development of psychic ability. He says: 'Even if such experiences are merely the product of imagination, they may serve to break down mental barriers that inhibit the use of psi.'[8] An anthropologist who has compared shamanism with magic, Nevill Drury (1982), also notes this relationship between the Jungian conception of the collective unconscious and its archetypes, and magical systems.[9]

Mishlove also notes that all the so-called pre-scientific training systems that he examines, namely shamanism, divination, yoga, Sufism, Buddhism, Judaism, and ceremonial magic, emphasise

> the value of personal instruction from one who is already experienced. They have all used some combination of concentration exercises, breathing exercises, diet regulation, behaviour regulation, solitude, secrecy, music, movement, altered states of consciousness, special clothing or jewellery, and particular mantras, prayers or spells. The combinations are different in each case, and vary with the progress of the student, but always the effect seems to be to focus many mental functions on the task of psi production, in a culturally acceptable manner. The intensity and length of training varies with each culture and undoubtedly also with individual circumstance.[10]

He also notes that some of the traditions that give psi training include an ethical code not permitting public demonstrations or scientific experiments of psi — except under rigorous conditions that would not violate the spirit of that tradition. We do have quite a strong taboo among certain members of the magical community in this country not to use psychic ability for money or other form of personal gain, or for public demonstration. I have a feeling that this taboo is breaking down, particularly when I see how many clairvoyants and fortune-tellers are selling their abilities in the marketplace, and I know of some parapsychologists and psychics who have specifically used precognition to make money from horse racing, bullion markets, stocks and shares and other legalised gambling ventures that govern our economy. Those people who initiated the highly successful remote viewing programme did this for some years before it eventually failed.[11] I am not yet sure whether this is wise or not. One part of me says that the time when that taboo had relevance is no longer. Another part of me feels that the 'selling' of psi is profanity, and that someone who is selling themself as a psychic is to be treated with caution. For me psychic ability is a sacred gift. Psi is a clear manifestation of our divinity. Through psi awareness we can move outside space and time to that place where it all just is. We are at our most spiritual when within the psi world. And to profane that space by using it to gain money, or power, or for public demonstration does not seem quite right. At another level, psi is just another talent, like a musical talent, and so why not use it to earn your daily bread? I think the essential ethic here is to use it and not abuse it; to use it so you have just enough for your needs.

In parapsychology this broader acceptance of the existence of psi in all its forms is permitting greater acceptance of disciplines that have previously been considered occult and, therefore, unacceptable to the scientific parapsychologist. For example, John Beloff, a psychologist at Edinburgh University who ran a Parapsychology Laboratory for a number of years — this laboratory now being the Koestler Parapsychhology Unit, since Arthur Koestler left his money to set up a chair in parapsychology — though one of the more conservative elements in parapsychology, is beginning to accept astrology as being worthy of study since the Gauquelin work in 1977. John Palmer, another parapsychologist from the USA, in his Presidential Address in 1979 to the Parapsychological Association,[12] encouraged careful exploration of formerly occult areas, and in the same year at the same conference David Read Barker reported on his exploration of the use of psi among Tibetans. Stanley Krippner has researched various cultural healing practices, some of them being shamanistic and magical.[13]

*Parapsychological anthropology*

Michael Winkelman has done a cross-cultural anthropological psi study in order to: 'determine the universals, and the cross-cultural differences, in magic and to investigate the social conditions associated with particular magical beliefs'. He considers that:

> Understanding the relationship between psi and culture requires examination of the nearly universal magical practices and beliefs. It appears that many magical practices use techniques or involve beliefs about magical functioning which show strong correspondence with psi-functioning principles well established in parapsychological research ... since many magical practices have evolved over long periods of time, in response to cultural needs, and with the intent of utilising psi, this suggests they may have evolved highly efficient psi-elicitation techniques, and could reveal important information about the functioning of psi.[14]

So this therefore is the attitude and perspective behind his work which is the most comprehensive survey of magic in relationship to parapsychology that I have come across. In 1983 he published an article in the *Parapsychology Review* from the perspective that the magical world is inherently the world of psychic happenings — whether at the level of synchronicity, wishes that come true, or the stuff of folklore and fantasy. Since then he has published a book on his work. [15]

In his work he has noted that different social conditions have fostered different adaptations of magical practices to the inherent psi potentials within the particular society. I have often noticed that while parapsychologists are happy to investigate the relationship between psi and magic in other cultures, they are very wary of doing so in our own and he has noticed the same thing in his discussion of previous correspondence between the two areas. He discusses the impact of an anthropologist called Tylor who worked in the last century.

> Tylor set the stage for subsequent anthropologists, suggesting that the disturbing parallels between the evidence of psychical research and primitive magical beliefs were best avoided by studying magical phenomena among culturally more distant peoples.[16]

Tylor argued that religion and magic were based in animism, the belief in spiritual entities or beings, but other early anthropologists suggest that a concept of supernatural power was more basic to magical actions than was animism, and they have taken the Melanesian word 'mana' to denote this power.

> Both psi and 'mana' conceptualise a power which transcends the physical laws or the ordinary course of nature and operates outside of normal

space—time constraints. They are characterised as both an action and a milieu as well as fundamental processes in nature and perhaps the basis of all phenomena of nature ... However, the concepts of psi and mana are not identical since conceptualisations of psi generally lack a personalised spirit conception, instead emphasising it as an impersonal force, reflecting the different assumptions made by many parapsychologists.[17]

Another point that Winkelman makes concerning complex and simple societies is that as a society becomes more complex so the magical methods change, and this change is only too apparent at the present time with the proliferation of pagan, neo-pagan, wiccan and other grass-root groups forming spontaneously all over the country. It seems as if magic is part of the inherent nature of certain people and the so-called 'Hippies', 'alternatives', 'New Agers', find in magical ceremony, ritual and beliefs an esoteric system that feels good and feels right. There are no formal rites, no formal laws, but there is mythology and lore, and people are actively searching back to neolithic and even paleolithic times, not to mention the Celtic Revival, for items of inspiration, for intuitive straws that will help to give some form to their need to become spiritual beings in this material world.

*Shamanic anthropology*

Nevill Drury is an anthropologist who relates ancient shamanic practices with modern magical methods. In his book *The Shaman and the Magician*[18] he has brought together the two themes 'to show that even in a modern urban context the mythic approaches to human consciousness continue to play a vital role alongside scientific technology'. In this book he discusses various forms of shamanism, e.g. the Shuar of the Amazon, together with their mythological aspects. Then he discusses various magical systems, e.g. The Golden Dawn, the Inner Guide meditation.

He concludes that 'both shamanism and magic offer techniques of approaching the visionary sources of our culture. Both systems of thought structure the universe in ways that are deeply and symbolically meaningful and which fully accommodate enlarged horizons of human consciousness.'[19]

Thus, we are beginning to get the complete circle where parapsychologists are synthesising their findings with those of anthropology, and vice versa; where anthropologists are synthesising their findings with those of modern magical disciplines; and where modern occultists are synthesising their experience and knowledge with parapsychologists. There are many, many more people who are bridging the gulf between science and magic at this time. I have mentioned these few as examples of the sort of work that is going on; more are mentioned in the rest of the book. This in no way denies the excellent work of those not mentioned either from my ignorance of their work or because of lack of space.

Now let us look at some of the findings within parapsychology which are relevant to the practice of magic.

## The Psychology of Occultism

All of the basic findings in parapsychology which have been mentioned are of course relevant to those who are interested in putting their magical ideas to practical use. However, there are some findings which seem to have even more relevance in that through them we can begin to understand why magic has traditionally been practised in certain ways.

### Principal requirements for opening to the subconscious

1. *Attitude* An attitude is the most powerful action we are capable of taking. It is of vital importance to believe totally, i.e. to know that the reality you are working towards really is. It is not an intellectual exercise — it is living reality. And once you start living that reality, your whole life becomes filled up with coincidences that are impossible, chance happenings that just aren't chance, hunches that invariably turn out, and all the other ways in which we become aware that psi is an everyday function of the subconscious. This importance of belief was re-iterated to me by many Tibetan people who I consulted in my recent research — they stated that if you ask for divination you must have complete faith and trust in the diviner — and the diviner must have complete faith in the deity they work with in their practice.[20]

2. *Relaxation* The second universal constant is that of relaxation. Every technique lays emphasis on total physical and mental relaxation — letting go is what I call this principle and it is a whole way of living in the world — letting go into the Universe and letting the Universe be in control.

3. *Concentration* Not the type that is associated with left-hemisphere tasks, but a holistic concentration in which the whole of your being

is totally engrossed in, encompassed by, the technique you are doing - as is a child at play. This is one of the Paramitas that are part of the Bodhisattva Way of Life – for enlightenment you need to have accomplished this art.

4. *Visualisation, or imagination* This aspect is far more important than a lot of people realise. It has even been noticed in biofeedback research. In order to develop voluntary control of normally involuntary behaviour it is necessary to become conscious of (or focus consciousness on) the present behaviour and at the same time visualise (imagine) the desired behaviour. Visualisation helps to enhance the self-suggestion. Many things may be an aid to effective visualisation, e.g. when healing someone, have a photo of them when healthy. Acting ability is another form of visualisation, a characteristic of good senders or telepathic agents. The essential requirements of acting and of effective visualisation are concentration, belief and imagination. *Effective* visualisation can take years to learn. The only problem is that in our ignorance we may unwittingly create ill effects. Once the heart chakra is free of greed, pride, ego and such like, then, however ignorant you may be, you will do no harm. And the first maxim in this field is: 'An ye harm no one, do what ye will.'

5. *Emotion* A *very* powerful force, as the example of poltergeists reveals. Rituals are not effective in themselves. Performed without 'power' they are empty words and actions. It really doesn't matter *what* is done if one is *confident* that it will work. Superstitions help because they inspire confidence, e.g. a word-perfect spell gets a charge around it which influences beliefs and emotions of the user.

6. *Zen action* The paradoxical Zen principle of action through non-action is to be found in every form of psi. As parapsychologist Rex Stanford says:

> Good ESP performance is not derived from non effort but from elimination of certain types of effort - specifically the conscious effort to try to force the psychic things ... to happen — combined with effective effort of other kinds (e.g. to concentrate on some inner mental image which is then dissolved and then is followed by the discipline of quietly but attentively waiting for (possibly) psi mediated impressions to emerge). Let us interpret receptive mode in this fashion. It is likely that certain altered states facilitate this mode by helping one to experience the world in a way in which psi events happen naturally and easily.[21]

In the above quote the ideas in brackets come from Rhea White's famous 1964 paper. These ideas are Stanford's interpretation and elaboration of William Braud's conception of the psi-conducive syndrome. This requirement of Zen action is vital to all aspects of working with the subliminal mind. An analogy could be that we are using the left hemisphere to bring to light the workings of the right hemisphere, and the older areas of the brain in which the subliminal mind has its foundation.

*Altered states of consciousness*

Alteration of consciousness is a central characteristic of magical practices noted by anthropologists in every culture past and present which have been studied. As mentioned earlier some cultures use psychotropic plants. Other cultures, notably the Tibetan, Sufi and Hindu, use discipline and mental techniques, as do the traditional Jewish (Cabbalistic), Egyptian and other formal ritual magical systems. The trend I see happening now, which I shall describe in great detail later in connection with people who are living the magical way, actually incorporates both psychotropic plants and mental training. In fact alteration of consciousness seems to be an essential for all magical workings, and this has also come to be one of the basic tenets in parapsychology.

Charles Honorton was an American parapsychologist who worked with altered states techniques since the early 1970s using a technique called the Ganzfeld. In a review[22] on the role of altered states in parapsychology experiments he found that a huge range of altered state procedures such as meditation, hypnosis, induced relaxation, and Ganzfeld stimulation were significantly effective in improving ESP performance. Recent meta-analyses of this line of research have been done by several people and there is an excellent overall description by Dean Radin.[23]

Let me give an example of this by describing the research using the Ganzfeld technique. The Ganzfeld is a very effective psychological technique for taking one into the hypnagogic state, and which is experienced by many people just as they are falling asleep. Just at that point between waking, sleeping and dreaming, many people experience intense dreamlike imagery which is brighter and stronger and realer than the real thing in waking life. This is precisely the state that many magical techniques aim for, and it is being induced in psychology laboratories around the world in the name of science and the pursuit of knowledge. As long as it is done with wisdom and care and love then that is fine, else we shall have to call these scientists 'black' magicians.

Research using the Ganzfeld technique has been going on in America and Britain and Holland since the early 1970s. In 1985 these were reviewed by a sceptic Ray Hyman, and one of the first people to have used the technique, Charles Honorton.[24] Forty-two experiments were reviewed of which 45 per cent could be considered to be 'significant', in that there was clear statistical evidence that people were picking up information psychically. Since then there have been several meta-analyses of Ganzfeld research with some people corroborating the overall success of this technique and others questioning its' effectiveness – par for the course in parapsychology.[25] Now for those who know how difficult it is to create a psi-conducive environment within laboratories, and who know how psi is so very shy under such situations which are in some ways the complete opposite of a magical environment, this finding that nearly half the studies produced clear evidence that people were picking up information psychically, is quite astounding. If one looks at the other experiments that were discarded for one reason or another, you will find that the reasons they were discarded is more because of statistical or scientific methodological reasons than because there was no suggestion of psi within the experiment. This 45 per cent acceptably successful evidence represents a really pretty tough line hardnosed scientific stance. Similar analyses looking at all the experiments using hypnosis as the altered state have found similar success rates. This is because within a laboratory environment creating an altered state is like creating just a little bit of a magical environment in which the people can relax and know that within this environment it's OK, they have permission to be psychic. Targ and Puthoff in their remote viewing experiments, and Ed May in the famous Stargate remote viewing work[26], did not use an altered-state technique but they did create an atmosphere in which it was OK to be psychic. This is the vital point.

Similarly John Palmer, whom I mentioned earlier as advocating more research into magical techniques, in a review in 1978[27] noted that it is not only experiencing an altered state that is important, but that the depth of the altered state significantly influences performance on ESP tests. So, those of you who are working magic to heal this planet, remember that the deeper you go, the better it is, because the deeper you go, the more you shift into the psychic realms of being in which all is one, and your wishes can become true. This has been strongly corroborated by my recent research with yogis and Tibetans into the effectiveness of meditation to shift our awareness so that we become more reliably receptive to psi impressions.[28] In my chapter on the pineal gland I will cover this area in much greater depth, including a glimmering of understanding into why it is that an altered state facilitates our entry into the psychic realms.

*Belief and attitude*

We create our thoughts and perceptions according to our attitudes and beliefs. We see what we expect to see, and if something is too novel we just won't see it — our reality in which we live is an artificial construct manufactured by a collaborative conspiracy of our conscious and subconscious minds. There is an excellent book which discusses this in great depth called *The Crack in the Cosmic Egg* by Joseph Chilton Pearce.[29] A lot of research in psychology and parapsychology endorses this concept.

The distinction between belief and attitude is a very fine one since attitude tends to harden into belief. In the old days it used to be called faith, and that is a deeper level of belief. The first person to bring this home to me was a writer on occult subjects,[30] although at the same time I was coming into contact with the scientific research into attitude and its paramount effect on the whole of the way in which we live our lives.

Belief controls all of our actions. It is at the basis of social conditioning and the basis of all culture. If we can conceive of something, then we can perceive it and — seeing is believing. Nowhere is this more true than in the psychic realms. This aspect of attitude is so strong a factor in psychic awareness that it has even been given a special name by parapsychologists — it is called 'the sheep—goat effect', sheep being believers and goats being dis-believers. This was first researched by Gertrude Schmeidler in 1946.[31] In this effect those who believe in the possibility of psi, and particularly those who have had psychic experiences and feel perfectly happy about taking part in experimental research, will show evidence for psi in laboratory tests. Those who think that there is no such thing as psi, have never had a psychic experience, or who think that it is impossible to demonstrate psi in the lab, will get chance results, or even significantly negative results.

One well-known medical example of this power of the subliminal mind is that of the placebo effect — anything can be healed if the doctor and patient believe totally in the method. When doubt creeps in then healing is prevented. The placebo effect is probably the strongest drug effect in modern medicine and there are numerous anecdotes showing the efficacy of placebos until the patient discovers that the medicine is a placebo when it immediately loses its potency. The doctor also must not know that it is a placebo or it doesn't work nearly so well. In psychosomatic illness the person creates their own sickness. Our minds directly affect our own bodies and other people's — our subliminal mind can not only heal but also harm if it is warped. The most extreme example of the power of the mind to harm is that of stigmatics who actually cause bleeding from their hands, feet and occasionally sides. Studies of stigmatics have found that it is often brought about by unacknowledged dreams, in other words our dreams can affect us physically. If the mind can cause bleeding in stigmatics then there really is no limit on its potential to heal or to hurt. Accident proneness is a well-known psychological example of this. So get it under conscious control - or at least become aware of your subliminal mind. It is vital to become more aware of our subliminal mind and to make it more healthy.

Now at a magical level this vital action of attitude or belief is even more important, because at the magical level we are working within the higher reaches of the mind. I know people who live their whole lives every moment of every day in a magical frame of mind; I call it living the synchronistic way.

Synchronicity is a term coined by Jung,[32] to describe meaningful coincidences that are more than coincidence - an acausal connecting principle is how Jung defines synchronicity. It's when things just happen to come together in a way that is like destiny or Lady Fate being pushy in your life. The more you become aware of the still small voice, the more you go somewhere not knowing why but knowing that you have to, the more you live on intuition and spontaneity, the more magically you are living your life, because you are allowing the psychic realms to interface with this earthly realm. And belief is the key. I return to this topic when discussing a parapsychological theory that has been proposed by Rex Stanford.[33] He calls his theory 'Conformance Theory' because he reckons that it is easiest for psi to manifest in a system that is flexible. Therefore, the more flexible you are the more you are likely to be living the synchronistic way because psi has more of a chance to act in your life.

In traditional magic, belief is the key because within a magical setting everyone believes the same world-view, whether it be an

Egyptian, a Cabbalistic, shamanic or a pagan view of the world; *and this world-view expressly permits the psychic to be an integral part* of it so that you flow in tune with the goddess, or you call up spirits typical of that particular culture, as in voodoo where you become possessed by the spirit you call up, or whatever other world-view the group holds. All these belief systems are just different ways of expressing the fact that there is more to this world than the see-touch realm that we see about us. Every culture throughout human history has had a belief system in some sort of spiritual realm. In our age it is emerging out of quantum mechanics which is about as bizarre as one could wish for, and I love it because it is so bizarre that our culture pursued its godless, mechanical way to the extreme limit and ended up finding a reality that is identical with the mystical, magical culture described in so many ways throughout the ages. I shall discuss this in greater depth in the next chapter.

Without belief in the whole magical world-view of that culture, one becomes an outsider and just cannot experience what all the others are experiencing. So belief is absolutely vital to magic just as disbelief has been vital to the creation of the soul-less, modern mechanistic, reductionist culture in which we live now. Two anthropologists, Malinowski and Mauss[34] suggested that *a priori* and collective belief was a characteristic of magic. Mauss also pointed out the widely held belief 'that the presence of non-believers rendered magical activities null and void'.[35] Both of these characteristics are also central to awareness of psychic phenomena as measured in laboratory experimental work. In fact the latter, the presence of disbelievers rendering magical or psychic activities null and void, has been the thorn in the side of parapsychologists, because whenever they do tests in the presence of sceptics nothing ever happens, and so they cannot get academic respectability and acceptance, which they would dearly like to do.

*Trickery — the trickster*

The widespread use of trickery and sleight-of-hand among various magical practitioners in different cultures may also be a psi-conducive technique as has been so clearly shown by the English researcher Batcheldor.[36] Apart from showing that psychic things happen better in groups, because people don't suffer so badly from the fear of psi if they can believe it was someone else who did it, he also showed that psychic things happen more easily if it has happened once before. In other words once the table has tipped then it's more likely to tip again. He found that this was true even when the first tipping was done by a trick.

The most obvious psychic aspect of this in real life is that of the psychic surgeons in the Philippines. Some of these people really do heal others. They do it in a very dramatic way with plenty of blood around the surgical incision, which has been done with a finger, and then bits of gall bladder or whatever are produced. Now it is possible that the various objects are psychically materialised, or it is possible that the blood and gore is a conjuring trick. It doesn't matter. What does matter is that the dramatic effect is such that onlookers, and the person being operated on, all believe that the surgeon has cured the person. And if the right atmosphere is created, the person will be healed. To me, this is the most important part of ritual. To create the right atmosphere in which the psychic realms can manifest. And whether it is done with pure high dramatics and conjuring, or whether the supposed psychic manifestations are truly psychic, is beside the point. As long as the end result of healing, in whatever form, is achieved then that is the only important thing. And what Batcheldor has shown is that though there might be some trickery at some point, this normally allows the atmosphere to become so charged that it allows the psychic realm to manifest, and this is when the real psychic, or magic, phenomena can begin to operate for what is really needed. A good recent book on this topic from the parapsychological perspective has been written by George Hansen.[37]

Richard Reichbart[38] devotes a whole article to this aspect of magic. He says:

> The glaring fact that magic [by which he means sleight-of-hand as in stage conjuring] in the distant past was always associated with apparently genuine psi phenomena and that the attempts to separate the two has been only a fairly recent activity in man's history, should prompt us to take a closer look at the relationship between magic and psi.

He considers that conjuring is used by shamans as a psi-conducive technique since it helps to build up the necessary degree of faith in the participants, and also to engender a suitably dramatic atmosphere. Shamans make no claim to omnipotence, and recognise that they aren't always in total control of their psi powers, and so conjuring was developed as a technique designed to tip the scales in favour of psi production. He also makes a pertinent point that psi is used by shamans only when it is absolutely necessary — for hunting, healing, etc. - and in these cases it is successful.

> Elkin[39] and Malinowski[40] noted the importance of the dramatic emotional states of anticipation and expectation of the desired results which were created by the practitioners. This suggests that the observations about the importance of motivation, positive mood and attitude and confidence inspiring expectation for successful psi manifestation which have been noted by Rhine *et al*, Kennedy and Taddonio and White[41] were also recognised by magical practitioners.[42.]

A similar psychological practice occurs in modern psychosomatic medicine, with the use of placebos, in which essentially a trick enables psychic healing to occur. This whole trickery aspect of magic in whatever form is yet another example of the power of belief.

*Visualisation*

All magical practices stress the importance of visualisation and imagination, but this has not yet been confirmed in parapsychology. Morris[43] has done a long series of experiments training visualisation and seeing if psi performance improves, with some measure of success, and Irwin and Cook[44] have correlated innate imagery ability with out-of-body experiences (astral projection), with rather confused results.

In an experiment by Child and Levi[45] the participants were asked to concentrate on the expected outcome of the PK task and to visualise it. Successful PK occurred. If, however, they visualised the process by which the PK machine worked and tried to influence the process itself, then no significant PK occurred. While other research has shown the effect of visualisation *per se*, this experiment is significant in elucidating the importance of where the attention is directed — namely to the outcome, leaving the actual process by which PK works to chance, coincidence, spirits or whatever force you believe in. Thus the important aspect of PK visualisation seems to be aiming for the goal and letting Nature fill in with the process. I discuss this principle again later in greater detail because it's an important one in magic.

Robert Morris *et al*.[46] also did an experiment with people using either a goal-oriented or a process-oriented imagery strategy. The participants looked at a ring of lights of which one light was on at a time and they had to move the on-light either in a clockwise or anti-clockwise direction. The lights moved at the rate of 16 times a second so it was a rapidly shifting circle of lights. In the process-oriented strategy the person was asked to visualise 'energy' building up inside their body, then flowing out to the circle of lights, and assisting the movement of the lights in the desired direction. In the goal-oriented strategy, the person was asked to point their finger at the light they wanted to light up next and vividly visualise it lighting up. Almost all of the PK occurred with the second strategy, and was particularly marked with those who had practised some form of mental training such as yoga or meditation.

Morris and Harnaday[47] repeated this experiment, but instead asked the participants to visualise being successful at influencing the lights for about four minutes, and then, after the visualisation period, actually seeing if they could. Then they went away for a week and spent time visualising moving the lights every day. At the end of the week they came back and had another go. Apart from general positive scoring nothing clear was found, which is to be expected because there was no really clear strong motivation behind the daily visualisations, and there could have been concomitant ego-problems because they did not do the actual influencing under PK conducive ritual circumstances. However, for real-life purposes, this procedure is highly recommended.

My feeling with regard to the rather patchy results in parapsychology is the mistake by the experimenters of assuming that a few training sessions would be sufficient. I know from their research into the effects of meditation on psi awareness that they have used people who have done only a few weeks' meditation at most. To me this is ridiculous. Having done yoga and meditation for thirty years I know how long it takes before one even begins to be fairly proficient at this level of mental art. And the same goes for visualisation ability of the sort used in magic. When I read books by Crowley or Dion Fortune, or others who consider themselves to be trained magicians I know that they have not just spent a few weeks training their visualisation ability, but rather they spent years of very disciplined work, to the point where they could mentally conjure up an image so strongly that it could be as if real to them. It's this level of visualisation that will then allow the image to be manifested into reality as Alexandra David-Neel so well explains in her book *Magic and Mystery in Tibet*[48] where she conjures up the image of a monk, who then becomes so real he can be seen by others, and then starts to take on a life of his own. I don't think that any parapsychologist has worked with people who have trained their imaging abilities to this sort of level, and only when they do will they find that imaging ability is related to the ability to manifest psychokinetically what one will.

What we are touching upon here is another aspect of the magical reality. It is to do with will, determination. Lots of people know the 'think positive' slogan. It is actually very effective. If you say 'I can't do that' then you won't be able to. You are blocking yourself at the first step. I teach my children to know that they can do whatever they wish to - and then they can. It might take years and a lot of hard slog, but they are not limiting themselves mentally before they've even tried. Well, the highly trained visualisation ability is an extension of this. And, if we really do want the 'New Aquarian Age' to be as beautiful as it could be, then we've all got to use our determination and will power, and train our visualisation ability so that we can truly envision with great clarity the way we wish society and people to live on this planet, including ourselves of course. Then we just might have a chance of realising our dreams.[49]

*Ritual in parapsychology*

Marilyn Schlitz is another parapsychologist who has done some research into magical techniques, particularly rituals. Marilyn is an experimenter and she is also psychic and has done some experiments with Elmer Gruber in Italy who visited places in Rome, while she in America used the remote viewing technique to successfully see the places he was visiting.[50]

She mentions four important features of rituals in so far as psi-induction is concerned.

> [Firstly, it is] a social process, deeply embedded within a set of symbols and beliefs that are shared by the actors in the procedure. Many ingredients are attributed to the successful outcome, including the dynamics of the group and the physical setting in which the ritual takes place. It may serve to unite the group, enhancing the cohesiveness of the participants. The intended outcome is sought after by the group and not just by an isolated individual.[51]

This echoes what Mishlove[52] noticed when studying all the different groups; that they each have a language unique to the system, which serves as a cohesive force creating a group bond as well as giving a philosophy that 'explains' the Mysteries to the group members. Even those who live magically and are isolated from a group of companions still tend on the whole to see themselves as part of a group, the extreme of this being the feeling that there is a magical reawakening happening on this planet, and that we are each of us finding a way forward to a new magical philosophy and way of being, that all is in flux at present, and that is why there is no clearly defined formula, why magical ritual is becoming more spontaneous, less formalised and intuitive. In this instance of the loosest of all possible groups there is still a feeling of kinship and community with unknown others who are felt to be walking the same path.

Marilyn's second point is that the leader of the ritual does not attempt to detach himself or herself from the induction process, but is considered to be of vital importance to the desired outcome. This process is in stark contrast to the traditional experimenter-subject relationship within parapsychology and science in general. And yet in science at present the experimenter is becoming a leader in a ritual, since the observer theories in quantum physics have emphasised that the experimenter is an integral part of the experiment, and cannot stand outside because by the very act of measurement the experimenter defines the outcome.

In parapsychology this is called the Experimenter Effect, because it has been observed so often that some experimenters always get clear evidence for psi in their research, while others with almost equal monotony do not. This same process has also been observed and commented on in subliminal perception and other psychological research. It seems that the most important factor in the experimenter effect is the attitude of the experimenter: an experimenter who is warm and open and friendly, and makes the participants feel comfortable and relaxed and secure and happy and know that they are in a space where it is OK to be psychic, will find that the participants are giving signs of picking up information psychically. An experimenter who is sceptical, or cold, aloof, tight and distant will almost certainly find no evidence for psi at all. I should say that this parallels a magical ritual, although it is highly unlikely that the leader of a ritual is ever likely to be the latter type of person.

Marilyn's third point is that the relationship between the leader and the participants is usually highly personal and individualised — again in strong contrast to normal experimental practice. Once again this is something that is changing in parapsychology, there being many more experimenters who are realising the importance of the personal relationship.

And, fourth, 'The ritual serves to reduce the amount of responsibility any one person must bear in the psi-elicitation process. Shifting the burden of success to the group procedure or outside forces accomplishes this goal.' This final aspect of the psychological process of ritual has been extensively studied in a table-tipping context by Batcheldor.[53] This theme of removing personal responsibility in order to allow strong PK phenomena to manifest recurs again and again. Batcheldor called it 'ownership resistance' and 'witness inhibition'. He defines the former as being a 'reluctance to possess paranormal powers oneself and the latter to a disturbance felt on witnessing displays of paranormality — whatever the source. Basically, the two forms probably spring from the same source - a fear of the unknown and the uncontrolled, and of what it might do.

> If belief is a positive factor in psi induction, then resistance is undoubtedly a negative factor. It can lead to all kinds of interference, such as explaining everything away in normal terms; continually making counter-suggestions (often thinking them but not verbalising) such as 'This won't work!'; distracting attention by talking about irrelevant matters; and, in extreme cases, complete refusal to continue with the experiment. Often subjects seem unaware of the emotional basis of this behaviour, but many openly admit that some aspects of the sitting make them apprehensive.[54]

Batcheldor is talking about table-tipping experiments, and factors that are conducive to PK. What he says is absolutely relevant to ritual and I recommend anyone who does or wishes to do magical ritual to read his report. Let me quote just one more paragraph:

> Resistance is another reason why validation may be difficult. In addition to provoking analysis and doubt, clear validation under watertight conditions will raise emotional resistance to a maximum. This tends to be foreseen, so that as the crucial moment approaches, our thoughts tend to sabotage the experiment. PK may still slip in, but on a target that is not being actively monitored. Or PK may jam the monitoring equipment. Sometimes there is an accidental success under tight control, but this is likely to arouse considerable retrospective resistance and so will be hard to repeat.[55]

Batcheldor recommends that the best way to overcome these two forms of resistance is to work in groups in a very ritualised way so that no one feels personal responsibility for producing magical effects — they can always feel it was someone else — so if your magic isn't working, look deep into your own psyche and see if you have any residual fear of the paranormal lurking there. We all know the person who always manages to bring the energy right down just when we are flying high, another example of what Batcheldor is talking about.

Marilyn Schlitz concludes that: 'Parapsychology may have much to gain from understanding the role that ritual can play in eliciting psi — thus providing fuel for the experimental fire.'[56] This attitude is one that is becoming more and more common in parapsychology now. And it is also true that research such as Batcheldor's can teach much to ritual magicians.

*Other aspects of goal-oriented strategies in PK*

Closely connected with the research into ritual and visualisation has been research into how the visualisation is directed in order to affect things outside oneself. In the course of research into psychokinesis several quite important principles have been formulated:

1. The ego must not become involved in the task. For example, Honorton and Barksdale[57] used a ritualistic exercise where the participants pointed dramatically with tensed arms at the PK machine which they were trying to affect. PK occurred only when they were asked *not* to try and exert any effort to affect the outcome. This is very Zen, you have the idea of what you want in mind and then you let go and let the Universe do it for you. According to Stanford this effect is analogous to his conception of one of the factors underlying magical ritual. This is what he says:

> The muscle tensing ritual was, in short, effective only when it was used in a way analogous to a true religious-magic ritual. In such a ritual one does not think one has, personally, to make something happen, and therefore one does not concentrate on trying to make something happen. One simply involves oneself with the ritual.[58]

Thus, one uses ritual as a direct psychological aid to permit psychokinesis to manifest. The final goal is visualised, then put out of mind and all one's concentration goes into performing the ritual impeccably, and let the goal look after itself.

2. PK occurs when the task is challenging, fun and motivating but one does not have a deadly serious attitude. If motivation is too high this results in anxiety which is self-defeating, and so a detached,

gamelike effortless intention to succeed is the best formula. As Stanford puts it:

> Perhaps conscious effort can be effective if it is expended on a ritual believed to aid or cause things to happen. Such a ritual, because it usually involves an appeal to powerful being(s) or force(s) outside oneself can be effective in actually reducing egocentric efforts to try to make things happen.[59]

There are a large number of experiments which concur with this view. My own research into telepathy also showed that too-high motivation is a problem. Batcheldor's work with table-tipping worked best when the participants were joking, singing, laughing and the table was 'accidentally' jogged — this then triggered real PK when the atmosphere was right. More and more experimenters are realising that a laboratory experiment is ritualistic in essence, certain formulae being rigidly followed, the experimenter being the high priest/ess, the participants and any assistants filling the other roles according to personality and function, and when the atmosphere is right psi can manifest.

3. PK seems to emerge after you have stopped trying - this is known as the lag or 'release of effort' effect. I call it the Cosmic Joker. You really are working hard at something doing all the right things and nothing at all happens, and after hours, weeks, months of really going for it hammer and tongs you just give up. You've had it, sod it, you don't care anyway, you'll do without — and what you were aiming for happens just like that.

Similarly, if there are two simultaneous objectives, one of which is more important to you than the other, the less important one will manifest. This relates to motivation and ego-involvement — or rather non-involvement of the ego, once again.

4. The strange thing with PK work is that significant results are achieved even when the target systems are hidden and/or unknown. Thus PK is totally independent of all normal sensory systems. In other words you do not need to know exactly what it is you are trying to influence in order to be able to affect it. You merely have to concentrate on affecting an anonymous target. As Stanford says:

> PK success does not depend on knowing the PK target, upon knowing the nature or existence of the REG, upon knowing one is in a PK study, upon the complexity or design of the REG, or upon the subjects knowing about the mechanics of the REG ... [This suggests] that PK somehow

occurs such that the favourable outcome ... is directly accomplished without mediation through sensory guidance and probably without any form of computation or information processing by the organism.[60]

Now isn't that a mind-blowing statement to read in a scientific textbook. Meditate upon it and the implications it has for Western science and the whole of our culture, since most people these days follow a world-view in which the scientists are the high priests, and they shift their attitudes and lifestyles in accordance with the prevailing scientific outlook.

*Conformance theory*

Another anthropologist, Evans Pritchard[61] has noted that:

> A common characteristic of magic is that it is used to influence chance or indeterminate events that are subject to change, or are likely to occur.

This is also true of psychic phenomena within the laboratory. Parapsychological research has indicated that things which are most likely to be influenced by PK within experimental situations are those which can be most easily altered. Rex Stanford[62] has come up with two interesting theories concerning psi functioning in everyday life. One I have mentioned already, the PMIR theory, is that psychic events are happening all the time, that they are subconscious so that we don't consciously know that they are there, but that they affect our behaviour. Then he brought out another idea about PK in which he says that PK happens easiest with things that are most flexible. For instance, if you want to learn how to affect matter it is best to start with a dice, or a candle, because you have a moving, spinning, rolling and tumbling object to influence. It's much more difficult to start with a table. The more sensitive a thing, the more unstable, and the easier to influence. Try playing with clouds.

Most parapsychologists agree with Stanford[63] and Kennedy[64] that psi can operate best when there is a situation in which an element of randomness or indeterminacy is involved. William Braud, and Marilyn Schlitz, will have done[65] a series of experiments which indicate just this. In some he measured the skin response of people, using an EDA meter, which measures slight changes in sweating on the hands, this sweating normally being linked to emotional response. He found that people's relaxation levels changed in accordance with the wishes of someone sitting in another room who had been told to try to influence the relaxation level as measured by EDA according to predetermined random instructions, i.e. on some occasions to make the EDA go down in activity, with control periods of no change. He did some experiments with fish, the person trying to influence the fish's movements. Once again the changes were in line with the desired outcome. Then he worked with gerbils influencing their activity levels. Several of these experiments were with Matthew Manning, and one Matthew did was with blood cells which he 'protected from haemolysis' (membrane permeability) to a significant extent, showing that he heals at a very direct and basic level right down to activity of blood cells. In all of these experiments the targets being influenced are essentially very mobile, very open, very random. The person influencing has a specific aim in mind, and merely attends to the end result. I call it the hitchhiking model.

When I am hitchhiking there is a strong element of 'randomness or indeterminacy'. So often though, when I am hitchhiking, things happen that are intensely meaningful and, even better, practical. For example, one other day I had been lecturing in London and needed to get home for the children coming out of school at 3.30 p.m. I left my friend's house at 9.30 a.m. I had plenty of time to get home, six hours in fact. However, on arriving at the underground station I discovered that there was a strike. It took me three hours to go from one part of London (Swiss Cottage) to my hitching place (Hammersmith) just a few miles away on the Western edge of London. When I got there I found the traffic so badly jammed that no cars could get on to the road leading out of London so I had a good two miles walk ahead of me to the next roundabout. While walking a car came out of a side road and dropped me at the next roundabout and there was a van waiting for me. In the van was a taxi driver who had the previous year given me a lift from my home to a nearby town. He had seen me walking and had seen the car pick me up, and had also seen the car drop me off by the roundabout, so he had known I was hitching out of town and had decided to wait for me. He took me home and I was in time for the children. It took me 3 hours to go the 130 miles from London to Glastonbury, the same length of time it took me to travel a few miles through London.

Now that is luck, or synchronicity, made possible by the fact that I had thrown myself completely open to the Universe. I was playing with chance, with randomicity, in that I was not getting a particular bus, but was letting whatever way of getting home come my way. And so magic could happen — and it did! This is not something that can be assessed by statistics and so scientifically speaking it is meaningless, but to me and my children the fact that I arrived home in time, when they were just arriving back from school, was very meaningful, very important to me and to them. The key to this happening in my life is that I held the important fact, to be home when they got back from school, as the time that I was aiming for, and let chance, luck, providence, the goddess, call it what you will, have the freedom to manifest my will, my aim, in whatever way it could. It is really very important to live one's life according to this tenet. For instance, one needs money to set a business project up. People who live in harmony with the Earth refuse to live in debt, so you cannot go to a bank for a loan or overdraft. In this case, you state your intent very clearly and then let Providence, or the goddess or synchronicity provide; remembering that 'God helps those who help themselves'.

I learnt this when playing backgammon. In the game of backgammon there are occasions when you need certain combinations of numbers in order to make the best move in the game. I learnt that one sussed the required numbers and then threw the dice without thinking. If I concentrated on the numbers it wouldn't work. The key was to forget, not to think what it was that I needed, but just to throw the dice with abandon and let Lady Luck have her chance to give me what I needed, to be flexible and 'let go' at every level. This principle of focusing on the end result and not on the manner or process by which to attain that end result is one of the findings of those who have done research into PK. And it works, I can assure you. It is also one of the tenets of magic. Visualise what is desired and then let the gods, goddesses, spirits, again call it what you will, bring that desired aim into being. People who work within a particular magical system have their own names. The word I use is the term Jung coined — synchronicity. It is the new way, the new 'in-word' that a lot of people are using now. The concept ties in with the new/old magical philosophy that is once again coming to the fore. I use it because I feel I don't need an anthropomorphised spirit or god or goddess to help me bring about magically what I need. I just need the right attitude of mind and a willingness to let go, and trust in the Universe to help me when I really need help. As Christ said 'See the lilies in the field, they toil not neither do they spin', yet they're alright. Become one with Mother Nature, live according to her lores, and you'll be alright too. It's just a question of letting go with perfect love and perfect trust. If there's the slightest shadow of doubt you'll blow everything.

Scientifically this principle of synchronicity has been shown by a lovely experiment by a German called Elmer Gruber.[66] He positioned a photoelectric cell by the door of a supermarket, and another by the entrance to a road tunnel. The 'cell' clicked every time someone walked into the supermarket, or every time a car went into the tunnel. He left them running for a long time so he had long stretches of tape with clicks on. He then cut the tapes up at random so he had short stretches taken at random from the complete tapes. Some weeks later, people came into the laboratory to ostensibly do a PK experiment in which they put on headphones, and tried to 'influence' the tape to produce lots of clicks, or very few clicks depending on the experimental condition, again decided on a random basis. He found that the number of clicks on the short stretches of tape correlated with the experimental condition of lots of, or very few, clicks.

Now the people doing the experiment did not influence the people walking into the supermarket or vice versa. By 'chance' the two coincided to a 'statistically significant' level. This, to me, is an experimental example of synchronicity, or my hitchhiking experiences, where 'by chance' just the right person was going my way when I most needed it. And it doesn't happen every time I hitch, and I try not to have time limits when I'm hitching because I'm superstitious and I don't like to push my luck. I feel one should not over-use one's luck, because then it will turn and one will be dreadfully unlucky. But if I really need it, then it will turn up trumps.

This can be conceived in conformance terms as follows: the ease with which a system can change from one state to another and the amount of 'free variability' in the system relates positively to the ability to influence the system, whether animate or inanimate. This suggests that magical practices were generally directed towards influencing systems which were in principle, and on the basis of their characteristics, amenable to PK influences.[67]

*Divination practices*

By divination is meant the attempt to elicit from some higher power, or supernatural being, or synchronicity, or the collective unconscious, etc., depending on your belief system, the answers to questions beyond the range of ordinary knowledge; things future or things past, hidden or removed in space; right conduct, and so on.

Michael Winkelman notes[68] that there appear to be three main types of divination procedures:

1. involving altered states of consciousness;
2. involving physiological or behavioural responses on the part of the practitioner or client, e.g. dowsing;
3. involving the creation or observation of an indeterminate random event which is interpreted in a symbolic framework to determine certain information, e.g. the Tarot, I Ching or runes.

As Winkelman says: 'in general it appears that most divination procedures integrate principles or practices which have been experimentally established as facilitating the manifestation of information via psi.'[69] As we have seen, altered states are psi-conducive, behavioural changes are often linked as in dowsing, and synchronicity plays a large part in our psi interactions with this universe.

In our culture we use the third method most often. Most people these days seem to feel happier playing with chance and letting synchronicity play its part in the divination, though there has recently been quite an upsurge of interest in pendulum dowsing which is an instance of the second method. I know people who dowse to find out which Bach Flower remedy is the most suitable one to use. In this way divination is being used for diagnosis. The altered-states method is best exemplified by 'scrying', in which you stare into a bowl of water, or crystal ball, until you get a vision or hallucination pertaining to the question being asked. Far fewer people in the West use this method nowadays although there are some people who 'channel' and give psychic readings, which are examples of this type of divination.

Culturally this is very interesting because the altered-states method is the most common form in ancient Western and other cultures. Everybody has heard of the oracle at Delphi to which for thousands of years kings, emperors, governors and other leaders used to go for advice on matters of state. This is the most direct form of altered-states divination.

In Tibet[70] various divination techniques were used for arranging marriages, planning births, undertaking journeys, business affairs and sites for houses, treating diseases, settling the outcome of legal disputes, finding lost objects, etc. The divination was normally done by a tulku, a lama reincarnated to continue their spiritual work, each tulku embodying a particular spiritual principle. For example,

His Holiness the Dalai Lama embodies the principle of compassion. Also wise women would be sought out because one rarely divined for oneself.

The methods used were primarily that of scrying in water, or a metal mirror or using the ball of the thumb which had been dipped in red wax and was then held up to a candle and stared at until it became very large and out of focus and the vision could manifest. Another method used the randomicity principle in that a mala (a string of 108 beads) was picked up by both hands at once and the number of beads between the thumbs was counted out in groups of four until between one and four remained. This was repeated three times and the resulting numbers interpreted. This is a bit like the yarrow-stalk method of divination in the I Ching in which the stalks are counted out in groups of four. Similarly three dice could be thrown and the numbers interpreted. All these methods are considered to be an expression of an underlying world order that we tune into through psi. As one works with symbolism and penetrates more deeply into its meaning one learns by its aid to arrive at an integrated view of the world, to see the one in the many.

In China three methods were used. Roasting of animal bones or turtle shells so that cracks appeared, which were then interpreted. Also the I Ching and Feng Shui, which is another name for geomancy, or dowsing out the Earth Energies.

> These were practised as a means of communication with sacred powers by methods that lie beyond the normal perception of the human sense, they were seen as a manifestation of eternal, universal truths that transcend the ephemeral attainments or mundane purposes of the human intellect; and they were believed to demonstrate and explain the regular cycles of birth, decay and rebirth in the created world of thought.[71]

Another form of divination is through omens or portents, such as the flight of birds, particularly crows which are sacred birds throughout the world. Omens or portents are derived from abnormal, irregular or violent occurrences of nature, which were believed to indicate a rupture of the normal, regular patterns. These events could be recognised as signs of cosmic upset which would reverberate in the different parts of the universe. In divination, scientific measurement, experiment and verification merges with intuition.

All of these different types are typical of cultures the world over, apart from our own which concentrates primarily on the randomicity method. I find this very intriguing because as always magical techniques reflect underlying beliefs and our world-view is very much the randomicity holographic world-view which I shall discuss in the next chapter.

*Social influences*

With regard to divination and other aspects of magic, Winkelman notes that various 'lines of research suggest that as societies become more complex, the use of ASCs in magical practice declines', although trance and possession states do occur in more complex societies. Thus, with divination simple societies use ASCs, while more complex ones like our own, use the randomisation techniques like the Tarot, where the practitioner appears to have less direct control over an apparently more objective outcome.

> The beliefs and practices in more complex societies tend to indicate a displacement of responsibility, repression of awareness and reduction of direct ego-control over information received or actions taken, while those associated with simpler societies indicate a more direct ego-contact with magical power, practices and experiences.[72]

He suggests that:

> These differences can be understood in terms of the 'filter' or 'barrier' model suggested by Bergson[73] ... Transposing this filter or barrier model to the social domain suggests that social conditions also create some filter or barrier which would affect the realisation of, or access to, psi abilities. This would be analogous to the apparent blocking of ESP information by people with high defensiveness[74] or with rigid response patterns or neurotic tendencies[75]. The available evidence cannot establish whether these differences between simple and complex societies with respect to magical practices indicate a reduction of access to psi abilities in more complex societies, different forms of access to psi abilities, or both. However, it is clear that we can expect social conditions to interact with psi potentials and that some of the cross-cultural differences in magical practices may reflect differences in access to or use of psi related abilities as a function of general tendencies or individual psychological dispositions fostered by particular social conditions.[76]

Mishlove, in his book also stresses this point, stating that:

> It is undeniable that the manifestation of psi training in each culture is affected by the dominant beliefs of the culture as well as by other

political, economic and historical factors.[77]

I discuss the historical factors in our culture in greater detail near the end of the book.

## Why it's so Important that the Bridge gets Built

While there are very encouraging signs of bridge-building going on on all sides, there is still a great void to be filled. Jeffrey Mishlove touches on this point when he notes that literally millions of people are currently participating in popular programmes for psi and spiritual development, programmes which are designed primarily for commercial reasons and which can do considerable harm. Public confusion regarding psi is rampant. Individuals who find themselves experiencing ostensible psi phenomena, often accompanied by altered states of consciousness, find that they are unable to discuss these experiences comfortably with anyone — the scientific world is too remote and generally too hostile to the psychic, the magical world is still too surrounded by fear and superstition for most people to feel comfortable about approaching it.

> A variety of subcultures, however, have provided a social format for explanation and discussion of such ostensible psi phenomena. These are often oriented toward new religions, pseudo-scientific cults, and/or hallucinogenic drug usage ... Thus, to an extent, the cultural information vacuum regarding psi has been filled. Nevertheless, the results leave much to be desired - from the perspective both of the public and of responsible peoples.
> The November 1978 suicides in Jonestown, Guyana, provide a striking example of innocent human beings being deceived and deeply betrayed by fraudulent psychic healing practices. There are many indications ... that a number of cults, involving hundreds of thousands of individuals, are not providing responsible guidance in terms of understanding psi, but are rather playing on popular credulity and thereby causing considerable psychological damage ... Dangerous cults are not likely to disappear merely from outrage or debunking; viable social alternatives are required to replace the positive functions of various cults.[78]

This problem of cult followings is one that I call 'flavour of the month' here in Glastonbury because they come and go so quickly: Rebirthing, a Course in Miracles, God Training, Immortality, Prosperity Consciousness, we've seen them all and more. Some have charged outrageously large sums of money to participate. They all promise outrageous things, they all tell you that you are incredibly special if you are one of them, and they are all basically the same in everything except name — bandwagon gurus in it for the money now and forgotten next week — or next year - or however long it takes for their groupies to start giving their energy, money and attention to the next cult. This is not to deny the good that they do for some of their groupies, and some good ideas, that can arise out of them. It is just that the cult rests solely on faith, and some are a rip-off, and some are positively dangerous. The void that Mishlove refers to is one that a bridge between parapsychology and some of the more responsible magical schools can usefully fill. The Western Magical schools have both an ethical and moral basis, and a long tradition of training people, while parapsychology can provide a modern cultural rationale and a firm basis for belief with which many Western people feel more comfortable. It is to this end that this book is being written so let us now explore a possible theoretical understanding for a world-view which incorporates psi as a necessary aspect — the Holographic Paradigm.

# 3. THE HOLOGRAPHIC UNIVERSE: A NEW/OLD WORLD-VIEW

## Introduction to the Looking-Glass Reality

The biggest implication of psi phenomena is that we are potentially aware of *everything* in the entire universe.

As mentioned in Chapter 1, experimental research over the past fifty years has confirmed that awareness of psi is dependent on one's state of consciousness, one's attitude, motivation, relaxation and other psychological (mental) aspects. We are separated from this potential omniscience by the thinnest of veils. In scientific terms this veil is called 'filter theory'. We would go totally crazy if we were aware of everything in the entire Universe even for only one second; so we need filters to block out all but the most essential items from our conscious awareness.

Unfortunately this trend has gone too far, and the pressures of living in a city have made most city people so blocked that they aren't even aware of what their close friends and family are feeling, let alone whether or not some distant relative is in need of help. We have as a race become insensitive - or desensitised, mainly as a result of living in cities where the pressure of so many people and all their thoughts and emotions is so great that we have to blank out all but the most immediate perceptions or go totally crazy. So now we have to reverse the trend and become more aware, more sensitive, open up the filters a little bit, and open our psychic, as well as our physical, eyes and ears. This of course is easiest in the country!

Some people call this process of increasing awareness 'becoming aware of the subconscious', since it is only in conscious awareness that we are so blocked, and the potential omniscience is all there in the subconscious. In becoming aware of the subconscious *per se*, we also tend to become aware of psychic information. This information is different from most subliminal information in that it was never *physically* present; we know something that we have never physically heard or seen or felt. The psychic information of which we become aware when we open up to the subconscious has come in by means other than our eyes or ears or other sense, possibly through the 'third eye'. In trying to understand this process by which psychic information is available I shall be travelling through some pretty strange waters.

## Thresholds and the subconscious

Conscious experience is a very limited channel. We can, it is now accepted, be conscious of only about 7 things at any one time. Yet there are 10 thousand million neurones and 100 thousand million million synapses in the brain.[1] Therefore, it is physically possible to process billions of bits of information at any one instant. This alone demonstrates how limited conscious experience is. And remember, nobody really understands what is happening when we become conscious, or aware, of something - be it a thought, a feeling, a beautiful view, another person, etc.

In 1903 Myers,[2] a Cambridge don and co-founder of the Society for Psychical Research, noted that the occurrence of psi is subconscious. We become aware of psychic information once it is 'in the system' so to speak, but cannot specify how it got there, although many people have offered various diverse theories. Psi phenomena appear to be ever-present subconsciously and occasionally break through into consciousness, but ordinarily 'press on' 'the stream of thought, emotion and behaviour in such a way as to steadily distort, modify, emphasize and deflect the ongoing processes of consciousness'.[3] In other words, the occasional breakthroughs into consciousness are less important than the constant interaction between the incoming psi information and our behaviour, emotions, thoughts, etc. *of which we are totally unaware.* Psi is a 'thorough-going part of the total behaviour of the individual',[4] and not only of humans but of the whole of nature.

Linked with this discussion of the subliminal, is a textbook psychological theory called Signal Detection Theory (SDT), in which the concept of a threshold as a fixed point of transition in sensory events is rejected, on the grounds that theoretically *all* signals contribute to the continuum of sensory activity on which decisions and behaviour is based. Theoretically, therefore, on a psychic level, one is stimulated by *everything* in the Universe. Every single thing is linked in some way to every other thing; every thought, every action affects every other thing in the Universe in some way or other. As this is impossible to cope with, the barriers blocking such awareness have to be proportionately stronger. Therefore, a psi stimulus has to be very strong to pass these barriers, which is possibly why spontaneous cases reported in surveys are primarily about death and disaster to one's nearest and dearest. Alternatively, a radical alteration to one's state of consciousness — so that what is normally subconscious may become conscious — will permit a certain amount of *relevant* psi material through, i.e. shamans, healers and mediums tend to concentrate their awareness on the task in hand rather than diffuse it through the Universe.

According to SDT the behaviour of *responding* to an external stimulus is really a two-stage process; the first is sensing, and the second is that of deciding on the appropriate response. With psi information only the first stage is not yet understood, i.e. we don't really understand how we can be potentially aware of everything in the Universe, or how there can be instantaneous information with no consideration for time and space. Physically, however, it *is* possible, and it is so at a quantum mechanical level.

*The quantum brain*

Let me introduce this topic with a quote from an excellent book by Danah Zohar[5] which is about precognition, and in which she attempts to understand how precognition works. This quote is about the brain, and the possibility that it works at the quantum level:

> A single photon of light will excite the optic nerve; the synapses of neurones are so tiny and so sensitive that the likelihood of their firing (and their stimulation threshold) varies according to quantum fluctuations in the surrounding ionic fluid. As these fluctuations are wholly random, it is no more possible to predict when any one neuron will fire than it is possible to predict when any one electron will become excited. Laboratory tests on isolated neurones have shown that their stimulation thresholds vary according to definite statistical law, just like any other quantum process.

> Not all cortical neurones are subject to quantum indeterminacy. Only those which are stimulated at or near their stimulation thresholds will be sensitive to the quantum level excitation of the surrounding fluids ... But of the $10^{10}$ neurones thought to exist in the brain, experimental data suggest that about 10 million are at any time being stimulated at or near the marginal threshold for quantum sensitivity.
>
> In states of reduced brain activity, such as during sleep or in meditative trance states, the proportion of neurones subject to marginal stimulation is increased — thus increasing the susceptibility to stimulation by quantum indeterminate phenomena in those states.[6]

It is very important to realise that some aspects of the brain appear to work at a quantum level. This point is key to the hypothesis. I am not too sure yet exactly how this link between the brain and the strange world of quantum mechanics works, but I feel it to be profoundly significant. Costa de Beauregard[7] mentions it in his writings, as does Pelletier.[8] Pelletier says that the synaptic cleft, that is, the space between the synapse of one neurone and the next neurone is in the range of 200-300 Angstroms. This is so tiny that it is in the range considered to be applicable to quantum effects. So, since the brain can be stimulated at the quantum level, in order to understand better how it is possible for us to sense everything in the Universe, we have to turn to quantum theory, and try to grapple with the reality and concepts in this realm of Nature and of ourselves, that is very different indeed from our familiar see-touch reality. As above, so below - as below, so above?

*From coincidence to synchronicity*

In Chilton Pearce's *The Crack in the Cosmic Egg* (1973)[9] a suggestion is made, based primarily on psychology, that we create our own reality through our beliefs about reality. Now I don't mean that if I am a beggar in India that it is 'my fault' or anything as fascistic as that. I am talking at a much more basic level. We can only perceive that which we can conceive. An example from psychology classes concerns pigmy peoples who had always lived in the jungle and who had never seen distances. When one went with the European researcher to the savannah and saw a herd of animals coming towards him he was terrified because to him they were getting larger and larger — transforming from insects to huge animals — not closer and closer as we would have perceived them. The level of belief about reality that I am talking about is at that level — a level so deeply ingrained that we don't even realise that we hold the belief that makes us perceive reality in that particular way.

Alan Vaughan[10] echoes this sentiment as a result of his experience of strange coincidences, or synchronicities. The same concept has repeatedly been formulated by physicists as a result of their work with the fundamental constituents of our universe. For instance, in everyday reality we function extensively within the Newtonian world of cause and effect - if I drop an apple it will fall to the ground. But if you start to notice coincidences in your life and then start to live by those coincidences, you will begin to get a glimmering that there is a way of living that is outside of the Newtonian cause-and-effect mechanical clockwork reality. Coincidences are the obvious aspect of a way of living in this world where one is living at the psychic level. Jung called this 'synchronicity'. Synchronicity is described as the acausal organising principle. When meaningful coincidences rather than cause and effect rule the way your life is going, then you are living synchronistically.

> If cause and effect are illusions — always not merely in strange situations — then synchronicity emerges as *the organising principle* of consciousness ... Consciousness overlays 'randomicity' to achieve synchronicity; it is the organizing principle of life opposing the disorganizing principles of entropy, i.e. the decay of matter and energy ... It is time to discard the constraints of our thinking and take a new look at the potentials of human beings. If we contain within us an image of all creation, then we also contain vast potentials of creative consciousness. Only beliefs confine our creativity ... Synchronicity suggests that there is an interconnection or unity of causally unrelated events, and thus postulates a unitary aspect of being.[11]

Let me give you a good example of synchronicity. I have very decided views on never incurring a debt, and certainly not basing one's life on debt, overdrafts, loans of any sort. I have a friend who started up a so-called ecological, green-oriented business, and is certainly a very aware person, but he started on loans, and never cleared them, and now his whole business is pricing itself out of the pockets of the people it was set up for because of all the interest charges he is having to pay off. Well, I wrote about my feelings on hearing of this and some time later I received a newsletter, in which the editor had a piece about living one's ideals in an ecological sense, so I sent her a copy of my ravings. Two days later I received a phone call from her. My letter had arrived just as she was about to incur a large loan for her newly fledged business and had tipped her, what must have been already uncertain, mind in the direction of not incurring the loan. The coincidence of my letter arriving at that particular moment, the very morning that she had an appointment to see her bank manager, was so striking as to be quite magical and very powerful — for her at least.

The quote I like best from the above is 'Consciousness overlays randomicity to achieve synchronicity.' I have always objected to the so-called Second Law of Thermodynamics which says that entropy is always increasing and the Universe is becoming more and more run down, chaotic and random, because it is so obviously totally untrue — the Universe is becoming more and more evolved and complex and growing more and more beautiful — in my eyes at least. I do not see the evolution on our particular planet as being the workings of the Second Law of Thermodynamics at all. I see it as the ever-evolving life force, consciousness, mind, spirit, call it what you will. Just where we are evolving to and why I'm not quite sure, but the universe is certainly not running down into chaos.

So what is the principle that is creating this evolving order — opposite to the entropy principle? Well, it's life, isn't it? But it's more than life — it's the life-force. Now I'm playing with words here, because we all know the difference between something that is alive and something that is dead, though whether the raw carrot I eat pulled fresh from the garden is alive or dead is a moot point. But that which is growing this beautiful planet in all her diverse forms is not just life, life is the outcome of it — so I call it the life-force, and this force is the organising force that is causing order and beauty — though if you look at recent work by 'chaos theorists' you begin to wonder if chaos is not more beautiful and ordered than order is! And many people reckon, me among them, that something which we call consciousness is this life-force — or is the aspect of this life-force that works against the entropy force. In Kashmiri Shaivism, this is described as Shiva in union with Shakti, Shiva being consciousness and Shakti the life force energy. Tibetan Buddhism symbolises this with the Yab-Yum statues, where for example Manjushri symbolises wisdom in union with his consort. And when we start to live with synchronicity so we are working in harmony with this principle of creation of form and harmony and beauty — the evolution of life principle always growing closer and closer to the purest expression of itself so that we are more self-conscious and self-aware and questing than are the plants around us who express their beings just as they are and do not ask questions all the time.

Now randomicity is also rather fun. When I first had to learn about statistics I groaned and muttered about lies and statistics. However, I am beginning to get a funny feeling about all this probability and possibility stuff. It's the stuff of which the Universe is made after all. All of quantum mechanics is based on probability.

It is a statistical world where there is a chance that Schrodinger's cat is alive or dead or both at once (and if you've never heard of Schrodinger's cat, don't worry because I will explain that later) but until we look — that is, until there is consciousness involved — all the different states exist simultaneously. Some theorists reckon that it is consciousness that is holding this world together in the way it is. That is why our thoughts are so powerful and magic can work. We live in a world of probability. That's why there is no contradiction between our knowing that we have free will and can determine the way our lives are, and the knowing that precognition, divination, astrology, etc. can at times be absolutely accurate.

A lot of philosophers have got stuck on this one, but if we understand exactly what a probabilistic world is then we see that the two are just different aspects of the whole. The future is a statistical probability that synchronises in a meaningful way with some aspects of the present — and we can definitely do something about it. The greater the likelihood (probability) that something will occur, the more likely that future will be realised, but there is always a possibility that the unlikely will happen — and when it does we go, 'Wow! what a coincidence!'

The *Hitchhiker's Guide to the Galaxy*[12] with its infinite improbability drive is a fine example of what I'm talking about here. Apart from their wonderful humour, I believe that the immense popularity of those books is due to the fact that Adams psyched into the underlying consciousness of our society and wrote something that is playing with concepts which we are all trying to grasp and to understand, because they are the new world-view by which we are beginning to live our lives, and so they are vitally important to us, and, like children, the best way to learn is through play and laughter.

Quantum mechanics is all about statistical probability. In the quantum world you cannot know everything about an individual particle, but you can know generalised laws that govern the whole. That is statistics; where the individual is unknowable, but the whole can be determined. And the whole is in every part, and every part contains the whole. 'Human know thyself.' 'As above, so below.' 'Humans are the microcosm of the macrocosm.' It all sounds very familiar really; it's just new words and a new way of seeing it, which gives a new perspective so that you know it better.

Einstein's Relativity Theory also plays havoc with causality, since it allows for the possibility of circular time and time reversibility. Within the tiny microcosm of the atom, and at the speed of light, neither time nor causality have any meaning in the accepted sense. The Quantum field is the most subtle level there is, at the level of the Planck scale, the most basic scale there is: $10^{-23}$ cm and $10^{-43}$ secs. Since our brains are susceptible to quantum influence, and therefore susceptible to this reality where synchronicity rules, we have to explore this world of the atom further. Time asymmetry in our see-touch world is really very weird because in quantum physics there is no time asymmetry – the possibility of going backwards in time exists.

## Heisenberg's Uncertainty Principle

1. *When observing something we choose what we want to see*

Let's start our quest here with the double slit experiment, in which I shall give a brief summary of what I shall then describe in a bit more detail, to hopefully help you to get a grasp on this looking glass world of quantum physics.[13]

If there is an emitter of electrons (or photons) and it sends out a beam through a single slit, then they will behave like particles so that on the screen behind the slit you get a straight line. However if you send the beam through two slits they will behave like a wave and you get an interference pattern of a strong line in the middle with weaker and weaker lines going out to each side. But in the double slit experiment if you send electrons one at a time, when by all accounts they shouldn't be able to interfere with one another to give an interference pattern, you still get an interference pattern even though there's no other electron for it to interfere with. This means that the electron is splitting itself into two and interfering with itself. But, and this is even stranger, mathematically and experimentally it is going through both slits, neither slit, one or the other slit and none of these is actually happening!

If you put an observer, a measuring device of some sort, by the slit then that effects the results, and the electrons behave like particles. The observation causes the electron to act differently, as if it was aware of being watched. The observer "collapses the wave function" simply by observing. In other words the electron is both particle and wave, in all possible states, until it is observed and the act of observation is called "collapsing the wave function." What we choose to examine it with changes the property of what is out there. Consciousness must be involved and is more fundamental than matter. There are two laws: one applies when you're not looking – one applies when you are looking. Looking at things involves interacting with them. We influence the observed; there is no separability. It's all created by consciousness: observer, observed and the process of observation. In what is called the virtual reality, the electron is in every possible place and not in any of those places; this is also called superposition of multiple possibilities. Superposition has now been observed in the real state, not just in virtual reality! In other words one electron has been observed in multiple places. Normally, when the observation takes place, the electron is found in one particular orbit. However, there is no movement from all possible places to one place; it just is in one or the other. The quantum jump is discrete; there is nothing between one level and the next. The particle is in a position in which questions about its position don't make sense, like asking about the political affiliation of a tuna sandwich doesn't make sense.

Particles are momentary manifestations of the implicate order, the quantum wavy function in which there is no particle but a waviness which can spontaneously pop out as particles. When you're not looking it's like a wave, when you are looking it's like a particle.

Let me introduce you now to an aspect of quantum mechanics called Heisenberg's Uncertainty Principle. This says that the harder one tries to scrutinise the movements of subatomic particles, the more elusive they become. Owing to the mechanics of quantum movement, the mere act of focusing on the particle is enough to disturb it, i.e. by observing something we change it.

David Bohm, who was a physicist at London University, noted that thought processes act in a manner akin to this uncertainty process. He states:

> If a person tries to observe what he is thinking about at the very moment that he is reflecting on a particular subject, it is generally agreed that he introduces unpredictable and uncontrollable changes in the *way* his thoughts proceed thereafter.[14]

This is something that anyone who has practised meditation knows all about. By observing our own minds we change what we observe. The simplest meditation exercise is to observe your breath. Easy. Try it for five minutes just watching every breath into and out of your body. Then forget about it; now watch it without changing the natural rhythm so that the breath you are watching is the same as the breath when you weren't watching it. Easy? Well, it wasn't for me. I found that the process of fixing my attention on my breath or my thoughts, changed them immediately, often subtly, but they were not the same as when my attention was elsewhere. I sussed this by learning to sneak up on them sideways, if you see what I mean. The quantum example goes something like this: when measuring an electron one has to choose between knowing its momentum (how fast it is moving) or knowing its position — one can never know both, i.e. one sees it as a particle or as a wave — as matter or as energy. One can never be certain about the whole of an individual particle, only that aspect which we decided to focus our attention on. This is very important to realise because it has implications as a profound philosophy underpinning a whole belief system. *We can never be certain about the whole of an individual, only about that aspect on which we have focused our attention.* Until we measure the electron it is both particle and wave, both matter and energy. The act of measurement of observation by consciousness is what makes it collapse into one form or the other. This is the Schrodinger Cat Paradox.

The electron has the potential of manifesting as either particle or wave — as matter or as energy - *and we choose which form it will adopt* by choosing how to measure it. According to Heisenberg this uncertainty is a built-in feature of the universe. Light too can be a particle (called a photon) or a wave — matter or energy.

*2. The individual can never be wholly known of itself*

Another aspect of Heisenberg's Uncertainty or Indeterminacy Principle is that *the individual can never be wholly known of itself.* Only as one of many can a generalised understanding of each be formed. Given a long enough run of measurements the approximations would, following the laws of probability, form a picture, but this exact picture would be the result of a statistical trend rather than an exact description of any one electron. So we have moved from a totally deterministic Universe where things are exactly as the Laws say they will be and there is no paradox or grey areas, where it's black and white and as the authority tells us it is, to a world where all is paradox and shades of grey and nothing is certain and all is ambiguous.

This is the new world-view philosophy that people are starting to live, and it is the essence of the magical world since it is attitude, belief and other psychological parameters, or lores of the mind, that determine how one sees the world, or reality, and hence how one lives in it. Since the universe, and you and I, are composed of these particles, are sensitive to a single particle, have neurones in our brains that are sensitive to quantum fluctuations, this implies that, at some level of our being, even if just the philosophical level which is ultimately important in its own right, you and I as individuals are probabilities with an infinite possibility for change.

We may define ourselves as material - as matter, as particle, but we are also non-material, energy, wave. The eighteenth and nineteenth centuries emphasised the material side of the universe. But nature, the universe, is dual, matter and energy, and when one is observed the other becomes potential and unknowable.

*The Schrödinger Cat Paradox*

As mentioned earlier, in choosing experiments the physicist determines which identity the electron or photon will manifest when it is measured, whether as particle or as wave, in other words, a physicist's choice of measurement of an event determines what he observes. In other words, we create our own reality — in a different sense this time! Consciousness apparently plays an active role in determining the outcome of any experiments conducted to study quantum phenomena. We get this same uncertainty with radioactive phenomena. We know precisely how long it takes for half of a certain quantity of radioactive material to disintegrate, but we cannot state exactly when any one atom will disintegrate, giving off its radioactive particles. Thus before the actual measurement the electron is in what is called a virtual state; it is in *every* possible state; and at the point of measurement it then collapses in to one particular state. This is called 'the collapse of the wave function' and in the mathematical equations which describe this process there are mathematical entities called 'hidden variables' that are essential to the equations, but nobody knows what they are. They know that they are essential for the electron to manifest as particle or wave in a particular state, i.e. for the collapse of the wave function, and all this has been proved experimentally. Some physicists have suggested that these hidden variables are 'will' and 'consciousness'. In other words, ultimately it is the will and consciousness of the observer which creates the reality that is observed. How's that for a worldview philosophy by which to live your life — and is there any actual difference from a magical philosophy? The reason for the cat in Schrödinger's Paradox is the awful thought that a cat in a box in this virtual state is both alive and dead at the same time because nobody knows whether or not the radioactive material in the box has or has not disintegrated. So it has both disintegrated and not done so at once — the virtual state — until the lid of the box is opened, the wave-function is collapsed and the cat emerges in one state only — either alive or dead.

To recapitulate briefly, matter and energy are the two sides of the same coin. While we can never be sure whether an electron or photon is a particle or a wave (matter or energy), the two possibilities complement each other in such a way that we can get an approximate picture of reality. This picture, though, is 'never more than a distribution of the probabilities that, under a given set of circumstances, the matter-wave will express itself in this way or in that, and until it does so, reality itself ... must be said to consist of probabilities'.[15] We have to change from an 'either-or' world view to a 'both-and' one.

## The Principle of Non-locality

The full implications of Heisenberg's Uncertainty Principle are that:

> Reality at its most primary level consists not of any fixed actualities that we *can* know, but rather of all the probabilities of the various fixed actualities that we *might* know ... When any physical process first starts, it sends out 'feelers' in all directions, feelers in which time and space may be reversed, normal rules are violated, and unexpected things may happen.[16]

As described above, an electron in this 'virtual' state, as it is called, will *simultaneously* situate itself in *all* of its possible new homes. *At once* it is everywhere, before eventually dropping back to its original stable orbit. Once again there are correlates at the mental, psychic, magical and philosophical levels. For example: if you can't decide whether to stay at home or go to the seaside - then in your mind you will be at *both* places at once - a most uncomfortable position — until you make your mind up. Then you will know where you are and feel much better. That is why in magic the intent must be clearly stated and clearly focused - if there is any wobbling in the mind then the energy gets dissipated into those wobblers — same thing at the mental level.

These 'virtual states', in which electrons, for example, find themselves, give rise to *the principle of non-locality*. In this principle something can be affected in the absence of any local cause. Sounds like magic, doesn't it? For instance, information can be transferred without any yet known means of transport, as is the case with psychic phenomena.

*The EPR paradox*

The prime example of non-locality is known as Bell's Theorem or the famous Einstein-Podolsky-Rosen (EPR) paradox, now known as quantum entanglement which has led to a book on parapsychology with the wonderful title of 'Entangled Minds.'[17] If a quantum particle is energised so that two related photons emerge, and these two photons zoom off in different directions, anything that happens to one of them instantaneously affects the other *so as to balance out the whole system, of which these are parts*. It's like *The Wizard of Earthsea*.[18] Thus, with quantum particles that were originally part of the same system, action-at-a-distance occurs, and information is transferred with no measurable means of transmission. Information is somehow different from anything else in the Universe and is closely linked to thought and hence to consciousness (hidden variables and collapse of the wave function again?). Freedman and Clauser and also Aspect and Grangier[19] verified the EPR paradox by measuring the change in the spin of polarised photons — light particles.

> They established beyond doubt that the mysterious correlations do happen exactly as quantum theory predicts, and their ability to do so by registering macroscopic effects, that is, effects visible on the everyday level of reality - on their laboratory apparatus — goes further to show that (this) has *implications* well beyond the sub-atomic level of reality.[20]

*Do remember while reading all this stuff about quantum phenomena that the neurones in the brain have several million synapses at critical threshold and these can be triggered by a quantum event — and psychic phenomena are perfectly valid at the quantum level. We are all part of a whole and information can be transmitted between us with no visible means of transmission — in order to maintain the balance of the whole system.*

In the past twenty years further research has shown that quantum entanglement occurs at the macroscopic level with effects in bulk properties of a magnetic material, so these ideas have been corroborated rather than negated by developments in physics.[21] However, critics of these ideas say that they do not show non-local signalling – for this we need to incorporate consciousness into the equations which is exactly what Bohm is proposing. Also in the past twenty years there has been a major development of superstring theory into M-theory. This postulates other dimensions which could also theoretically give a physical understanding of psi phenomena. Bernard Carr is the main exponent of these ideas.[22] What I like about these ideas is that they link with the Yogic Tantric philosophy, which states that the universe is created by the union of consciousness with vibration (spanda), emanating at various different levels from the subtlest through to the grossest, this last being our physical universe. Each of these levels can be understood as different dimensions, with superstring vibration as one of these subtle levels. I actually see both of these different theories as informing our understanding of how psi can happen, they complement each other.

Thus, in quantum entanglement, different aspects of a system cannot be broken down into separate bits affecting each other via causal laws, but rather the whole system must be seen in terms of its indivisible inter-connectedness, each part making sense in terms of the whole. 'In broad conceptual terms, at least, most of the once insurmountable philosophic and material objections to the mere possibility of things like telepathy and precognition have been turned over by the new physics'.[23] If you accept the theory of the Big Bang, then this means that the whole Universe is interconnected and that something affected in one place has reverberations across the whole Universe — picking a flower affects the twinkling of a star. This is the true meaning of karma, the knowledge that every thought, word and action reverberates across the whole universe, and that everything I do to others comes home to roost in me. We come across this reference to indivisible interconnectedness again when we encounter the strange world of David Bohm.

As I mentioned earlier, thought is akin to quantum mechanical processes in that it cannot be pinned down since the act of observing it changes it. And, just as in quantum entanglement one must take systems as a whole since there is instantaneous information transmission between different parts of the same system, so thought processes appear to have an indivisibility of a sort with information transmission from person to person occurring instantaneously through psi.

For example, I can think about being on a certain beach and I am immediately, in my mind, on that beach. Furthermore, a friend of mine who just happened to be on the beach that day might at that moment think about me. This can be called telepathy, or understood in terms of synchronicity, or in the quantum mechanical ideas I have been elucidating here. It doesn't really matter other than that the more ways in which we understand why, the better we understand how. A thought needs to be seen in its whole context. The central theme of physicist David Bohm's book *Wholeness and the Implicate Order* is this theme of 'The unbroken wholeness of the totality of existence as an undivided flowing movement without borders'.[24]

Thus, quantum theory indicates that there are no such things as separate parts in reality, but instead only intimately related phenomena so bound up with each other as to be inseparable. In fact this aspect of quantum theory is the crux of magical philosophy and there is no difference. In a lovely book entitled *The Dancing Wu Li Masters* which, like Fritjof Capra's *The Tao of Physics*,[25] describes the findings of quantum physics and relates them to the ageless mystical truths, I came across this wonderful sentence:

> This view holds that our physical world is not a structure built out of independently existing unanalysable entities, but rather a web of relationships between elements whose meanings arise wholly from their relationships to the whole. [26]

I find it just so amazing that the most basic of the material world physics gives a world-view that is indistinguishable from the magical.

## The Implicate Order

I have discussed the notion that we are theoretically capable of being aware of everything in the Universe, albeit at a subliminal level. I have also discussed how it is possible through quantum mechanical principles for that information of which we become aware to be eternally everywhere. Since the brain is affected at a quantum mechanical level this means that we can link into the universe in this manner. Now I want to discuss the implications of all this as formulated by David Bohm, with special reference to what we do in magic. Bohm states that:

> Ultimately the entire universe has to be understood as a single undivided

whole, in which analysis into separately and independently existent parts has no fundamental status.[27]

We really need to realise the implications of such a world-view. Anything we do affects the whole, and in turn we ourselves are affected. Thus the dictum that those who do harm have that harm return to them makes immediate sense, because in doing harm, they are affecting the whole — of which they are an integral part.

This single undivided whole he calls 'the implicate order', that is, that which is implied, which stands behind. When you imply something you don't state it explicitly, it's there in the background of what you're saying or doing. The see-touch world of our external reality he calls the 'explicate order'.

This implicate order is the realm of spirits, angels, archetypes, call it (them) what you will. It is the psychic realm, the magical realm, the synchronistic realm. We have so many names for this level of reality, because each culture has known that it is there, but it is difficult to conceptualise because it is implicit only; and so each culture frames it according to their conceptions, since we can perceive only that which we can conceive. In our time, we are growing up and moving beyond the need for concrete conceptualisations, and so we can conceive of it as an implicate realm, or the synchronistic way, or however else you like. At present there are no rigid formalised dogmas: all is in the cauldron.

Bohm considers that the implicate order is the primary, independently existing, universal and autonomously active order. In other words, it is the basis for everything, the ground of all being, the prime mover, and other words taken from religious philosophies. The explicate order flows out of the implicate and is secondary, derivative and appropriate only in certain contexts. We get this same idea in the Hindu concept of maya as well. In fact his concept seems to me to be a far better conceptualisation of maya than the usual interpretation of 'illusion'. The magic idea of times when the veils are thin and one can move between the worlds seems to me to be describing the interconnections between the two realities described here by Bohm as implicate and explicate.

The relationship between the two is described by him as 'enfolded structures that interweave and interpenetrate each other, throughout the whole of space, rather than between the abstracted and separated forms that are manifest to our senses'.[28] This is pure poetry — and he claims that he is interpreting mathematical equations arising out of quantum theory.

Bohm explains his ideas about the implicate order with the analogy of a spot of ink dropped into a beaker of glycerine. If this glycerine is stirred, the ink spot gets gradually spread out so thinly that it disappears from sight, i.e. it becomes 'enfolded' into the glycerine. However, and here's the magic, if the glycerine is stirred back an equal amount the spot of ink reappears! When the ink drop is enfolded in the glycerine it is present, but only in an implicate state - when one can see the drop it is explicit. The drop is always there — it is just changing its state. It is possible to see an analogy here with life and death if you wish to.

This can also be done with a series of ink drops, first one which is slightly stirred, then another which is slightly stirred, then another and so on. After a while all will be enfolded in the glycerine and invisible. On stirring back, each drop will reappear one after the other, disappearing again in turn. If done right, this gives the appearance of a single *moving* spot. This is an illusion — as is so much in the explicate order. When stirred slowly we see each spot manifest as itself, separate from any other spot. But think about those spots when they are all invisible. Each of them is spread out in the glycerine, all interpenetrating and interweaving one another.

> Ordinarily we think of each point of space and time as distinct and separate, and that all relationships are between contiguous points in space and time. In the enfolded order we see, first of all, that when the droplet is enfolded, it's in the *whole* thing, and every part of the whole thing contributes to that droplet. When two droplets are present they are in different positions in the explicit state, but when they are enfolded they are distributed through the whole and are interspersed (interpenetrated) with each other. In the enfolded order every element has direct connections with each other — even with distant elements — unlike the Cartesian model where the field connects only with field elements very near to it in space and time.[29]

In the series of droplets that manifest as a moving spot, we have to remember that there is an aspect (the moving spot) which is perceptible by us, but the whole of the order is not just what is manifest. The moving spot is only an abstraction and secondary to the primary reality of the large number of separate droplets enfolded in the glycerine which are all interspersed and in contact with one another. The fundamental movement is unperceived by us, and is the folding and unfolding. The traditional movement is a crossing across space and time and we see it as a moving spot — a localised entity moving from one place to another, but in this case that is just illusion and the actual movement is that of the different spots coming into view and then being enfolded back into invisibility in the glycerine again — i.e. the degree of implication or enfoldment is the basic movement or parameter. The ink droplet that has turned N times differs from one that has turned '2N' times. That difference is of no significance in the Cartesian view, but here is the *fundamental* thing: those drops that are most closely connected have the closest degree of implication. Therefore, the primary order is not that of time and space, but of degree of implication.

> If you take this picture of the ink droplets converging to form a particle and going out, the particles are actually all over space. If you were to put obstacles in the way of the particle it would converge differently, like a wave. It would begin to show a wave-like property. So you see, all the properties of the particle are in the whole order.[30]

Now think of living beings. Here in the explicate realm we all seem to be separate entities from one another (although as a group we manifest something else altogether). But in the implicate realm, that which is primary and stands behind this one, we are all part of one another and part of the whole, interpenetrating and interweaving together. This is the basis for what is being called the 'Holographic Paradigm'.

*The holographic universe*

A hologram is like a photograph that is made out of two laser light beams which interfere with each other. This creates ripples like you get when throwing stones into a pond, and the resulting interference makes a picture of the object that was in the path of one of the light beams. The hologram is an excellent example of the enfolded or implicate order, since the information is not directly accessible to our senses, but must be 'read out', or unfolded, using light. This is because a hologram is composed of light patterns and light is wholly within the frequency domain when in its wave form, and it is only by stimulating these light patterns in a certain way that we can see the image that is embedded in them. Otherwise it is just a blank sheet.

In a hologram the whole picture is contained in every little bit. That is, if you cut a hologram in two you will get two whole pictures. Some of the fine detail will have been lost but the picture

# The Holographic Universe: A New/Old World-View

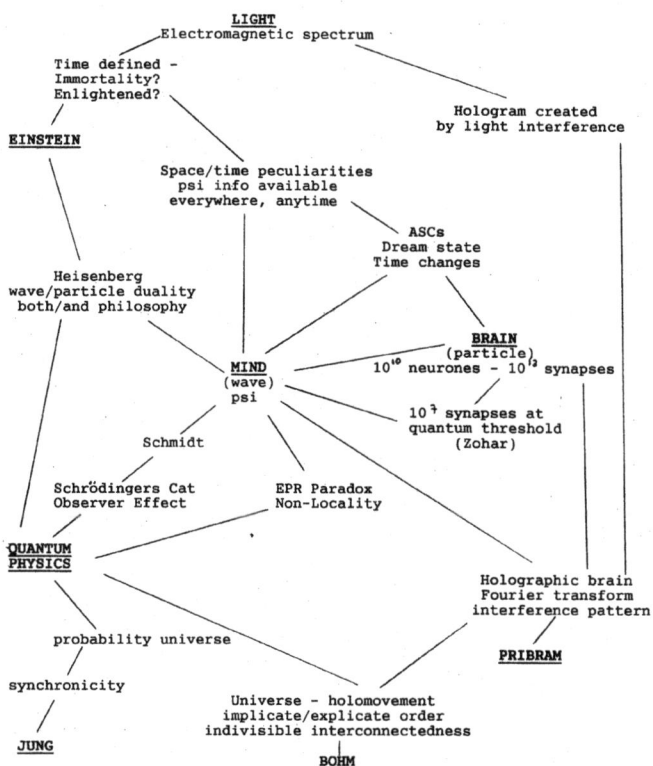

*The Holographic Paradigm (illustration courtesy of Serena Roney-Dougal)*

is all there in both bits. Whichever bit of the original you cut off you get the whole picture, but the less of the whole you have the less detail you get. Thus, we as humans contain the whole within us - everything is the microcosm of the macrocosm, it's just the fine detail that gets lost as the parts get smaller and smaller. And there's something special about human consciousness that links in directly with the whole - with the implicate order, the Akashic record or the Collective Unconscious.

Pribram[31] has been looking at the way memory works in the brain and he thinks that it works on holographic principles. You can take one bit of the brain and find one memory, but if that bit of the brain is removed the memory is not lost. In people who have operations, even if quite large chunks of their brain are removed, they still keep their memories. In other words, memory seems to be spread right across the neo-cortex, as well as specific memories being located in specific places. This is not to deny Penfield's work[32] in which stimulation of a specific area causes specific memories to be re-run as if one were really living them. We seem here to have a brain memory equivalent of the particle-wave paradox of quantum physics.

In Pribram's holographic model of the brain each part contains information about the whole, i.e. the form and structure of the entire body and mind may be said to be *enfolded* within each region of the brain. In taking a hologram, light waves are encoded and the resulting hologram that's projected then decodes, or deblurs, the image. The brain may similarly decode its stored memory traces, these being stored according to a mathematical principle known as a Fourier transform. It occurred to Pribram that the brain's mathematics may also comprise a lens. These mathematical transforms act as a lens to make objects out of blurs or frequencies, making them into sounds, colours, etc. (Remember that all sounds and colours are frequencies, wavelengths of energy in different forms.) If we didn't have that lens (the mathematics performed by the brain) maybe we would know a world organised in the frequency domain — no space, no time, just events.

It should be noted that Pribram's speculations incorporate the functioning of the synapses, which Bohm and Pelletier suggest work according to quantum mechanical principles. It is at the synapses that the processing of the incoming sensory information in the frequency domain is accomplished.

Not only memory and sensory experience have been incorporated into this holographic model of the brain but also characteristics of the experience of imaging and visualisation have been explored using these ideas. Now we have already seen that visualisation is a key aspect of spell-working and other magical practices, such as PK, in which we use our mind to affect directly this see-touch realm in which we live. It seems as if our seeing of images results from a brain process akin to stereo loudspeaker systems called phase relations. When computer simulations are done of image processing they have found that the holographic method provides a rich texture of scenes similar to what we experience in our heads.

The holographic domain however, not only describes brain

processing rather well but also physical reality as Bohm describes so well with his ideas about the implicate order.

> In the holographic domain, each organism represents in some manner the universe, and each portion of the universe represents in some manner the organisms within it.[33]

This holographic principle is a vital component of the new/old world view that is emerging in the magical religious peoples of our time. Once again we have the conception that everything in the Universe is in some sense linked to everything else — there is no there, only here — everywhere. Time is also lost in the frequency (holographic) domain, since frequency by definition deals with how densely things occur. So, as with psychic phenomena, we are back to a situation where the usual space-time world we know about does not exist, and so the usual understanding of causality does not exist either. Instead:

> Complementarities, synchronicities, symmetries, and dualities must be called upon as explanatory principles.[34]

Let us conclude this section with some excellent quotes from a book called *The Holographic Paradigm and Other Paradoxes*, edited by Ken Wilber, because these quotes say it all as far as I am concerned.

> There are intriguing implications in a paradigm that says the brain employs a holographic process to abstract from a holographic domain. Parapsychologists have searched in vain for an energy that might transmit telepathy, psychokinesis, healing, etc. If these events emerge from frequencies transcending time and space, they don't have to be transmitted. *They are potentially simultaneous and everywhere*[35]
> One need not speculate how information can travel from point A to point B if that information is already at point B ... If the brain is a hologram interpreting a holographic universe, ESP and PK are necessary components of that universe. Indeed holographic theorists would have to hypothesise the existence of ESP and PK had not parapsychologists carefully documented their existence over the years.[36]
> Psychic phenomena are only by-products of the simultaneous - everywhere matrix. Individual brains are bits of the greater hologram. They have access under certain circumstances to all the information in the total system. Synchronicity ... also fits in with the holographic model. Such meaningful coincidences derive from the purposeful, patterned, organising nature of the matrix. Psychokinesis, mind affecting matter, may be a natural result of interaction at the primary level. The

holographic model resolves one long-standing riddle of psi: the inability of instrumentation to track the apparent energy transfer in telepathy, healing, clairvoyance. If these events occur in a dimension transcending time and space, there is no need for energy to travel from here to there ... [Further] implicit in the theory is the assumption that harmonious, coherent states of consciousness are more nearly attuned to the primary level of reality.[37]

## Bohm's implicate order — a description of psychic space

Bohm discusses in depth the various quantum principles that I have already introduced like the Uncertainty Principle and Non-locality and concludes that:

> One discovers ... both from consideration of the meaning of the mathematical equations and from the results of the actual experiments, that the various particles have to be taken literally as projections of a higher dimensional reality which cannot be accounted for in terms of any force of interaction between them. [In other words:] We may regard each of the 'particles' constituting a system as a projection of a 'higher dimensional reality' rather than as a separate particle.[38]

Thus, the example I gave earlier of my thinking of being on a beach, and a friend coincidentally thinking of me, can be seen in these terms as being two outer manifestations of one and the same thing — at some level, 'a higher dimensional reality' as Bohm calls it, these two thoughts are not two but one, being projected out into this external world of ours. They only appear as separate thoughts to us on this side of the veil of maya.

Another common-sense notion that Bohm demolishes (while we're about it) is that of empty space — internal as well as external.

> What we call empty space contains an immense background of energy, and ... matter as we know it is a small 'quantized', wave-like excitation on top of this background, rather like a ripple on a vast sea.[39]

In other words, space is *full*, not empty, and our solar system and all the stars of our galaxy, and all the thousands of other galaxies are but specks of dust on this vast sea of energy that we call space. And in the so-called vacuum of space there is an enormous amount of energy available.

> What we perceive through the senses as empty space is actually the 'plenum', which is the ground for the existence of everything, including ourselves.

> The things that appear to our senses are derivative forms and their true meaning can be seen only when we consider the plenum, in which they are generated and sustained, and into which they must ultimately vanish ... It may indeed be said that life is enfolded in the totality and that, even when it is not manifest, it is somehow implicit in what we generally consider to be lifeless.[40]

Let me explain that in terms of breath. We breathe in oxygen and we breathe out carbon dioxide. When within us, the air is part and parcel of our aliveness; without the air and our breath we do not live, yet on being released from our lungs and rejoining the rest of the air from which it came it does not 'die'. It may be taken in by a plant and so be part of the plant's aliveness, and so on. Therefore, life itself has to be regarded as belonging in some sense to a totality, including plant, person and the whole environment — which Bohm extends to include the whole Universe. The Gaia Hypothesis[41] comes to mind at this point, since the world-view engendered by that hypothesis is identical in essence to these thoughts of Bohm, which in turn are identical with old and new magical philosophies.

Bohm proposes that this implicate order, which stands behind the external world, applies both to matter (living and non-living) and to consciousness, in which he includes thought, feeling, desire, and will. Consciousness does not have substance and so can be understood in terms of a dimension that is closer to the implicate than to the explicate. The way in which consciousness works constitutes a striking parallel to the activity proposed for the implicate order in general. For example, when listening to music, we do not perceive each note separately but we hear the tone, the feeling, the mood, etc. of the music as a whole. We hear the implications that the notes as a whole convey. Each moment of consciousness has a certain explicit content, which is the foreground, and an implicit content, which is the corresponding background. The actual structure, function and activation of thought is in the implicate order. We aren't aware of *how we think,* we are only aware of the end-result, the thoughts themselves. A really skilled meditator can start to become aware of a thought beginning, but even then it seems to bubble up as if from some pool below, which is seething with activity but of which we can barely begin to have any conscious idea. This pool below is the subliminal mind, which contains personal and collective unconscious, the subconscious and the preconscious levels of mind. And the subliminal mind is closer still to the implicate order than is the conscious mind. Anyone who has done any dream work, or pathworking, or worked with myth and archetypes, is well aware of the feeling tone of the symbolism which is the 'language' of this level of mind, and how whole enormous aspects of life can be implied by one symbol. Conscious thought has its origin in this realm of mind. Thought therefore is the manifestation of some deeper mind, and the relation between thought and the deeper mind may be like the relation between matter and the great sea of energy we call space. Thus each individual manifests in their own way the consciousness of the whole. The non-manifest is the subtler aspect, and the subtler has power to transform the gross.

We are however, aware of the implicate in some aspects of our lives. Movement is sensed primarily in the implicate order, as the example with the series of inkdrops shows. If you remember, you put a whole series of drops into the glycerine and on rotating the glycerine back again you appear to get a single moving spot, whereas there is in fact a series of static spots — a bit like a film which is a series of pictures which run together fast enough to give the appearance of continuous movement. In a similar way electrons in an electric current do not really move forwards any more than do waves in the sea — they only appear to. Thus waves in a sea are really moving up and down, up and down — the forward motion is an illusion as different bits of water move up and down in a sequence. In an electric current, *it is the energy which moves transmitted from particle to particle* — the movement is implicit in that each electron receives energy in turn, in sequence just as with the waves in the sea. In seeing movement we are seeing the implicate order, rather than the explicate which is the up-down vibration of the particle. Perhaps this is why sitting by the sea is so peaceful and refreshing to the spirit.

The psychologist Jean Piaget[42] who has worked extensively with small children, supports the notion that experiencing the implicate order is fundamentally much more immediate and direct than is that of the explicate which requires a complex construction that has to be learned. One reason why we do not generally notice the primacy of the implicate order is that we have become so habituated to the see-touch realm, and have emphasised it so much in our thoughts and language, that we tend to feel strongly that our primary experience is of that which is explicit and manifest. Gary Zukav[43] explains it this way. He shows that a particle such as an electron is not a particle in and of itself, but only in so far as it relates to other particles. *Matter is a web of relationships. You and I exist by reason of our relationships with the world.* Things exist because of their interaction with each other, just as movement of waves, or electrons in an electric current, or the moving ink spot in the glycerine, occur by reason of the particles' *relationship* to one another. When we focus our attention on the energy side of nature rather than on the matter side we begin to regain that childhood view of the world which focuses primarily on the implicate. I think that this is another incredibly important principle to come out of this game of linking modern physics with magical/mystical philosophies. What I keep coming back to is that our world-view is essentially determined by our understanding of how the world functions. The mechanistic, clockwork universe derived primarily from Newton, Darwin and the other scientists and philosophers of the seventeenth to the nineteenth centuries, who were in full flight from the magical world-view of the Middle Ages. It reached its high point in terms of complete denial of the numinous in the 20$^{th}$ century, when the new physics had already been born. The philosophy engendered by the old physics was probably no more directly relevant as a philosophy on which to base our lives than the new one is. The interesting thing is that our establishment culture is basing its view of life on an out-of-date scientific theory. The big advantage of the new physics is that the philosophy it gives rise to is totally in tune with that of the mystics, sages and spiritual teachers from all the ages, and totally in line with the magical philosophy of the Western peoples.

## Bohm — the implications of his ideas

Bohm then goes on to propose that the more comprehensive, deeper and more inward actuality is neither mind nor body but rather a yet higher-dimensional actuality, which is their common ground and which is of a nature beyond both. In this dimension, mind and body are ultimately one, as is past, present and future, and all of space. The implications of this is that it would be

> misleading and indeed wrong to suppose that each human being is an independent actuality who interacts with other human beings and with nature. Rather, all these are projections of a single totality.[44]

My thoughts, then, are connected with those of others. I am connected with everything else in this Universe in a manner directly analogous to the fundamental particles of the Universe, of which I am composed and the study of which led David Bohm to propose this view of reality — a view which corresponds closely to that taught by mystics and sages both East and West for millennia. It does mean, if we accept it, and I think most of the newly growing pagans, witches, shamans, druids, and all the plethora of spiritual groups and paths that are so fecund at present, do accept this world-view, that we have to be ever aware of the interlinking of everything, and view ourselves, our thoughts, our actions from a far wider perspective. When we scream 'Why me?' we realise that there are levels of answer undreamt of before. And we begin to be ever so careful of our thoughts, and what we think, because if we want this planet to be healed of her dreadful ills inflicted on her by unthinking and uncaring people, then we have to think the best thoughts possible so that we too do not add to the pain our mother, the earth, is suffering.

The implications of Bohm's theory are that the easily accessible content of consciousness is included within a much greater implicit background - the personal subconscious. This in turn is contained within a greater background that includes not only all body processes of which we are not generally conscious (the unconscious), but also a yet greater background of unknown depths of inwardness that may be analogous to the 'sea' of energy that fills so-called 'empty space' — cosmic consciousness or perhaps the collective unconscious of Jung. Whatever may be the nature of these inward depths of consciousness they are the very ground of our being, and it is these depths that we plumb when we work magic. Psychic phenomena and synchronicities are manifest aspects of this level of consciousness, they are manifestations of a reality that exists outside of the space and time of our see-touch realm. Just as the vast 'sea' of energy in space is present to our perception as a sense of emptiness or nothingness, so the vast 'unconscious' background of explicit consciousness is normally totally imperceptible to us — that is to say, it is *sensed* as an emptiness, a nothingness, within which the usual content of consciousness is only a vanishingly small set of facets. The void of inner space is as well documented a place as the void of outer space. I am sure that most people who have done any meditation or magical work will recognise exactly what it is that Bohm is saying here.

Telepathy, clairvoyance, precognition and the various forms of 'mind over matter' arise from this conception of consciousness as the explicit modes of action and perception of this implicate higher dimensional ground of being. That is why they have no known forms of energy transmission, because they arise from a realm that is everywhere at all times — outside of our time and space.

The surprise is that we, in our culture, are so little aware of psi occurrences, since psi is everywhere if only we know how to link in to that dimension. It is probably our increasing identification with the explicit order (the see-touch realm), and the increasing intellectualisation of our thoughts, and blocking of the subconscious, as we grow older that partially contribute to this lack of awareness of the prime manifestations of this higher dimensional order. It is often noted that babies and children and less intellectual adults experience psi far more readily. Also dreams, and other ASCs, where our awareness is more inward turned and therefore closer to the 'background sea of nothingness', are very psi-conducive states. Advanced yogis, mystics and magicians, who have specifically trained their minds to turn inwards to the deeper levels of consciousness, the Void as they call it, also tend to become proficient in psychic awareness, although many consider this blossoming of talent to be a distraction on the path of full spiritual growth, and from what I've seen of the magical world and its pitfalls of glamour and ego, I'm inclined to agree with this. However, psi ability is definitely a milestone marker along the path of development of our divine nature, and if one *lives* by psi — as opposed to *practising* it — then one is actually living in conformity with the harmony of the Universe — at its implicate level.

It seems that as one becomes more aware of this higher-dimensional reality in general, so one becomes more aware of the web of psychic relationships (synchronicities) that surround us — the implicate phenomena or order of the universe. I think that all this is summed up very well by Paul Davies:

> Over 50 years ago something strange happened in physical science; bizarre and stunning new ideas about space and time, mind and matter erupted among the scientific community. Only now are these ideas beginning to reach the general public. Concepts that have intrigued and inspired physicists themselves for two generations are at last gaining the attention of ordinary people, who never suspected that a major revolution in human thought had occurred. The new physics has come of age.[45]

> In the first quarter of this century two momentous theories were proposed: the theory of relativity and the quantum theory. From them sprang most of 20th century physics. But the new physics soon revealed more than simply a better model of the physical world.

Physicists began to realize that their discoveries demanded a radical reformulation of the most fundamental aspects of reality. They learned to approach their subject in totally unexpected and novel ways that seemed to turn commonsense on its head and find closer accord with mysticism than materialism.[46]

# 4. THE PINEAL GLAND: THIRD EYE AND PSYCHIC CHAKRA

From all that I have said so far, it is clear that we are slowly but surely becoming increasingly certain that a relaxed meditative state of consciousness is psi-conducive. The best examples of this are the various relaxation, dream, hypnotic, Ganzfeld, and remote-viewing experiments.[1] This corroborates traditional teachings and practices both East and West, e.g. Patanjali's teachings on yoga. Since this aspect of traditional lore is yielding to experimental research, it is worth examining other items of traditional knowledge.

I am very interested in the link between psychic phenomena and our physical state. The mind and body are inextricably linked while we are alive and living on this earth. Everything we do to the body affects the mind as anyone who needs a cup of coffee in the morning knows.

In my first year of doing my PhD, I was invited to go to America to deliver a paper outlining my ideas for research for my PhD. I hadn't done any research at this point — I just had some good ideas — and I was given £400 to fly to St Louis to tell all the leading researchers in parapsychology what my ideas were. This still strikes me as being outrageous, since they are normally really strict about accepting only the best of the year's research for this particular conference. Anyway, there I was flying into Chicago, catching a Greyhound bus down to St Louis, and arriving on the campus one day early. Everything was shut — so I went into the student coffee room, got a drink, sat down at a table where there were some students and after a while plucked up courage and asked if anyone could give me floor space on which to sleep that night. Yes, said this long-haired man, he and his lady would be delighted to have me to stay. So off I went. We got to talking and when he realised I was interested in psychic matters he got out his collection of slides because he was doing research into psychotropic plants and he was just back from South America where he had been staying with a tribe in the Amazon area where they used a particular vine, called Ayahuasca, specifically for psychic purposes. It was speculated by a Chilean psychotherapist, Claudio Naranjo, who had used Ayahuasca in his practice, that the active ingredients of the vine were chemically similar to a substance made in the pineal gland. And so I was introduced to the subject which I have been investigating ever since. Now let me draw back a little and give you the details of this slowly.

*Folklore: East and West*

While there are doubts as to whether one can consider psi equivalent to a form of 'perception', there is a considerable body of folklore which does so: the idea of a 'third eye'; the 'second sight' as it is called in Scotland. In Eastern (Indian) terminology the 'third eye' is called the 'ajna chakra', about which there is a considerable body of information. Some modern teachers of yoga, who also keep aware of Western research, equate the 'ajna chakra' with the pineal gland, while Descartes called the pineal gland 'the seat of the soul'. Let me quote, as a brief introduction, from a booklet written by one such teacher, Swami Satyananda Saraswati:

> All psychic systems have their physical aspects in the body ... With ajna chakra the physical equivalent is the pineal gland, which has long baffled doctors and scientists as to its precise function [He then goes on to discuss recent scientific research on the pineal gland and notes:] Yogis, who are scientists of the subtle mind, have always spoken of telepathy as a 'siddhi', a psychic power for thought communication and clairaudience, etc. The medium of such siddhis is ajna chakra, and its physical terminus is the pineal gland, which is connected to the brain. It has been stated by great yogis ... that the pineal gland is the receptor and sender of the subtle vibrations which carry thoughts and psychic phenomena throughout the cosmos.[2]

Thus, folklore and yogic teaching suggest that the 'third eye' (pineal gland) plays some part in the process of becoming aware of psi information, but what evidence is there that this is actually the case?

## The anthropological evidence

The folklore surrounding the pineal gains greater physical reality from certain specific divinatory practices among a large number of South America Indian tribes. These tribes all use a vine of the genus Banisteriopsis mixed with other plants, which contain di-methyltryptamine (DMT, a potent hallucinogen), in order to induce visions specifically for a variety of psychic purposes. The Amazonian Indians call this vine 'the Sacred Vine.' There are hundreds of psychoactive plants in the Amazon basin, yet all the tribes scattered over this huge area use these plants.[3]

Early chemical investigations of Banisteriopsis indicated the presence of an alkaloid which was actually given the name 'telepathine' in 1905 by Zerda Barron[4] because of its supposed telepathic properties. This alkaloid was later called yagein, banisterine, and was finally identified as harmine which had been independently isolated from seeds of Syrian rue *(Peganum harmala)* more than a century ago. Peganum harmala is used in Morocco to counteract harmful psychic influences. More recently Rivier and Lindgren[5] have analysed the drink prepared by the Sharanahua and Cuhina Indians from Banisteriopsis, and found it to contain five different harmala alkaloids. Hochstein and Paradies[6] isolated three harmala alkaloids (harmine, harmaline and d-tetrahydroharmine) from *Banisteriopsis caapi* and concluded that probably the most psychoactive ingredients were the harmaline or the d-tetrahydroharmine.[7]

In the 1960s the Chilean psychotherapist, Claudio Naranjo,[8] whom I mentioned earlier, used a variety of hallucinogens, including harmaline, in the psychotherapeutic setting, and came to the conclusion that harmaline is possibly more hallucinogenic than mescaline, creating a large number of incredibly realistic images so that you feel you really are experiencing the visions, as portrayed so well in the film *The Emerald Forest,* which depicts the tribal use of Ayahuasca. Some of the people who took harmaline with him felt that certain scenes which they saw had really happened, and that they had been out-of-body witnesses. Vomiting is a very common side effect of taking Ayahuasca.

The harmala alkaloids are extracted by shamans from *Banisteriopsis (Banisteria) caapi* in Colombia under the name of Yage, in Ecuador and Peru by the name of Ayahuasca, in Brazil by the name of Caapi, by the Shuar people who call it Natema, and by the Cashinahua who call it Nixipae. Banisteriopsis is usually mixed with other plants such as *Prestonia amazonica* and *Psychotria viridis* (Cawa), which have hallucinogenic properties, the active ingredient being dimethoxytryptamine (DMT), that are essential in order to trigger the full effects.[9] The anthropological evidence, however, points to Ayahuasca being more than merely hallucinogenic. The original name 'telepathine' was not inappropriate as the following anthropologist's reports suggest:

> Among the Jivaro, it is felt that part of the soul may leave the body, with the subject having the sensation of flying, returning when the effects of the drug wear off ... The Conibo-Shipibo Indians ... report that a common function of Ayahuasca taking by shamans is to permit the shaman's soul to leave his body in the form of a bird ... Among the Amahuaca 'a man's soul may leave his body when he drinks Ayahuasca'.[10]

These experiences are also reported by the Ziparo, the Tukano and the Siona, and can be considered to be fairly typical reports of what are now called out-of-the-body experiences. (This use of a psychotropic plant for out-of-body experiences matches that of the witches' flying ointment.) But experience of out-of-body effects does not necessarily mean that the vine is psi-conducive, although it is a good indication that psi may be close by.

The Amahuaca report not only separation of the soul from the body, but that after the sorcerer has drunk Ayahuasca, his yoshi — a jaguar spirit — will appear to him and tell him everything he wants to know, including the whereabouts of the intended victim. The Conibo Indians believe that the taking of Ayahuasca permits them to see the supernatural aspect of nature, and the Jivaro shamans believe that they are seeing distant persons and what they are doing. Normally these are people and places that the shaman knows, but he frequently has the experience of travelling to distant and unfamiliar villages, towns and cities of the whites which he cannot identify but whose reality can readily be ascertained. These experiences can best be compared to clairvoyance and remote viewing.

Divination is, however, the most important aspect of the rite among those who use Ayahuasca for healing — or murder. To 'see' the shaman who has bewitched the patient the Ayahuasca drink is used, since it is considered to allow one better vision while curing and to allow for better diagnosis. It is also used to identify personal enemies and to locate the resting place of stolen or lost articles. Shamans also drink Ayahuasca:

> when called upon to adjudicate in a dispute or quarrel; to give the proper answer to an embassy; to discover plans of an enemy; to tell if strangers are coming; to ascertain if wives are faithful; in the case of a sick man to tell who has bewitched him.[11]

Possibly the most revealing evidence comes from a footnote in an article concerning the Cashinahua by the anthropologist K. M. Kensinger:

> Hallucinations generally involve scenes which are a part of the Cashinahuas' daily experience. However, informants have described hallucinations far removed, both geographically and from their own experience.
>
> Several informants who have never been to, or seen pictures of, Pucallpa ... have described their visits, under the influence of Ayahuasca, to the town with sufficient detail for me to be able to recognise specific shops and sights. On the day following one Ayahuasca party, six of nine men informed me of seeing the death of my chai, 'my mother's father'. This occurred two days before I was informed by radio of his death.[12]
>
> [More generally] the Cashinahua drink Ayahuasca in order to learn about things, persons and events removed from them by time and/or space which would affect either the society as a whole or its individual members ... Although in most cases little can be done to alter events foreseen in visions, some precautions can be taken ... Rarely, however, would decisions based on information gained through Ayahuasca affect an entire village, and never the whole society ... In conclusion, the Cashinahua use Banisteriopsis as a means of gaining information not available through the normal channels of communication, which, in addition to other information, forms the basis for personal action.[13]

Of course, this anthropological evidence needs testing within controlled laboratory conditions before we can judge the extent, if any, of the psi-conducive properties of harmaline or the other harmala alkaloids present in Banisteriopsis, with or without the DMT normally present in the tribes' drink. So let us now look at the pineal gland or 'third eye' which produces a chemical that is almost identical in structure to the harmala alkaloids present in Ayahuasca.

## The Neurochemistry of Psi

Remember when I was discussing the parts of the brain considered to be connected with the subliminal mind and the processing of psychic and subliminal information, and I mentioned that the limbic system or temporal cortex was felt to be the prime area. Well, once again the limbic system is the place to which we must turn our attention since the pineal neurohormones act upon the hypothalamus and limbic system; the midbrain. Let me remind you: the midbrain is the mammal part of the brain, prior to language; the emotional, symbolic, archetypal mind; the area where our hormones affect our brain/mind.

The other intriguing item of information I have come across that links in at this point, since it concerns the functioning of the neurones themselves, is the fact that at any one time approximately a million neurones in the brain are at what is called 'quantum threshold of excitation,' and in certain states of consciousness more synapses will be in this state that is susceptible to quantum phenomena, i.e. more synapses will be working within that strange reality outside space and time, the psychic space. Needless to say, ASCs such as are

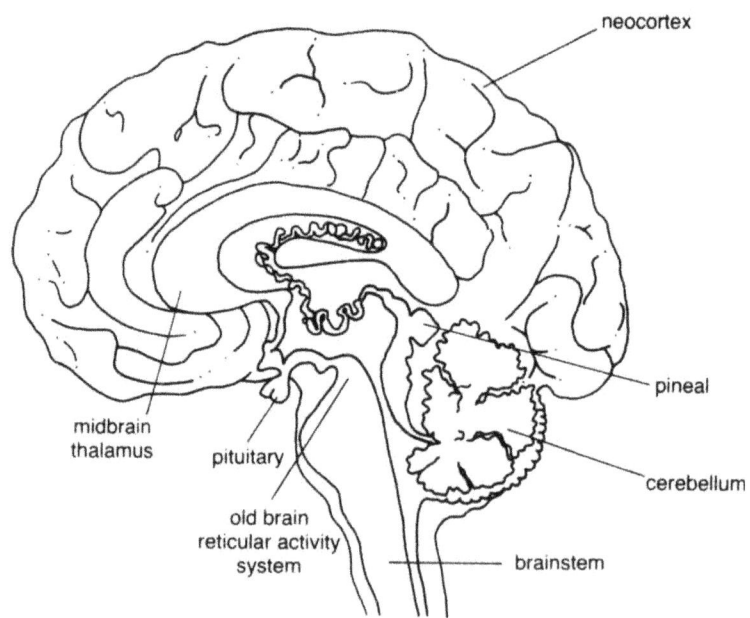

*Some relevant parts of the brain (diagram adapted from Most, 1986).*

experienced with Ayahuasca are precisely the best states for this.

## The pineal gland

The pineal gland is found right in the centre of our brain. It is tiny, about the size of the tip of the nail of our little finger and is shaped like a pine cone, which I think is where its name comes from. It shows extreme variability in size, form and internal structure from one individual to the next.[14] In general the pineal is a very active organ, having the second highest blood flow after the kidneys and equal in volume to the pituitary.

A unique anatomical feature is that it is an unpaired midline organ in the brain which, alone of all equivalent organs, has resisted encroachment by the corpus callosum, the part that connects the right side of our brain with the left.[15]

While being right in the centre of the brain, by the ventricles, it is actually outside the blood-brain barrier and so is theoretically not part of the brain. The blood-brain barrier is a skin or membrane which goes right around the brain and protects it from unwanted chemicals in the blood stream. It has the highest absorption of

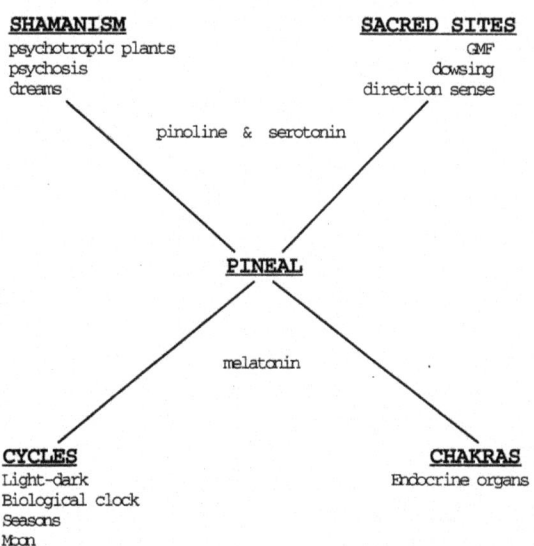

***Pineal gland connections (Illustration courtesy of Serena Roney-Dougal)***

phosphorus in the whole body and the second highest absorption of iodine, after the thyroid.

No other part of the brain contains so much serotonin, a neurotransmitter that works at the synapses, and is capable of making melatonin which is a neurohormone. (This means it works both as a hormone and as a neurochemical.)

A unique anatomical feature is that it is an unpaired midline organ in the brain which, alone of all equivalent organs, has resisted encroachment by the corpus callosum, the part that connects the right side of our brain with the left.15

While being right in the centre of the brain, by the ventricles, it is actually outside the blood-brain barrier and so is theoretically not part of the brain. The blood-brain barrier is a skin or membrane which goes right around the brain and protects it from unwanted chemicals in the blood stream.

The pineal only has nerves from the autonomic nervous system going to and from it. (Remember where I discussed the autonomic nervous system when talking about healing and other PK effects, and one of the ways by which we can detect whether someone has picked up a psi message.)[16] These autonomic nerves use noradrenaline as their transmitter. The pineal synthesises and releases melatonin and other neurochemicals in response to noradrenaline.

There seems to be considerable functional connection between the pineal and the pituitary, in that their actions tend to be antagonistic, the pineal being inhibitory in relation to the pituitary i.e. it is the off-switch for hormones activated by the pituitary. The pituitary used to be called the master gland because it was in charge of switching on the hormones. In similar vein the pineal could be called the mistress gland because it is responsible for switching them off!

When LSD level is measured in the brain, it turns out that it concentrates mostly in the pineal and pituitary glands; secondarily in the limbic system structures such as the hippocampus, amygdala and fornix; and thirdly in the hypothalamus.[17]

The rate at which noradrenaline is released declines when light activates retinal photoreceptors and increases when the sympathetic nervous system is stimulated, for example by severe stress.[18] Thus the amount of chemicals the pineal gland releases is determined by light and by stress. This has all sorts of implications, so now let us look at those chemicals themselves in a bit more detail.

It took me years to even begin to understand all these long names used in neurochemistry. In fact I still find neurochemistry an intensely confusing science and I am beginning to suspect that this is because the neurochemists are intensely confused. Our brains are immensely complex. We are dealing with a most wonderful creation of nature that contains billions of millions of cells each of which can have up to a thousand connections, all involved with our body processes, our emotions, our thoughts, our moods, our creativity, music, language, and so on. This works at a physical level through electricity and chemistry. I am finding that gaining understanding at this level is helping me to understand our psychic nature more clearly. It earths the knowledge, so to speak. We are spiritual beings in physical bodies here on the earth plane, and by understanding the physical facet better it is possible to understand the emotional, mental, psychic and spiritual facets of our multi-faceted being more clearly; so please bear with me as I delve into this world of the neurochemist. I can assure you that it is worth it.

*Serotonin and melatonin*

It is known that melatonin is linked with the neurochemical serotonin, which is stored in synaptic vesicles (just the place speculated by Zohar for possible quantum effects) together with noradrenaline, the other neurotransmitter involved. Those synapses in which serotonin is found form what are called serotonergic pathways. Serotonin's chemical name is 5-hydroxy-tryptamine (5HT) – another sort of tryptamine, closely related to the one found in ayahuasca. Think of them as different sorts of T – a bit like PG Tips, or Twinings! These are concentrated in the mid-brain structures of the pons, medulla oblongata, hypothalamus, limbic system and median forebrain bundle which ascends to the forebrain, with other connections to the reticular activating system and the spinal cord. These are the very same mid-brain areas specified by Dixon[19] as being the most likely interface point for psi to link in with the brain, from his understanding of the psychology of the situation. So here we have physiology corroborating a psychological hypothesis.

The concentration and rate of turnover of serotonin in the pineal is more than 50 times greater than in any other area of the brain. The pineal contains a pair of enzymes which are able to convert serotonin into hallucinogens.[20] Normally, when serotonin has done its job of transmitting across the synapse it is inactivated by the mitochondrial enzyme, monoamine oxidase (MAO), which converts it to an inactive metabolite. MAO is the major enzyme involved in the breakdown and inactivation of the neurotransmitters serotonin, dopamine, epinephrine and noradrenaline. Thus any enzyme which interferes with MAO will cause a build-up in serotonin levels, which has been found to lead to the formation of various endogenous hallucinogens, for example, 5-methoxy-N,N-dimethyltryptamine (5-MeDMT), a hallucinogen similar to the DMT found in cawa, which is a vital ingredient the Amazonian Indians add to their Banisteriopsis brew.[21] Harmala alkaloids such as harmaline are serotonin antagonists, CNS (central nervous system) stimulants, and extremely potent, short-term MAO inhibitors.

Comparison of the chemical structures of various hallucinogenic agents and tranquillising drugs, such as DMT, with the structures of noradrenaline and serotonin show incredibly close similarities, so it is not surprising that hallucinogenic drugs have a profound influence on the transmission of nerve impulses and, as a consequence, on mental and emotional states.[22]

Serotonin is a neurotransmitter which has been implicated in a wide range of mental phenomena from sleep cycles to psychosis and psychedelics. Of more importance here is the fact that it is a chemical precursor of melatonin with which it alternates on a day-night basis. Serotonin is found in greatest concentrations in the pineal gland and melatonin is synthesised in the pineal gland. Both serotonin and melatonin exhibit a circadian rhythm, serotonin concentration being greatest during the day and melatonin at night. This rhythm is free-running if one is in constant darkness, but is severely disrupted if one stays in constant light. This is possibly the chemical mechanism behind the severe disorientation experienced by people who stay awake for several days, and is almost certainly implicated in the manic-depressive syndrome when this follows a seasonal cycle of mania in the summer and depression in the winter. In a milder form it has even been given a special name — Seasonal Affective Disorder — or SAD,[23] because most people who suffer from this find they go for medical help when suffering from the depression in the winter. No one much minds the extra energy and bounce in the summer.

Melatonin is produced by the pineal gland, but is found also in the retina, the Harderian gland, and in the gut. This latter could possibly be because serotonin causes contraction of the intestinal muscles. (This is a typical autonomic nervous system function.) Is this what is behind butterflies in the stomach when we are nervous and that wonderful tingling when something really magical is happening, or our gut reactions to people and events? Melatonin inhibits serotonin's action, so melatonin calms us down again.

Melatonin also controls eye pigmentation and thus regulates the amount of light reaching the photoreceptors in the retina, where it is also to be found. There are possibly very important implications here with regard to colour and intensity of light in order to induce specific states of consciousness. For example, some mediums always used to work either in the dark or in red light because they found that the strange phenomena they produced, such as ectoplasm and phantom limbs, would happen more readily in dim light. Many people consider that magical rituals need specific light colours for specific effect. No one has researched this yet, but it is well worth looking in to. The Ganzfeld used by some parapsychologists to help induce a psychic state of consciousness specifically uses a red light to induce the hypnagogic state. The pineal gland has been well named 'the third eye' as melatonin is active both in the pineal gland and in the eyes. Melatonin is also implicated in the production of melanin, that which tans us in the summer — more interesting implications here.

Melatonin is called a neuroendocrine transducer, which means that it is a hormone which also has an effect on neurones. At a neural level the single clearest effect of melatonin is that it induces drowsiness during darkness. Peak production is three to six hours after sunset; it is a creature of the night. This could well shed light on other strange folklores surrounding psi phenomena. For example, many spiritual groups such as Catholic monks, Buddhists and Yogis all recommend rising at 3 a.m. to meditate, or to chant matins, or some other practice which is primarily aimed at personal development but which seems to bring enhanced psi effects in its wake. And what about our own saying that 'Midnight is the magic or witching hour, when witches ride abroad'? Not to be forgotten in this context is the research by Ullman, Krippner and Vaughan[24] concerning the psi-conducive nature of dreams, which of course was done at night. All of these practices, however trivial, become more meaningful when linked with our slowly emerging knowledge concerning the pineal gland.

## Beta-carbolines: harmaline and the pineal gland

This section is the lynchpin of the whole of my hypothesis concerning the pineal gland: namely that, together with serotonin and melatonin in the pineal gland and retina, there is another class of compounds called beta-carbolines, which are produced by the pineal gland, our third eye, and which are chemically very similar to the harmala alkaloids. There is a suggestion that the pineal effect on psi functions through the action of the neurotransmitter serotonin, which is known to be most active in the pineal gland where it is converted at night into melatonin (5-methoxy tryptamine) and the beta-carboline called pinoline (6-methoxy tetrahydro-beta-carboline). Pinoline can be formed in the mammalian body under physiological conditions from serotonin (5HT) or as a tricyclic metabolite of melatonin.[26] Callaway[27] suggests that this may regulate dream cycles. Strassman[28] has suggested that MAO (mono-amine oxidase) inhibitors, such as pinoline, could be involved in

6-methoxytetra-         6-methoxyharmalan         harmaline
hydrobetacarboline
(6-MeOTHBC)

*The chemical structure of Harmaline, 6-Methoxyharmalan and 6-Methoxytetra-hydrobetacarboline(illustration courtesy of Serena Roney-Dougal).[25]*

converting serotonin into di-methyl-tryptamine (DMT). This is the visionary hallucinogen found in certain ingredients (e.g., *Psychotria viridis*) of the Amazonian shamanic brew, Ayahuasca. Endogenous biosynthesis of DMT might also occur through the conversion of the common, nutritionally essential, amino acid tryptophan.[29] In other words our pineal gland makes our own endogenous Ayahuasca. The pineal gland is therefore possibly involved in altering our state of consciousness to a potentially psi-conducive state. Could this be the neurochemical trigger that stimulates neuronal thresholds to a psi-conducive quantum state, whatever that may be?

Neurochemical terminology can be very confusing so it must be understood clearly that harmaline and the harmala alkaloids are all beta-carbolines, of which there are many varieties with very similar properties. See the illustration above which shows three alkaloids which all look very similar. The first is found in the pineal gland; the other two are found in Ayahuasca.

W. B. Quay[30] in his textbook on pineal chemistry says that the pineal gland may synthesise small amounts of beta-carbolines. He believes that the pineal would make these beta-carbolines from either serotonin or melatonin. McIsaac[31] demonstrated that a particular beta-carboline called 6-MeOTHBC (this is a shortened version of the full name which you can see in the diagram; from now on I am going to call it 'pinoline' for simplicity) can be formed under certain specific conditions. It was also found that, at least in cell culture, 6-methoxyharmalan can be formed from serotonin.[32] This has been confirmed by several researchers. For example, Barker *et al.*[33] report

103

## The Pineal Gland: Third Eye and Psychic Chakra

that two different sorts of beta-carboline, of which pinoline is one, are found in extracts of rat brains and adrenal glands. Then in 1984, Langer *et al.*[34] found that pinoline was present in the human pineal gland. They proposed that it might act to modulate the uptake of serotonin in the synapses — and remember what I said about serotonin and its potential conversion to the hallucinogen DMT, and the quantum threshold properties of synapses. They also found that the pineal contains as much pinoline as it does melatonin.

Thus, they are suggesting that pinoline, which is found in our brains in the pineal gland, works by preventing the breakdown of serotonin. Prozialeck *et al*[35] have observed the apparent precursor for pinoline to be located mainly in mid- and hind-brain structures and to the greatest extent in the pineal. They have also shown that THBC and pinoline are potent inhibitors of serotonin neuronal uptake and so elevate plasma and brain levels of serotonin.[36]

Beta-carbolines are neuromodulators in the sense of playing an important role in the fine-tuning of the actions of neurotransmitters. Their main action is inhibition of MAO-A, which breaks down serotonin and noradrenaline. That is, they prevent the breakdown of these neurotransmitters and so cause a build-up of them in the synapses. It is this action that is the chemical concomitant of hallucinogens.[37]

Beta-carbolines also inhibit the transport of Na+ ions, and so affect the transport of nervous impulses.[38] Some beta-carbolines

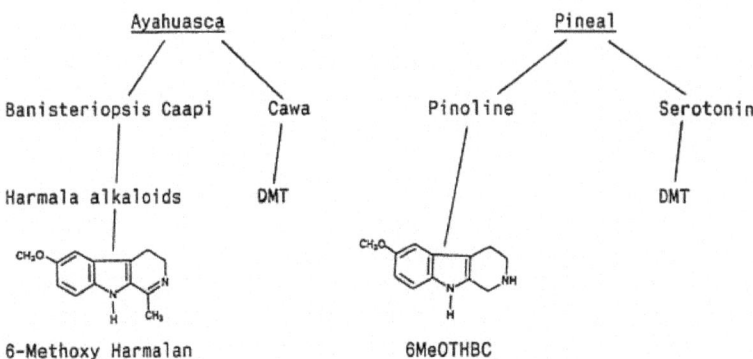

***Shamanic-pineal link (Illustration courtesy of Serena Roney-Dougal)***

bind to benzodiazepine receptors (benzodiazepines are best known by the name of Vallium — tranquillisers which have become the biggest prescribed addictive drugs used in the Western world by stressed and unhappy women). Beta-carbolines, such as pinoline, bind to the same receptors as Vallium. In other words they may have a similar stress-reducing function, albeit in a natural way as they are produced by the brain anyway and are non-addictive.

This also means that they affect circulating plasma vasopressin concentration.[39] Vasopressin has been linked with psi abilities in certain psychics.[40] It is linked with people who suffer from 'water retention' which often makes them unduly fat. This sort of fat person is not one who overeats but whose metabolic level and vasopressin levels combine to cause them to be fat. It has been suggested that a large number of mediums and psychics have this sort of metabolism.

Beta-carbolines seem to have an effect on temperature regulation, which could be connected with moulting, hibernation and other cyclic-seasonal effects noted in connection with melatonin.

Some beta-carbolines have analgesic effects, probably because of binding to the opiate receptors. It has been suggested that the visual symptoms have a retinal origin since beta-carbolines occur in the retina, and in cats given harmaline, changes in the electro-retinagram are apparent before changes in the brain cortex are observed, these changes being increased alpha and decreased beta waves.[41] Thus beta-carbolines appear to have quite wide-ranging neuro-endocrinological functions and effects.[42]

There is one specific beta-carboline, though, which I feel is of particular interest in this context and that is pinoline because its structure is so similar to the harmala alkaloids found in the sacred vine used by shamans. The hormonal effects of this beta-carboline, in so far as it has only recently been identified and its effects are only beginning to be deduced, appear to resemble those of melatonin. It inhibits the development of the genital organs, and in females it makes the dioestrus period longer or even totally abolishes the oestrus.

More recent research has found that pinoline is involved in the neurotransmitter glutamate pathways that regulate several neuroendocrine functions;[43] that pinoline may be an antioxidant that prevents membrane rigidity mediated by lipid peroxidation, and this ability is enhanced by melatonin. The membranes of cells are crucial with regard to protection from getting cancer. Thus pinoline like melatonin is possibly involved in protection from cancer.[44] Because pinoline and melatonin are antioxidants this means they also protect against free radical damage, to which the brain is particularly susceptible. Lipid peroxidation products have been accepted as biomarkers for oxidative stress in biological systems.[45] Pinoline acts in cells and nuclei as an antioxidant, particularly in the adrenal glands and cerebral cortex. Direct genomic effects of pinoline cannot be excluded.[46]

Thus pinoline has effects on the kidneys and on neurochemicals such as our endogenous opioids, and hence is involved in relief from pain.[47] As pinoline has been shown to exert an antidepressant-like effect in behavioral experiments and has been reported to have a low toxicity, this compound should be further studied as a potential antidepressant with pronounced antioxidative effects. These results further support the importance of pineal gland in antioxidative protection.[48]

When harmala alkaloids (the form of beta-carboline found in the Amazonian vine Banisteriopsis) are taken orally by humans, at the highest doses they cause visions, hallucinations, vomiting, tremor, buzzing in the ears, cold sweating, dysphoria, and a drop in the heart rate. At lower doses they cause bradycardia (irregular heart beat), difficulty in focusing the eyes, tingling, hypotension, cold extremities and light-headedness. All of these are physiological effects caused by MAO inhibition. The major psychological effect of 6-methoxyharmalan is said to be akin to a state of inspiration and heightened introspection. There is less effect on the emotions and thought processes than with LSD, although there is a change in perception of colours, increased sensitivity to sound and taste, passivity and withdrawal.[49]

My personal experiences have been different according to dosage. I used Syrian rue because in those days the sacred vine wasn't available in Britain. At a very low dose I floated into the hypnagogic state where I was aware of what was going on around me but I was very 'spaced'. At the next dosage I had incredible energy and clarity of perception, everything became crystal clear, shining, and I had so much strength and energy that I felt I could achieve anything physical. At the next level, which was just below that recommended for full psychic effects, I felt very still and peaceful and as if I could see through things; in other words, I could see the wall but felt as if I could see the room beyond. And if I moved my head it was as if all the world dissolved into flashes of light — everything material became energy, which I know to be true from physics and which was a practical confirmation for me of what had always before been just theory. However I have never had it at full dosage and so cannot confirm the anthropologists' reports. And also the full effects occur only when it is taken with the DMT, which is the hallucinogen. Sadly, at the time I tried out the Syrian rue, there was no DMT available.

*Light/dark cycles - the biological clock*

It is known that melatonin, the major pineal neurohormone, makes one drowsy and that one's circadian (daily) rhythm of sleeping and waking can be altered by taking melatonin. This is now being used to help with jet-lag and for people who suffer from insomnia. Light therapy is another way of influencing the natural light/dark rhythm of the pineal and it has been found that normal domestic lighting is not enough to affect the pineal, so those people who use light boxes to help them with winter blues need to make sure they are getting at least 1000 lux.[50] So let us first investigate the role of melatonin, serotonin and pinoline in sleep, since it is during sleep and dreaming that the majority of spontaneous psychic experiences are reported.[51]

There is a nocturnal increase in electrical activity in the pineal. Pineal activities are driven by the hypothalamic suprachiasmatic nuclei (SCN), which are linked to the pineal via the sympathetic nerves of the ANS. Reppert *et al.*[52] have found that most of the melatonin is found in the SCN which is located very close to the optic nerves, at the front of the hypothalamus near the base of the brain. There are specific nerves from the retina to the SCN, so that the pineal has its own supply of nerves, separate from the optic nerves, transmitting information about whether or not it is night or day. When these nerves are activated by light, the pineal is dormant, sleeping. When it gets dark the pineal becomes active. *In other words when it gets dark the pineal gland starts making melatonin in earnest.*

Peak levels of melatonin occur at 2 a.m., when cortisol and serotonin concentrations are at their lowest. I have found only one reference to a circadian rhythm of pinoline, which is not very specific, but I assume that it too has its peak at night.[53] This assumption is possibly supported by research with hamsters which has found that there is a time-dependency for melatonin inhibiting running behaviour in hamsters, both in the day and for most of the night, similar to that known to occur for other effects of melatonin. This effect is attenuated by giving a central type benzo-diazepine (BZP) antagonist (of which pinoline is possibly one form). [54]

Callaway[55] and Whitehouse[56] have both connected this rise with REM (rapid eye movement) dreaming sleep. Whitehouse notes that babies, foetuses and small children dream and sleep a lot *and* have higher melatonin levels. They are also considered to be more psychic.[57]

Whitehouse suggests that melatonin influences the serotonergic cells of the reticular activating system (RAS), which is considered by Norman Dixon to be central to the processing of subliminal perception (see brain diagram on p.97), and so could also be linked with the awareness of psi information. The RAS is right at the top of the brain stem and is the oldest part of our brain, equivalent to the earliest animals like earthworms. It is the basic awareness mechanism — one is either conscious or unconscious (as in coma), and this is the network that switches us on or off, so to speak.

Most people who sleep will have four dreams every night of their lives. The dreams occur every ninety minutes through the night. Callaway[58] has suggested that the periodicity of REM sleep is due to the interaction between pinoline and serotonin; that pinoline is functional in inducing REM dreaming, lucid dreaming and other naturally occurring ASCs, such as some meditative states. He proposes that the melatonin production beginning with the onset of dark affects the RAS to cause sleep onset. Melatonin production continues to a point where it reaches sufficient concentration for production of pinoline to begin. At a certain concentration this triggers dreaming because of its interaction with the serotonin creating the hallucinogenic DMT. At first there are only small amounts of DMT so the dreams are very short. This cycle recurs through the night, the increasing concentrations of pinoline acting to increase dream duration progressively until morning arrives.

Pavel *et al*[59] in an experiment with thirteen young boys found that administration of melatonin and of arginine vasotocin (also synthesised by the pineal) caused sleep onset within 15 minutes at 9 a.m.! Both also reduced the onset of dreaming (REM sleep latency) to approximately 20 minutes after sleep onset, whereas normally the first dream doesn't occur for about two hours.

*Psychosis, psychedelics and the pineal*

This is probably the most insubstantial part of my hypothesis, and I include it more for the sake of completeness than from satisfaction that I have researched it sufficiently, since I like the way in which it links with the rest. Anthropological studies suggest that some epileptics, and certain very sensitive people who in our society sometimes end up being labelled psychotic, are often chosen for training as shamans, since they are considered, as a result of their spiritual emergence experience, to be linked in some special way with the world of spirits and psi abilities. People in possession trance also sometimes appear to be in a state very similar to certain phases of epilepsy or of, what is called in the West, psychosis. And folklore of psychosis links it with the psychic.

Some research has found that admission to mental hospital varies with season and time of the month, and also that mental illness is more common the further north you go — i.e., into long light summers, and long dark winters. Arendt[60] has shown that those who suffer from depression can be helped by sun-lamp treatment — or alternatively anti-melatonin treatment in pill form — since melatonin levels rise in winter with the long nights, there being more hours of darkness during which melatonin is made. In Holland a long term study of elderly people found that daily treatment with whole-day bright light helped maintain good cognitive faculties and helped ameliorate depression. Giving melatonin in the evening helped create better sleep patterns, but made the emotional state worse, though this was mitigated to some extent if the person was getting bright light.[61]

Since one of the primary effects of melatonin is to make us drowsy, this makes us want to sleep more during the winter. If we insist on working the same hours as we do in the summer, this goes against the natural inclinations of the body, which affects us emotionally and this then leads to mental depression. Similarly, manic-depressives whose mental states alter with the seasons are being affected by a similar imbalance. Seasonal affective disorder (SAD) is really an exaggerated form of natural seasonal variation that we all experience. It is part of nature's way that we should have more energy in the summer and wish to hibernate in the winter. Those of us who are becoming more sensitive to the cycles in nature are becoming more and more aware of this variation over the year.[62] I suggest that we alter our work patterns to mega-flexi time!

Reiter[63] found that manic-depressives have a low melatonin concentration during suicidal episodes and a high melatonin concentration during manic episodes. Also when melatonin is given to people suffering severe depression this can exacerbate symptoms.[64] With chronic psychotics, the neurochemistry of their brains may be severely malfunctioning and is the physical level of what is often a complex of physical, psychological, family, social, psychic and spiritual disorders.

Melatonin has been shown to have an anti-epileptic effect, epileptics often being considered to be seers in olden times; and, together with thyrotropin, is involved in coping with persistent long-term stress. It is well-known that many shamans and mystics will undergo a rigorous training that puts them under persistent long-term stress, and during this time they tend to have mystical visions, out-of-body and psychic experiences.

Another interesting snippet of information is that the highest concentrations of serotonin have been found in the pineal glands of schizophrenics. A dysfunction of central serotonin metabolism in schizophrenia has been repeatedly suggested, and this would fit in with the cyclic seasonal nature of manic-depressives. Melatonin elevates cerebral serotonin, particularly in the mid-brain. Schizophrenics often exhibit sleep disturbances; insomnia may be the first symptom of a psychotic episode, while changes in REM sleep and in EEG patterns during sleep have been observed. Also arousal level, thus implicating the RAS, is altered in psychotics. All of these are connected with serotonin and melatonin.

Another possible link is that psychosis such as schizophrenia and manic-depression rarely occur in childhood, often first manifesting at adolescence. There may be a delay in sexual maturation, though in some patients an increase in sexual activity may occur during the acute phase. Onset of puberty is controlled by the pineal gland. At puberty there are enormous hormonal changes which affect us at every level of our being: physical, emotional, mental, and psychic. I look at this in greater detail below.

Smith[65] tested two groups of psychiatric patients: one showed lower than normal melatonin levels with a shallow circadian rhythm, while the second group was being treated with chlorpromazine (an anti-psychotic drug) and they exhibited very high melatonin levels, especially during the day. Chlorpromazine and haloperidol, which are both used in the treatment of psychosis, both decrease the activity of the pineal enzyme which makes the melatonin from serotonin.

Several people have suggested that the beta-carbolines may play some sort of role in psychosis since they have hallucinogenic effects because of the interaction with serotonin creating DMT, and appear to be synthesised together with melatonin, but tests of serum and CSF levels of pinoline were not significantly related to the psychopathological symptoms, nor did the values correlate with such variables as age, sex, subtype of schizophrenia or duration of illness.[66] In fact, Airaksinen and Kari consider that 'If the dopamine hypothesis is valid for any type of schizophrenia, the beta-carbolines may protect against such mental disorders rather than aggravate or induce them.'[67] But, diazepam, which is used to treat psychotics, has an antagonistic function to beta-carbolines. It really is too early yet to know for sure what is happening with regard to the beta-carbolines as they have only recently been identified as a normal constituent of the pineal.

Correlations between suicides and psychotic disturbances have been reported on several occasions to coincide with geomagnetic storms, which links this area with the section on geomagnetism[68] in the next chapter. The geomagnetic field has been found to affect pineal functioning.

More generally the pineal plays an anti-stress function and 'forms part of a broader neurohormonal feedback mechanism linking the stress response of the hypothalamus-hypophysis neuroendocrine complex'.[69]

Basically, the more sensitive you are the more unstable you are — as a candle flame is exquisitely sensitive to minute draughts and is inherently unstable, it is always flickering and wobbling about. In tribal societies sometimes a person who had an epileptic fit or psychotic breakdown as an adolescent was taken in by the shaman of the tribe and taught how to use the talents without becoming unstable. They were given a strict diet, daily regime, discipline and training to ground and stabilise them, so that they did not end up chronically ill and the psychic gifts that come with the hallucinations could be used for the benefit of the nation. In this way some people, who in our society are labelled as mad and fed pills that help not one bit in terms of learning to control the mind, through right training could become vitally useful members of the society. Let us hope that there are moves in this direction beginning now in our society. I do know that some psychiatrists, clinical psychologists and other people within the mental health world are beginning to be aware and to consider the psychic and spiritual aspects in relationship to mental illness. There are now web sites that can be accessed if in need.[70]

I also think that ashrams and monasteries may well enable sensitive people to find a safe harbour where the regular daily regime, simple diet and relatively stress free environment and spiritual training could benefit the person. For more detail about this topic please go to my web site where there is an article called "Walking Between the Worlds" which can be downloaded.[71]

As yet there is no clear brain functioning linked with psychological aspects of psychic functioning. However, the pieces of the puzzle are turning up and the picture is becoming clearer, and this validates a lot of folk and magical lore concerning psychic functioning.

Although it is at present unclear as to the exact mechanism within the brain of the pineal chemicals, and their exact effects on our state of consciousness and behaviour, this information could lead to a greater understanding of a physiological process underlying certain psi-conducive states of consciousness. This neurochemistry also links in with other aspects of yogic lore in a most intriguing way.

## The Kundalini — Is It Real?

Over the past few years there has been increasing interest in 'translating' the knowledge of one system into the language of another. This same process has been occurring in psychology with Charlie Tart's *Transpersonal Psychologies*[72] and Paranjpe's *Theoretical Psychology*[73] both examining Eastern philosophies and religions from a Western psychological standpoint.

Much of this translation has, of necessity, been in very general terms, since we have to clarify the overall picture first. By chance, I seem to be involved in this process from a rather different perspective. I have been researching a specific topic, the pineal gland, which seems to be generalising to a whole system, the chakras. The chakras and kundalini are considered to be inextricably linked. I must stress that what follows is still very much in an early speculative and exploratory stage. John Davidson also discusses this linking between scientific knowledge concerning the body and the Eastern concept of the chakras in his books.[74] His ideas about the chakras are different from mine. Which is the more correct conception, or whether they are just different ways of perceiving the same thing, I don't know. All I know is that I was led to the following information from studying the pineal gland, and it was only after several years of gathering knowledge on the pineal that I realised I was seeing what seemed to be an endocrinological chakra system. So this is a pattern that emerged, rather than my fitting the pieces into a preconceived theoretical system.

Most of my knowledge concerning kundalini and the associated chakra system comes from Swami Satyananda Saraswati.[75] He teaches the traditional northern Indian Yogic Tantric system of the chakras, a body of knowledge originating in the Vedas and Upanishads which are thought to be at least 4,000 years old. He states that the flowering of kundalini is in Sahasrara chakra which he equates with the cerebral cortex of the brain. He describes the chakras as manipulating centres similar to switches, each chakra being directly linked to Sahasrara, and that in the brain there are layers and layers of potential functions which are dormant. Awakening the kundalini is the awakening of all these potential psychological, psychic and spiritual functions.

Satyananda is concerned to teach kundalini in a very practical sense so that we in the West can find the teachings more accessible. I have found that his teachings link kundalini and the chakras with psycho-neuro-endocrinological knowledge of the West, which links in with everything I have just said about the pineal gland.

*The yogic chakra system*

The yogic chakra system as explained by Swami Satyananda consists of seven chakras which are normally depicted as a sort of 'spinal column' with three channels which interweave, the crossing-points being the sites of the chakras. These three channels are called sushumna, which seems to be analogous to the central nervous system; ida, which seems to be like the sympathetic aspect of the ANS; and pingala, which links with the parasympathetic aspect of the ANS.

The root chakra (Muladhara) is situated at the tip end of the tail bone in the perineum and is linked with certain aspects of the urino-genital system. The second chakra (Swadhistana) is sited in the sacrum at the root of the spinal cord which is just below the small of the back, and is linked with other aspects of the genital system. The third chakra (Manipura) is sited in the spine behind the navel and is linked with the solar plexus. The fourth chakra (Anahata) is sited in the spine behind the heart and is linked with the cardiac plexus. The fifth chakra (Vishuddi) is sited in the neck and is linked with the throat. The sixth chakra (Ajna) is sited in the pineal gland and is considered to be the 'command' chakra, and the seventh chakra (Sahasrara) is the crown chakra at the crown of the head. These chakras are considered to be important points for the channelling of consciousness, energy nodes linking the physical with the spiritual. They have been adopted quite widely into popular usage in the West, partly through the Theosophists (whose teachings differ from those of Satyananda), partly because of their obvious correspondence with the Kabbalistic Tree of Life, and partly because of the intense interest in Eastern spirituality that emerged in the 1960s.

Each of these chakra points can be understood not only in physiological terms, but more importantly in terms of emotional, mental and psychic energy. By awakening the chakra we not only become aware of the emotional energy connected with that part of our beings, but by transmuting the emotional energy we become more spiritually aware beings.

## The pineal gland: Ajna Chakra

I have found that Swami Satyananda's concept of the pineal gland as both the psychic chakra and the command chakra has a sound neurochemical basis. I have just gone into the psychic chakra aspect in some detail, which is linked with the neural function of the pineal. The command chakra function of the pineal gland is connected with its hormonal aspect maintaining the biological clock, both on a daily basis according to the sun, on an annual basis according to length of day, and on a lunar basis as well.[76] In doing this it also controls, together with the pituitary gland, the regulation of some of the hormones from the endocrine glands throughout our body. And each of these endocrine glands is located at a chakra point. Most people have heard of the pituitary gland, which is often known as the 'master gland' in that the hormones it makes exert a controlling effect on all the other glands. As said earlier, we can think of the pituitary as being an 'on switch' and the pineal as being an 'off switch' in that it works with the pituitary in keeping the hormone output at the level needed.[77] Therefore it is the command chakra.

## The thyroid glands: Vishuddhi Chakra

According to Satyananda, Vishuddhi chakra is located in the throat and is the centre of 'the nectar of immortality'. It is connected with the sense of hearing and thus with the ears, and of course with the vocal cords and with self-expression.

The thyroid and parathyroid glands are found in the throat. Neurochemically, the thyroid is under the inhibitory control of the pineal gland, so that removal of the pineal results in thyroid enlargement and increased hormonal secretion rate. Synthetic melatonin has the effect of inhibiting iodine uptake and the secretion of thyroid hormones, and, given at the correct times, can reproduce the daily and annual light-dark cycles, or circadian rhythms, since iodine uptake decreases during the night. Evening injections of melatonin are more effective than morning ones, because the time of day when hormone supplementation is given is a significant factor. Also worthy of note is the finding that the effect of synthetic melatonin on the secretion of thyroid hormones decreases after puberty.[78]

Experimental evidence indicates that the pineal is under feedback control by the glands which it influences. Pineal cells respond to thyroxine, which is one of the hormones made by the thyroid, the response being particularly strong at night, showing the influence of the circadian rhythm once again.[79] Too much thyroxine reduces pineal activity so that less melatonin is made so that the person has, for example, difficulty in sleeping.

The thyroid regulates the metabolic rate via thyroxine and tri-iodothyronine, which means that it controls how fast the body runs. Thus an overactive thyroid causes the heart to beat fast, one becomes thin, sexual desire increases, and the mind works overtime; while an underactive thyroid has the opposite effect, in severe cases resulting in mental subnormality. In other words hormones work at a physical level with emotional and mental consequences. Thyrotropin (TSH) made by the pituitary under the control of thyrotropin releasing factor from the hypothalamus is, together with melatonin, involved in coping with long-term stress.

Stress is intimately connected with metabolic rate, heart rate, an overactive mind, and also with age, as an older person cannot cope with stress as well as a younger person. Long-term stress is very different from short-term stress (which is dealt with by the adrenals) and it is interesting that Ajna, Vishuddi and Manipura are all concerned with stress — which also affects the heart. Relaxation is the first step in meditation; slowing down, letting go, releasing the stress, stilling the endless internal chatter - when the mind *just won't stop* wittering away in circles (the beta-rhythm mental chatter) —

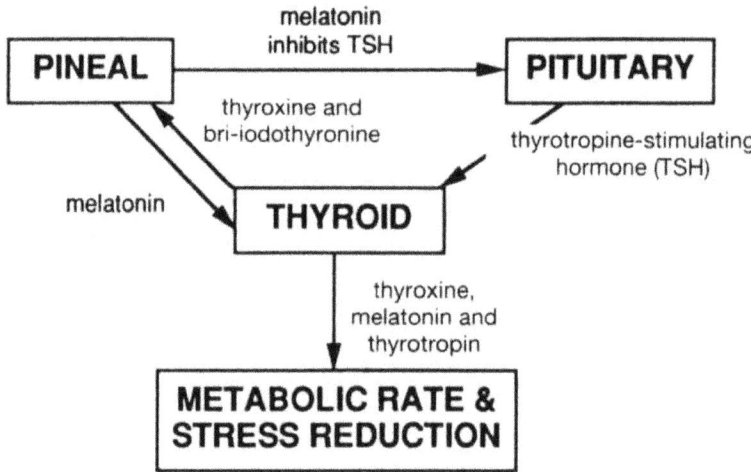

***Pineal and thyroid links (Illustration courtesy of Serena Roney-Dougal)***

which is one of the worst aspects of long-term stress. These are all the negative aspects of Vishuddi, and we learn through meditation to overcome these aspects and so to become peaceful, still, calm and to live to a ripe old age. To me this is merely another way of saying that the thyroid is connected with immortality. I say this because, as is exemplified so well by the Chinese symbol of immortality, the tortoise, the slower you go, the longer you live. Also, statues of Chinese immortalists depict men who are immensely fat and jovial, suggesting that they are totally unstressed with very low metabolic levels. And the yogis say that it is perfectly possible to regulate the functioning of the endocrine system, thus learning how to control one's metabolic rate. This has been confirmed by biofeedback research.

Stress plays an important part in dictating whether or not people display symptoms from an allergy reaction. The overloading, or overstress, may come about because there are too many germs about, or too much pollen, or it may arise from emotional stress, overwork, over-indulgence, or even a reaction to the electromagnetic environment, which I discuss in greater detail next chapter. Allergy sufferers have a lower threshold of tolerance and may be very sensitive to electric fields. The response of patients to electromagnetic fields can be very similar to their chemical or biological allergy response. Water can be used to treat allergies by exposing it to electromagnetic fields at certain frequencies specific for the patient. Water holds potency for effectively treating the allergy for up to two months.

Most people seem to have a threshold level of tolerance to stress. In women who are just about coping, staying healthy, unstressed, etc., the onset of the menstrual period may tip them over the threshold and they become ill. If this is the case, then these people are well advised to take a day off from work or from the children, the day before the bleeding starts, and so prevent illness or freakout.

*The heart centre: Anahata Chakra*

According to Swami Satyananda, Anahata chakra is concerned with will and with feeling, touch, the skin especially the hands, manifesting in such arts as painting, poetry and music, which are obviously aspects of heart energy. He also states: 'That in the present age, at this moment, we, the people of this world, are passing through a phase of Anahata chakra'[80] This means that this chakra is beginning to awaken in us, whether we are working on it or not. Anahata chakra is the lowest of the three non-earthly or spiritual chakras: Anahata, Vishuddi and Ajna.

*Where Science and Magic Meet*

To me, the feeding of a new-born baby is the essence of the meaning of the heart chakra, the nurturing, giving, loving, touching element. Such blessed joy to love, to give, to care, to share, to touch with others — that wonderful soaring emotion of the heart. Prana, the breath of life, is love. When the heart centre is open then one can do no harm to others any more, because love is the base of all one's actions. It is very important, at this time, to learn to love yourself, because then you no longer feel depressed and negative and upset and bitchy and mythering and moaning and churning away inside with misery — because if you love yourself, you will learn to love the earth and all upon her — so the negative aspects of the heart chakra, such as jealousy, can finally be fully transmuted.

Touch is connected with the hands, and the hands have a really interesting nervous system in that they sweat according to emotional state rather than temperature. They are connected to the autonomic nervous system and the limbic system in the brain, the emotional centre, and if you are tense the sympathetic nervous system makes your hands sweat, and if you are relaxed the parasympathetic nervous system stops the sweating. If you are severely emotionally withdrawn then the hands dry up completely and this is as bad as being continuously emotional and nervously sweaty. In other words, our hands are connected with our hearts,

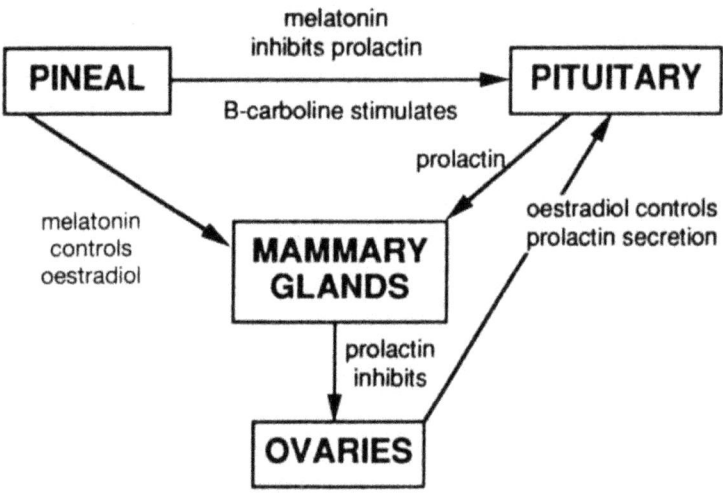

***Pineal and heart centre (Illustration courtesy of Serena Roney-Dougal)***

with our emotions.

Some have suggested that the endocrine gland connected with this heart energy is the thymus. The thymus is most active in children and is concerned with the immune system, which is under such stress in these polluted times. Recently it has been found that the pineal is implicated in regulating the immune system,[81] so the thymus could thus be seen as the heart chakra protection system. However, what I have noticed in my research into the pineal is that it is linked with a hormone called prolactin which is connected with pregnancy, the mammary glands and the mother's production of milk. Men too have prolactin, in fact all beings right from a 3-month foetus in the womb make prolactin so it is involved in something as well as lactation.[82] What I have found is that it is connected with manic-depressive disorder, often known as affective disorder. In this case affective is a technical term for emotional. So prolactin is involved in keeping our moods, our emotions regulated. Enough said! This is the physical aspect of heart chakra! Prolactin is linked both with puberty and the circadian rhythm and is felt by men as well as women — to keep both men and women in tune with the cycles that women are so obviously in tune with through menstruation.[83]

Most of the research with prolactin has been with animals. All ruminants (e.g. cows, sheep) show a marked seasonal fluctuation in plasma prolactin concentration, i.e. high in summer and low in winter, this fluctuation being controlled by the pineal gland, even though prolactin is made by the pituitary, because it is the off switch which regulates how much is made. (This is an important point for many things in life – a 'no' is often far stronger than a 'yes' in determining if something can happen or not. One person being negative can taint everything in a group endeavour, or in the family home.) This inhibition of prolactin secretion in ruminants inhibits implantation of the blastocyst during the winter, so that the foetus does not implant into the womb until spring time, even though mating and fertilisation occurred in autumn.[84] Prolactin secretion is also controlled by the ovarian steroids, its level being modified by the fluctuating oestradiol levels, which in women is a lunar cycle, so if any of you have tender breasts once a month this is possibly the reason why. Prolactin production normally peaks in women just prior to ovulation, when the secretion of oestrogen is at its maximum.

Most of the above-mentioned connections are more related to melatonin than to the beta-carbolines, since melatonin inhibits prolactin secretion, but plasma prolactin levels increase when the

*Where Science and Magic Meet*

***The Chakras (by Monica Sjoo, illustration courtesy of Serena Roney-Dougal)***

pineal beta-carboline is given.

Thus heart energy — love — fluctuates with the seasons (as so many of us know!) and is intimately linked with our sexuality, not just in terms of actual lovemaking, but more particularly in relation to lunar menstrual cycle and to pregnancy. I feel sure that every woman who has been pregnant will resonate with this. I know that during my pregnancy, after about the sixth month, I became so soft and open and flowing with pure heart energy.

It is important to remember here that hormones have psychological and emotional effects as well as their more obvious physiological effects. Just because men do not bear children does not mean that they cannot feel pure heart energy; they can and they must if they are to start caring for themselves, others and this beautiful planet in the truly loving way that is vital for life at this critical time. When a man hurts a woman it is the same action in essence as when men chop down the rain forests or damage the planet in other ways. The essence is that of damage, hurt, attempt to have power over, to control, to use and to abuse. This is the essence of a man who is denying his heart energy, with regard to woman and to our planet at this moment in time. It manifests in many varied ways that differ with each person, but the essence is the same — a lack of loving, of pure open heart.

*The solar plexus: Manipura Chakra*

Manipura chakra causes old age, decay and emaciation by burning up what Swami Satyananda calls the nectar of immortality. It is also connected with the sense of sight and the eyes and it is the organ of action and hence is also connected with the feet. The solar plexus is the locus for our 'gut feelings' about people and situations, and is connected with digestion. Manipura is the uppermost of the 'earthly' or base chakras.

The adrenals are the endocrine glands related to the solar plexus. Most people know these as the 'fight or flight' glands in that adrenaline is produced when we are in a stressful situation and we burn up our body energy in order to cope with a crisis; adrenalin is the hormone of action.

The pineal is connected with the adrenals, and in particular with their hormones adrenalin and noradrenalin in many ways. The adrenals comprise two parts: the cortex and the medulla. The cortex secretes glucocorticoids such as cortisol and corticosterone, on a rhythmic light-dark cycle stimulated from the pituitary which is stimulated by CRF from the hypothalamus. These are stress-related hormones. The cortex also secretes mineral corticoids, such as aldosterone, which controls salt and water balance by stimulating conservation of salt by the kidneys. The cortex also produces small amounts of the sex hormones, oestrogen and androgen. The medulla secretes adrenalin. Melatonin from the pineal inhibits the release of all of these hormones, thus controlling our physical level of immediate short-term stress — as it does with the thyroids on a long-term basis. Thus melatonin is *the* hormone of relaxation.

Melatonin is actually found in the gut as are the beta-carbolines.[85] Beta-carbolines interact with adrenalin and noradrenalin uptake and outputs as well as with corticosterone secretion, thus interacting closely with the adrenal functions. Removal of the pineal gland causes enlarged adrenals.[86] The significance of this enlargement of adrenals — as with the thyroids, when for some reason or other there is no pineal — is that the inhibitory effect on these glands

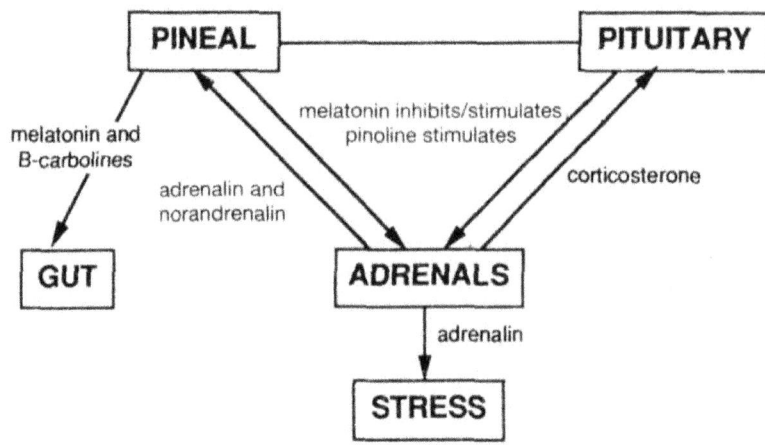

*Pineal – Adrenal Link (Illustration courtesy of Serena Roney-Dougal)*

has been removed so that they work overtime. And, as a result, one burns up! A child born with no pineal will be sexually active by the age of four or five and normally dies before reaching the teens. This can be understood in the spiritual as well as in the physical sense.

Chronic administration of small doses of beta-carbolines causes the weights of the pituitary, ovaries and uterus to decrease, and the adrenals to increase. Plasma corticotrophin, aldosterone, beta-endorphin and prolactin concentration all increase with administration of pinoline.[87]

Of interest here is that not only melatonin is found in the gut, but also pinoline, though why we need a beta-carboline in our gut I cannot imagine, unless it is a chemical link with the psychic aspect of our solar plexus. Perhaps this is a link with our gut-reactions to events, our basic, instinctive, intuitive level of reaction, one that can often be trusted over and above the more intellectual 'head' reaction. It also perhaps explains 'indigestion-type' dreams and nightmares and possibly has an earthing function in our response to psychic events.

### The root of the spinal cord: Swadhistana Chakra

Swami Satyananda states that Swadhistana is connected with all 'the phases of the unconscious', like subliminal perception. 'Swadhistana is made up of all the rubbish which you never wanted, which you never needed, which you never desired but which got in'.[88] It is the generative aspect of sexuality embodied by the womb in women, and by the hormones progesterone and testosterone. These hormones are central to the development of the secondary sexual characteristics — that which is the essence of man or woman; they define our gender, ourselves as sexual beings in terms of mother or father, boy or girl. This essence of masculinity or femininity becomes easily distorted, as the patriarchal system demonstrates, because it is so unconscious, so instinctive, such a basic part of our selves. We are defined as male or female from the moment of birth and are programmed with all of our society's attitudes. It is impossible to escape this conditioning and it is so deep that we do not realise its effect on us, although living in another culture which has very different gender behaviour can help us to become aware of this deeply unconscious part of our psyche.

The glandular connection of Swadhistana is with the testes, ovaries and related systems so that to some extent it overlaps with Muladhara Chakra. I feel strongly that we lump together too much in our understanding of sexuality and that is why I am not clear in my separation of the two lower chakras in endocrinological terms. In *The Wise Wound,* Penelope Shuttle and Peter Redgrove[89] distinguish two clear cycles in women and I think that the menstrual cycle and progesterone is connected with Muladhara and the ovarian and oestrogen with Swadhistana, but how the male hormones link with the two lower chakras I am not too sure.

Among its hormonal effects, melatonin mediates the effects of light on reproductive events in seasonally breeding mammals, by inhibiting ovarian function and preventing blastocyst implantation during winter. It is also implicated in puberty onset in humans.

*The coccygeal plexus: Muladhara Chakra*

Muladhara chakra is the root chakra, intimately connected in the male with the prostate gland and testes, and in the female with the uterus. This chakra is connected with the sense of smell, the nose and the earth element. Working on this chakra releases suppressed emotions and unconscious memories, and causes extreme swings in mood. It is the seat of kundalini, and has obvious and direct connections with sexual energy in its most earthy aspect. This is the

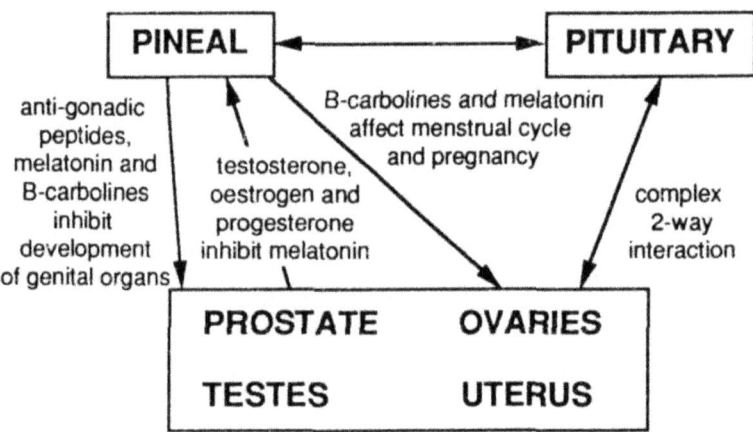

*Pineal – Gonads Link (Illustration courtesy of Serena Roney-Dougal)*

most base of all chakras, the one most closely connected with our instinctive animal self.

The pineal and the gonadal system interact extensively. Satyananda considers that there is a special connection between Ajna chakra and Muladhara, and there are certainly extensive connections between the pineal gland and the gonads. The pineal synthesises antigonadotropic peptides. In their turn the gonadal hormones testosterone, oestrogen and progesterone inhibit the biosynthesis of the pineal hormone melatonin. As we have already seen prolactin secretion is inhibited by ovarian steroids.

Thus, the pineal makes a hormone which inhibits gonadal development and regulates the onset of sexuality, both at puberty for humans, and on a seasonal basis in animals. There is a fall in plasma melatonin associated with human pubertal development.[90] In fact it is thought that it is this fall in melatonin which triggers puberty. So the pineal is responsible for maintaining our child status for so long which is linked with the extensive period of learning that we humans enjoy, compared with animals who become sexually active much earlier and rely more on instinct and less on learning. And the pineal normally becomes calcified at puberty.

This is a rather intriguing aspect of the pineal which I merely note here. By 'calcified' I mean that crystals of calcite solidify in the pineal. This is no way inhibits the functioning of the pineal. But it does mean we have a crystal in the centre of our head which is piezoelectric, and could possibly be involved in the exquisite sensitivity of the pineal to the earth's magnetic field, of which more in the next chapter. Our bones and teeth are made partially of calcium, and there are suggestions that this may be some sort of electromagnetic detector.

The pineal night-time melatonin concentration decreases progressively during the oestrus cycles, with a massive increase at ovulation and peak values during menstruation. Women also show a 28-day melatonin rhythm. Melatonin seems to be 'taken up' by the ovaries, testes and uterus. Continuous light, which causes a decrease in melatonin production, also causes a decrease in ovarian melatonin concentration, while injections of melatonin result in smaller testes.[91] And some are now becoming aware of variations with the moon cycle over each month as well.[92] Another link with lunar cycles is shown by the 25-hour circadian rhythm established if one is kept in constant darkness above ground. This is discussed in the next chapter.

Since continuous light decreases melatonin production, and melatonin regulates the onset of puberty, it is possible that continuous light causes earlier onset of puberty. However, our house lighting is, in most cases, below the level that inhibits melatonin production as is shown so well by the use of artificial sun lamps to help those suffering from depression. This is not to say that we are not affected by our artificial lights because I think we are, but exactly how is not yet clear.

Cyclicity in mammals is seasonal rather than monthly, according to sun cycles rather than moon cycles, and certain animals become impregnated in autumn at the end of the long daylight hours. As mentioned with regard to Anahata chakra, through the shortened days of winter, melatonin inhibits pituitary prolactin secretion and hence implantation of the blastocyst which is held in a sort of suspended animation until the spring and the return of longer days. While few clinicians would accept a seasonal basis for reproduction in humans, older epidemiological data, and data more recently derived from conditions of borderline fertility, both support a seasonal change. The exact link to melatonin is as yet unestablished but seasonal changes in plasma melatonin have been described.[93]

It has been shown that the foetus can synthesise and store melatonin. Reppert *et al.*[94] believe that melatonin from the mother

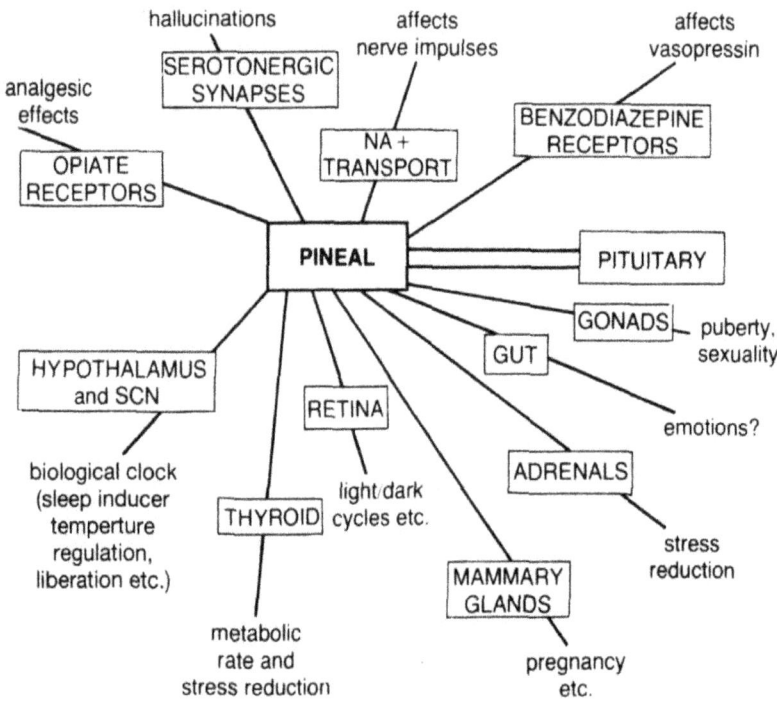

*Pineal – neurochemicals and endocrine links (Illustration courtesy of Serena Roney-Dougal)*

crosses the placenta and drives foetal rhythms, since the biological clock works rhythmically even before the optic nerves are fully developed (as I mentioned earlier the pineal has its own nerves independent of the optic nerves). During pregnancy the diurnal rise in plasma melatonin appears enhanced as pregnancy progresses, supporting the idea for a role in the maternal pineal entraining foetal rhythms. Melatonin levels in both mother and father are exceptionally high at birth.

There are similar findings between the gonads and beta-carbolines, such as pinoline. Thus concentrations of beta-carbolines seem to decrease with age. Beta-carbolines fed to young male animals inhibited the development of the genital organs. In female rats in high doses it made the dioestrus longer or totally abolished the oestrus period. In smaller doses the weights of ovaries, uterus and pituitary decreased.[95] All these research findings indicate that beta-carbolines are also connected with certain aspects of sexuality.

Formal laboratory studies of the relationship between sexuality and psi are virtually nonexistent. Apart from studies with children versus those with adults, there are two that I have found that look at psi in relation to the menstrual cycle — with very inconclusive results.[96] There is, though, a widespread folklore on the relationship. Children and celibates are almost universally those chosen as temple seers and prophets, the oracle at Delphi being an excellent example of this. Yet Tantra suggests certain sexual practices enhance spiritual development. Also poltergeist studies seem to suggest a relationship between puberty and adolescence and the chaotic psychic effects called poltergeist.[97]

The wavelength and intensity of light which influences pineal melatonin biosynthesis is not known for humans, but there is a suggestion that the pineal reacts differently to different wavelengths of light.[98] There must be some deep organic reason for our association of sexuality with red light, which has been the whore's colour since the time of Babylon and is the colour worn for marriage by women in India. Animals kept in red light show increased gonadal development, whereas those kept in green light don't. Light and the colour of light may have a pronounced effect upon sexuality, possibly through some impact upon the pineal (since it is the only hormonal gland directly affected by light). And why is the red light used in the Ganzfeld, which appears to be a method for inducing the hypnagogic state (drowsiness) and is psi-conducive.[99]

## Conclusions

My knowledge of the endocrine system is still meagre. The neurochemists have only recently isolated the beta-carbolines from the pineal and are still learning about melatonin and serotonin. Thus our knowledge about the chemistry of our body-mind system is still rudimentary. However, partial as our knowledge may be, it does fit together with what the yogis, 'scientists of the subtle mind', tell us about the yogic chakra system. Our disciplines, apparently so different in language and method, appear to corroborate each other.

I must admit that I find it absolutely fascinating that a spiritual system as exact as that of Kundalini Yoga can be found to correspond so precisely with a science as complex as psycho-neuro-endocrinology. They talk in different languages, but the translation from one to the other seems to give greater depth to each. To understand that my root chakra is connected both physically and psychically with my deepest emotional and sexual urges; that Swadhistana is connected both physically and spiritually with my deepest level of self as a woman; that Muladhara is my gut reaction, and so on, helps me to awaken and to develop those aspects of myself. My body, my emotions, my mind and my spirit all become clearer as a result.

# 5. EARTH MAGIC: SENSITIVITY TO THE EARTH'S MAGNETIC FIELD

> Any form of explanation that ignores some of the evidence, that leaves out or rejects those things that do not fit in, ends up by explaining things away, which is no pathway to a better understanding.[1]

## Electricity

Most of the information which follows is culled from several fascinating books: *Cycles of Heaven* by Guy Lyon Playfair and Scott Hill, *The Electric Shock Book* by Michael Shallis, *Needles of Stone* by Tom Graves, *Spiritual Dowsing* by Sig Lonegren, *The Opening Eye* by Frank McGillion, *Earth Lights* by Paul Devereux and *The Secret Language of Stone* by Don Robins. All of these books deal with our interaction with this earth in an electromagnetic and/or psychic sense.[2] These books are now 20 years old, but the information in them is still essentially correct and sadly our knowledge has increased only in the area of the effect of the magnetic field on our bodies and minds in the intervening time.

Matter is a complex interweaving of charges and fields. Moving electric charges induce a magnetic field and inside all materials are moving electrons - if the atoms are aligned as in metals then the magnetic fields become aligned and you get magnetic effects. *All matter is both electric and magnetic.*

Magnets get their properties from electric charges in the atoms and electrons moving in circles inside them (like whirling dervishes!). Similarly a moving magnetic field creates an electric current; thus we get such useful items as the dynamo for bicycle lamps, the generator, the electric motor and the transformer. In all of these the electricity is generated by rotating a coil of metal between the poles of a magnet, or rotating a magnet around a stationary coil. This means that around the wire leading to the lamp in your room by which you are reading this, or to the stereo, or to any other item of electrical equipment which is on at the moment, there will be a magnetic field that changes its polarity fifty times a second.

Light is an electromagnetic phenomenon. Light is formed when electric charge moves — it is a moving wave of electric charge. Different colours are just light moving at different frequencies, a frequency being how frequently the wave of energy cycles up and down each second, and on out of the visible spectrum into ultra violet, radio waves, microwaves, X-rays, etc. Any change in the motion of electrons gives rise to radiation, for example the light from your light bulb; photons of light are created by the electrons. Light and matter are different aspects of the same energy.

> After all light is simply the propagation of changing electromagnetic fields, so maybe the electron is nothing more than a compact and self-contained electromagnetic field which we observe as a particle containing an electric charge.[3]

In the photoelectric effect, ultraviolet light releases electrons from the surface of a metal; in other words, the energy of a single photon of light releases an electron. Thus, light interacting with materials can release an electric current (this effect underpins TV screens and phtotovoltaic solar panels). We shall come across this again when we turn our attention to stone circles and the effect of the rising sun on the stones.

*Some basic electrostatics*

Let us look more closely at these strange properties of electricity and magnetism since they are linked with many a magical idea for which we have as yet had no down-to-earth explanation, and my contention is that all magical ideas have very real psychological and physical earth-based reasons.

For instance, could the supposed magical properties of silicon, amber, quartz and other crystals be related to electrostatic properties? Amber and ebony attract feathers, wool, silk, etc., when rubbed. Rubbing amber creates static electricity. The Greek word for amber is elektron and Gilbert, physician to Queen Elizabeth 1, coined the word electricity for this 'power of attraction'. Ebony and amber which do not transmit the electric energy but retain it are called insulators.

Electricity flows through a conductor like a copper wire but the material resists the flow and so you get heat as in an electric fire, or light as in a light bulb. In an insulator the resistance is so high that the electricity cannot flow and becomes static. The electric charge on amber or ebony can be transmitted through the human body, through water and through metals because these are conductors of electricity. Metal will attract when rubbed by fur provided it is held by a glass or amber handle.

Silica, of which glass is made and which is the basis of computers and other modern tools that use microchips, is a semiconductor — half-way between an insulator and a conductor — and so current in it can easily be switched on and off, and electrical circuits can be created within the material itself. Silica is sand and is found in nearly all stones. In old crystalline granites such as make up many of the stone circles, the stone is formed of a mixture of different types of quartz, which is crystalline silicon dioxide.

Very high charges can be built up in amber and ebony which, when discharged, give off a lightning-type spark and crack (light and sound). Those of you who have gas cookers and have 'piezoelectric lighters', or those who have electronic lighters, are using this same effect of static electricity from a quartz crystal giving off a spark and crackle when discharged by pressure, when you squeeze the trigger. Hair becomes charged if touched by an electrostatic item and stands on end. It also stands on end if you go into a haunted house, suggesting that the place has some unusual electrostatic properties.

This is the physical, electrical basis behind the magical use of amber and quartz. Those people who are into using crystals would do well to learn about their strange electrical properties and how these interact with humans and the earth. Gemstones and crystals have electrostatic properties. I do not wish to discuss them too much here because insufficient research into the properties of crystals has been done, and I thoroughly dislike all the New Age crystal mania. If anyone knows of any good research being done into crystals I would love to hear from them because I am very intrigued by them and would love to know more. Many of the so-called crystal therapies are based more on hot air than anything else, and so I totally distrust them. However, there is a basis to all of this excitement about crystal therapy and that basis is that crystals are piezoelectric, which means that any slight mechanical pressure or stress on the crystal will cause it to discharge electricity.

Thus, crystals are very sensitive to pressure and also to psychokinetic effects as has been shown so well in research by John Hasted and David Robertson.[4] In this research which started off as metal-bending research with young children, they found that crystals can be used as 'strain gauges' measuring the amount of psychokinetic energy exerted on a piece of metal. These piezoelectric strain gauges pick up the stress in the metal from the psychic energy by which the children are trying to bend the metal. The piezoelectric crystals translate this mechanical stress into electrical energy.

New Age crystal therapy, and the age-old magical ideas surrounding gemstones has always attributed different properties to the different gems. Much of this is linked with their colour. Colour is electromagnetic in nature, since it is defined by the wavelength of light. Very little research has been done with colour although I know that some research into the healing properties of colour is being done, but have not heard yet of any clear results. I do know though that colour affects us psychologically very profoundly and I do know that the chakras are associated with energy and with colours. If anyone knows of good research in this area please let me know because I am very interested.

Lightning is another electrostatic phenomenon and occurs between electrically charged clouds or between clouds and earth even up as far as the ionosphere. Lightning represents a discharge of static electricity and is enveloped by its own magnetic field. Other associated sounds, such as clicks, swishing, hisses, are more likely to be the effects of electromagnetic energy impinging on the eardrum and being heard as sounds. These sounds are very similar to the noises some people hear in their ears when in a house with electricity, often heard as a constant high pitched humming.

Lightning always follows the route of lowest electrical resistance, and so can follow some very strange paths. It often strikes preferred trees and places. For instance, there is ten times more lightning over land; and the electrical resistance of oaks is very low and so this is why they are often struck by lightning. Could this be at least partially one of the reasons why oaks are held in such honour by pagan peoples such as Druids? Our psychic sensitivity appears to be strongly linked to electromagnetism. Static electricity is also attracted by pointed objects, and there is a suggestion that church steeples were specifically constructed so as to enhance the electrostatic properties of that place. There is a connection here with the wizard's pointed hat, and with the wearing of horns since they are pointed also.

Lightning is also frequently connected with Fortean type events[5] such as inexplicable accounts of so-called ball lightning and what is known as St Elmo's fire, which is a glowing discharge around, for example, ships' masts when surrounded by a concentrated and intense electric field. These strange electrical anomalies have been connected with both UFOs and fairy lore. Lightning has also been connected with areas that are associated with UFOs, the correlation being a negative one, which means that an area which has a lot of lightning is unlikely to have reported UFO sightings. Why, I do not know, but it is clearly something to do with the underlying electromagnetic energy of the area. Sometimes people who have an electric shock, e.g. are struck by lightning, develop psychic abilities afterwards.[6]

### The negative effects of electricity

Extremely low frequency electromagnetic fields (ELF-EMFs) induce electric fields inside the body. These fields represent the internal exposure one receives.[7] A whole range of endogenous electric fields exist within the body arising from normal physiological activity. These fields will combine with any field induced by external exposure to electromagnetic fields.

Some people are supersensitive to electricity - both static and current. They get headaches, fatigue, irritation, allergies, plus more bizarre things. Some people disrupt electrical equipment, pop light bulbs, destroy insulation, and are liable to be struck by lightning.

It is mainly women who are sensitives and there are often psychic occurrences connected with this type of sensitivity. Several poltergeist cases have involved electrical craziness. For example, in Rosenheim a young girl psychically affected the telephone, the light bulbs and other electrical equipment in the office where she worked, so that eventually she had to leave work.[8]

Overhead power cables can seriously damage your health. Common complaints are dizziness, feelings of weakness, poor concentration, headaches, nausea, insomnia, asthma and skin rashes, blackouts, heart attacks, leukaemia - wet and windy weather makes it worse as does pre-thunderstorm weather. This could well be due to an accumulation of positive ions, static electricity.

For example, Shallis refers to the fact that Dr Perry has correlated suicide cases with unusual electromagnetic fields in the home, and found that the incidence of suicide was 40 per cent higher in high-field environments, composed not just of overhead power lines, but also underground cables, household wiring, domestic appliances and the type of construction of the dwelling. Stone helps to shield us from the negative effects of electricity whereas concrete tends to enhance electric and magnetic fields.[9] This finding is of crucial importance when we come to consider tumuli and barrows, orgone and Faraday chambers.

Chromosome development abnormalities occur with animals in laboratories which have conditions of increased magnetic and electric fields. It is being suggested by some researchers that people working in microwave electric field environments have lower sperm count, have children with Downs syndrome, have lowered immune response systems, have altered insulin and other hormone production.

More recent studies have suggested that exposure to ELF-EMF may affect human health, because of the increased incidence of certain types of cancer, depression, and miscarriage amongst people living or working in places exposed to these fields,[10] and laboratory findings of in vivo and in vitro bioeffects such as increased oxidative stress, using cultured cells and human erythrocytes.[11] There is also evidence suggesting that environmental and artificial magnetic fields have significant impact on cardiovascular systems of animals and humans.[12] However the findings are contradictory, some showing an association between exposure to power-frequency (50–60 Hz) magnetic fields and increased health risks, but others showing a beneficial effect. Lower cholesterol level have been found in studies with rabbits, rats and mice.[13]

Electromagnetic fields act as biologically stressful agents particularly when one is already suffering from emotional or mental or psychic stress. I have already discussed the integral relationship between hormones, stress levels and the pineal gland. Later I shall show exactly how sensitive the pineal gland is to magnetic fields. Thus it is no wonder that we are so strongly affected by electromagnetic fields when we are feeling stressed. Dr Cyril Smith who has researched this area extensively considers that the electrical effects on the body are mediated by the endocrine system — the chakra link with the body.

And yet electricity has created the global village in which the whole planet is linked as one. Never before have we been so closely linked together on a planet-wide scale. The day that the photograph of planet Earth as seen from space was given to us all was a key moment in the shift of our human mentality on a global scale. We are now all of us intimately interconnected — through electromagnetic means. The magical means by which I write these words is a piece of silicon chip called a computer. So we benefit highly from our electromagnetic society. What we need to do is minimise the negative side effects and we can only do that by understanding fully what is going on — at every level.

## The healing properties of electromagnetic energy

Shallis quotes research that suggests that artificially generated bioelectric currents can help new limbs to regenerate in rats, as they do in salamanders. He considers that rhythmically pulsing magnetic fields, such as those that are used by physiotherapists, encourage the body's own electrical energy to operate, regenerate the necessary nerve impulses and restore the organ to its proper action. More recent research with stressed rats has found that extremely low frequency electric fields (ELF-EMF) might have some influence on lipid peroxide metabolism. We have already seen in the previous chapter that this mechanism is related to melatonin's cancer prevention system.[14]

Thus magnetic fields can help to treat cancer because cancer is a result of the bioelectric field being disturbed and cells growing out of control. Normally cells grow according to the specific body pattern but in cancer the pattern breaks down. Some electrical sensitives and allergy sufferers find magnetism helps them recover. The suggestion here is that our electric fields define our body pattern and when this malfunctions so we have diseases like cancer. I wonder if the growing increase in our use of electricity since Victorian times is, together with the ever-increasing pollution of our air, earth and water, related with the rise in cancer since then. Thus, electrical energy can both heal and harm, depending on various factors such as intensity or frequency. Please note this point as it is of vital importance in relation to recent discoveries concerning sacred sites.

In general, sensitivity levels seem to relate to gender in that boys are more allergy-sensitive than girls, and women more so than men. In fact there are changes dependent on hormonal changes such as during the menstrual cycle. Pregnancy and childbirth are times of critical sensitivity, which is analogous to the findings with regard to the pineal gland and melatonin production. With men times of stress are critical times. As old age approaches, allergy sensitivity dies down. Allergy, and electrical, sensitivity seems to be a partially inherited trait. When some people are suffering from their allergy they often seem to exhibit psi abilities: clairvoyance, telepathy, and healing abilities. Michael Shallis considers that their lowered thresholds somehow make them open to another level:

> As the body's thresholds become lowered, so the person seems to open up to subtle, low-level electromagnetic changes around them ... the more the EMF [Earth's Magnetic Field] is modified and complicated by local factors - such as minerals in the rocks, water and additions to the environment by man-made structures - so it becomes less puzzling that places affect the ... reactions of these highly sensitive people.[15]

In general, the electrical field strengths that affect people are very low. The electric human being presents an intricate, complex and subtle set of electric potentials that interact with each other and with external fields. In general the weaker the field the more likely it is delicately to alter the fields in and around a human subject. The direct result of such an interaction will depend on the nature of the electric fields and the state of health of the individual. If a person's thresholds are reduced, through stress or imbalance, then they are more likely to be affected by the electrical environment.[16](Cf. chakras.)

Michael Shallis also states that burying food and water in earth helps to remove the electrical component. I feel slightly uncertain whether this applies to all objects. Kirlian photography shows a corona discharge or electrical field inherent in most objects, so it is possible that this aspect of burying something is equivalent to the old folklore of 'earthing' objects such as magical implements in order to 'cleanse them' of unwanted magical energies. He says that electrically treated water remains unaltered by freezing; that water retains electromagnetic 'information'; it has some sort of memory of its earlier electrical exposure and reproduces the effect even when diluted. This information links with the findings of homeopathy in which dilutions past a certain point seem to be without any molecules of the actual medicine, and the healing effect is held in the water.[17] Also Douglas Dean,[18] an American parapsychologist, found that water held by healers and water from holy wells shows a different infrared absorption peak. He used a technique called infrared spectroscopy in which a beam of infrared light — that

***Chalice Well at Equinox (photograph courtesy of Lesley Delamont)***

is, light which is just past the red end of the visible spectrum — is shone through the water. The water absorbs some of it and a spectrum, a bit like a rainbow, can be measured which shows the precise wavelengths of light and how much they have been absorbed by the water. Ordinary tap water and holy water, or water held by a healer, absorb the light at different frequencies, which shows that the holy water is different from ordinary water.

## The Sun

Everyone knows that the Sun has spots. The full solar sunspot cycle is 22.6 years: 11.3 years from peak numbers of spots to the next peak number, but the first peak is in the negative phase, and the second peak is in the positive so it takes 22.6 years for a complete positive-to-positive, or negative-to-negative cycle. In other words, the magnetic field of the Sun undergoes an 11-year cycle building up in complexity and strength producing sunspots and then fading away again. After 11 years the cycle is over and is then repeated with a reversed polarity. But this cycle is very irregular, at times sunspots disappearing for decades. The sunspot periodicity has been linked to many terrestial effects described clearly in *The Cycles of Heaven*,[19] such as disturbance of the Earth's magnetic field; extreme weather conditions; general climate (when sunspots disappear weather is colder; when active it is warmer); the length of the crop-growing season; the growth pattern of trees; admission to mental hospitals; the blood-clotting rate in humans; and political and social change.

Sunspots are connected with the solar wind and affect the Earth as follows: The Sun's magnetic field rotates at different rates at different latitudes because it's a gas. This causes distortions to erupt at the solar surface giving rise to visible sunspots and solar storms where electrons and ions are ejected from the Sun with enormous energies. These storms result in bursts of interference to the electromagnetic properties of the Earth's atmosphere causing disruption of terrestrial radio signals and disruption in the ionosphere. The Sun's magnetic field is divided into four sections, each one charged opposite to the next. The rotation of the Sun spins the magnetic structure every 27 days so that the boundary of each section passes across the Earth every four weeks. The solar wind and the resulting effect on the Earth's Magnetic Field (GMF) could be the mechanism through which changes in the Sun can affect Earth, because each time a solar magnetic section passes across our planet 'it produces a sharp jolt in geomagnetic activity one or two days later'.[20]

Above the ionosphere of this planet are belts of magnetically trapped charged solar particles, van Allen belts, in the magnetosphere. The limits of the magnetosphere are influenced heavily by its interaction with the solar wind. The Aurora Borealis (Northern Lights) is formed from glowing ions, charged by the solar wind high in the atmosphere, trapped in the convolutions of the Earth's magnetic field. Sounds and smells associated with auroras have characteristics associated with static electrical discharges: sounds vary from swishing noises to sharp cracks; from ozone to sulphurous smells. There is a possible connection between auroras and thunderstorms and auroras and earthquakes - the earth movement resulting in electrical discharge causing ionised air in auroral regions of sky. Auroras produce measurable electric effects on the ground, immersing us in an electric field; fluorescent minerals glow from the strength of the electric field induced in the crystals.

In 1989 we had one of the biggest solar flares ever recorded; the Sun had enormous sunspots both in size and quantity with all sorts of incredible world events coincident. The last peak of the same phase was in 1967! In fact, even as far south as Glastonbury I saw the first Northern Lights I have ever seen. It was wonderful: streamers of colour stretching over the sky and then for hours after an eerie green glow in the northern sky so bright that the stars were eclipsed. The Northern Lights were seen that night as far south as Southampton, an unprecedented occurrence. I found that I actually had a severe stomach ache for several hours afterwards. I suspect that the intense magnetic field associated with the solar flare that causes the Northern Lights affected me very strongly, and my stomach has become very sensitive. We are well advised to watch our body reactions to events because they, and our associated language, have a lot to tell us. Exactly what is someone doing to you who is being 'a pain in the neck'. The neck is associated with the thyroids and stress at the communication, hearing, mental wittering level.

*An electromagnetic basis behind the aura?*

In a liquid, like water, it is not the electrons that carry the current, but the positive ions caused by the electric field that move from one electrode to the other in a process called electrolysis — this is an electrochemical reaction and is a very important electrical effect in living bodies, since all living things are primarily composed of water. This means that electricity is a life force — as it can be the death force in the electric chair. In fact some doctors define death as the absence of electrical activity in the brain. Therefore light from the stars can be seen as their life force reaching through the universe. I feel that the electromagnetic is the physical aspect of the life force, not the whole of it. It is how this aspect manifests in this physical universe.

Without electricity there would be no matter: electricity is the force that holds everything together, equivalent to the Chinese chi energy which represents the flow of cosmic energy that holds all things together and maintains the world in its present form. In India it is called prana, the breath of life. These ideas all appear to have some level of similarity in that they are talking about the underlying force that gives this world its form. We are all electric creatures as are the stars, the Sun, the planets and space. Electrical signals are the means our bodies use to move information around the organs and limbs, senses and brains to provide the basis for our thought processes. Harold Saxton Burr called this electrical energy L-fields. He found that the overall body fields in all living things are electric:

> The organisation of a living thing, its pattern, is established by an electrodynamic field, which determines the arrangement of the components of the organism and is determined by them. This electric field ... will be a characteristic of a particular thing; it determines its growth and pattern and maintains the pattern and structure of the organism during its life. The electrodynamic field must be the mechanism by which the wholeness, organisation and continuity of life are maintained.[21]

This finding by Burr that our form is defined by an overall electric field, a physical equivalent of the aura, is very similar in conception to Rupert Sheldrake's more recent hypothesis of what he calls a 'morphogenetic field'. I shall not go into it in detail now; if you're interested I suggest you read Sheldrake's books *A New Science of Life* and *The Presence of the Past*,[22] but, for example, frogs' eggs show a voltage axis, and the nervous system develops along the axis. Even an unfertilised salamander egg shows this voltage axis, the positive part of the field coinciding with the mark on the egg, which later develops into the heart. The L-field is independent of and remains unchanged by fertilisation — the future possible form of the embryonic animal is already mapped out in the unfertilised egg. L-fields extend out into space for a distance of several millimetres — a rotating salamander can act as a dynamo. It seems as if the electrical L-field is the physical aspect of Sheldrake's non-physical form-fields. If something is to have effect on this earth plane then at some point it has to 'shift through' into physical form, and the electromagnetic seems to be that linking form.

Burr showed the 'auric' effects of our bio-electrical field by using a voltmeter. He measured the voltage difference between the index fingers of right and left hand. This difference is individual — some people have high potentials and some low. Women's voltages change over the menstrual cycle, voltages rising dramatically at ovulation. Unusual changes in voltages are also associated with disease, the change in voltage often occurring before the disease is clinically visible. This sounds surprisingly similar to some of the claims concerning Kirlian photography which is another measure of our electric bodies, which some people liken to the aura. In measuring the voltages of sweet corn, there were marked differences between different hybrids, even a difference of one gene showing up. They found they could also distinguish strong healthy seeds, from others.

L-fields are strongest at maturity and fade as death approaches. Dr Ravitz hypnotised a person and measured the L-fields. He then suggested a grief memory and found that this severely affected the L-field. In other words emotional instability is marked by a high L-field. It is now well-known from psychosomatic medicine that mental stress results in physical disorder and Dr Ravitz has shown that mental states affect the L-field; the abnormal L-field could then affect the body, inducing symptoms of ill health. Also emotional states can be equated with energy. I have already mentioned this with regard to the GSR which is used so extensively in parapsychological research and in subliminal perception — electrical changes in the skin due to emotion, thoughts and level of stress, tension/relaxation.

In a type of Kirlian photography technique called electronography you get images of the whole body showing internal organs. The patterns that emerge on films are pictures of the L-fields modified by the state of the subject. In the space around our bodies is a region containing air ions, resulting from our own electrodynamic fields and from the chemical changes associated with biological processes. Tumours themselves radiate electromagnetic waves. There are clear parallels between L-fields, Kirlian photos and the aura. All are affected by physical and mental condition, state of health and emotions, and the spiritual state of the person.[23]

Harold Saxton Burr monitored trees' L-fields over a thirty-year period. He found that trees display a daily L-field rhythm, peaking at midday with a minimum at night. Tree growth follows this pattern. Changes in electric field were found to be due to fluctuations in the local environment and followed a monthly and a seasonal pattern.[24] The sunspot cycle is evident in electrical voltage changes of the earth, air and trees. This is true for all living things.

> The electric fields in and around living things necessarily interact with the electric fields of the natural environment, in which they are embedded. Not only is 'no man an island' but electrically all things are interrelated; the sun, the moon, the earth and all its inhabitants. We are electric creatures living in an electric world - naturally.[25]

In conclusion, all things are electric at a fundamental level - stones, metal, plants, animals — and the electric fields that surround all matter, that reach out into the space around things, connect and affect all other things. 'The cosmic web is spun from electricity.' The electric air around us works on our susceptibilities.

> This extraordinary interconnectedness, at an electromagnetic level, explains why we are affected by places, times, colours, sounds, atmospheres, objects and people. We talk about being tuned in to the atmosphere, of connecting with the vibrations of different people, and we mean what we say. At the level of the electric field the world is a massive, complicated and interwoven array of patterns of electrical energy. These affect us physically, mentally, emotionally and spiritually. The spark of life rises and falls within us as we move through this dazzling array, reflecting in our moods, our thoughts, our whole beings.[26]

Just as Burr's trees interacted with the changing state of the electric fields in the air and in the Earth, so too do we. A good example of this is psychometry in which we pick up energy from an object, and if we are sufficiently psychically sensitive we will be able to talk about certain aspects of the history of the object: to whom it has belonged, what they are like, and so on. This is the same sort of energy interaction with psychic sensitivity as occurs in hauntings of certain places. As Shallis says:

> There is clearly an electrical component to these energy fields and the sensitive person is able to read the message imprinted on that field. It could also be that the field around objects, living or inert, enables the psychic to direct his telepathic ability in the direction towards which the object is focused. If there is a direct electrical component in ESP, then it operates through the life fields that pervade the space around us. Healers also operate in this psychometric fashion. Radionics is an example of that. Those healers who work through the laying on of hands also seem to be invoking an influence on the life fields of their patients.[27]

## The Earth is a Giant Bar Magnet

The Earth's Magnetic Field (GMF) is linked with the electrical properties of the space of the solar system — Sun, Moon and planets. Picture Earth as a bar magnet with magnetism coming out of North Pole and going vertically to the South Pole. The actual magnetic poles move around and there are occasional reversals of polarity, the last one about 34 million years ago. The Sun and planets and stars all have a similar magnetic field. The GMF is due to the spinning of the planet, the interior being molten metal alloy of iron and nickel, which creates a dynamo effect. There is an electric current running through the Earth's crust which is linked to the magnetic field.

Geologists can measure the extent to which rocks contain a record of the GMF when they were formed — rocks were formed from the Earth's molten magma which cools when it erupts on to the surface, solidifying with any alloys of iron present within it becoming aligned with the GMF as it is at that time and place, and so forming a record of our planet's magnetic history. Since the GMF is constantly changing the rocks also have an induced magnetism due to the GMF on that day.

Rocks with a high proportion of magnetic minerals, primarily those of iron, will have a higher magnetic susceptibility to those with few magnetic minerals. Igneous and metamorphic rocks, found in areas where tectonic activity has been, or still is, important, generally tend to exhibit higher magnetic susceptibility than those of most sedimentary rocks. Don Robins[28] has suggested that the highly crystalline stones in standing stones and stone circles act as condensers. In a condenser the electric charge will get stored until something happens, when the electricity will discharge with a cracking spark — as with amber. Condensers consist of two conducting regions separated by an insulator, known as a dielectric. The electric effect is spread in a field around the charged object. Contour lines can be drawn around a charge showing the intensity of the field dropping off with increasing distance.

It was from permanently magnetic rocks that the lodestones were discovered. Lodestones are naturally magnetic rocks. A piece of iron rubbed with a lodestone becomes magnetic and points north and, if free, will dip downwards according to latitude. This is how a compass works. From research on the pineal it appears that we have a sort of internal compass. It is worth noting here that an electric current creates a magnetic field which will both attract a compass and magnetise iron.

The important point to remember from this is that some rocks hold unusual magnetic strength, others hold very little, and they are all oriented in different directions. Thus we live on a magnetic planet; the magnetism permeates the rocks of the crust on which we walk and thereby affects the surrounding atmosphere, and this affects both us and the site where these rocks are placed.

*Our sixth sense*

In the previous chapter I have shown how research over the past ten years into the pineal gland, our psychic chakra and third eye, has provided evidence that it makes a hallucinogen which could well take us into an altered state that is psi-conducive and which affects us at physical, emotional, mental and psychic levels.[29] Shifts in intensity of the GMF appear to be one of the environmental conditions which affect pineal activity and psychic sensitivity.[30]

There has been a lot of research into our sensitivity to electricity and magnetism recently and an intriguing picture is emerging, which suggests that the old idea of a 'sixth sense' may be just that, a sensitivity to the GMF that is in some way connected with our psychic sensitivity. This sixth sense is so subliminal that most people aren't even aware they possess it. It is not a psychic sense although it may be connected with a psychic sense, as seen in dowsing. Because it is a subliminal perception, the psychological process by which we become aware of this information is similar to that by which we become aware of psi inputs.[31] From a practical, subjective level therefore, awareness of the GMF and awareness of psi information will manifest in very similar ways. Thus map dowsing, a psi activity, and field dowsing, which is probably sensitivity to magnetic field change, manifest in similar ways. This suggests, as with my research into more traditional subliminal perceptions, that the line between subliminal sense perceptions and psi perceptions is a very fuzzy one, the one merging imperceptibly into the other.

The main point that I want you to remember from this is that we are being affected subliminally at all times by changes in the Earth's magnetic field and this has enormous implications for astrology, magic, witchcraft and other Earth-based systems of knowledge. We are seeing here a generalised sensitivity or awareness which is concomitant with spiritual development, and easier by far for the country person than for the city person.

This magnetic sense as a direction finder has been extensively researched in both animals and humans, and has been shown for bacteria, planarians, molluscs, insects, some types of fish, and salamanders.[32] For example, Mather and Baker[33] found that the woodmouse uses magnetic cues for direction finding; magnetic fields effect spatial discrimination learning in mice;[34] and there is the well-researched migratory birds and homing pigeons' magnetic direction sense. When preparing for migration, birds will assemble, even in a laboratory, in the direction they want to migrate. If the magnetic field is artificially changed, so that magnetic north is in a different direction, birds will orient to this new direction.[35]

Baker[36] in his book on human navigation calls this magnetic sense the 'sixth sense'. He blindfolded people and then took them on a coach trip that twisted and turned. At the end of the trip they were asked to point to where they had come from. He found that a significant number of people could do this, but if they wore magnets on their heads they lost their ability. He found that cues such as sun, wind direction and, when the blindfold was taken off, distant landmarks also helped people.

Baker's work has been taken a step further by Gai Murphy[37] who worked with children sitting in a swivel chair blindfolded. The chair was spun and the children then had to say which direction they were facing, north or south, etc. She tested 1,279 children and found that young children have no sense of direction until about age 11 and then girls become really remarkable in their accuracy while boys remain inept. When the girls wore a magnet at eye level the result was a significant decline in accuracy. Murphy reckoned the boys' insensitivity was because they tried to deduce logically which direction they were facing rather than 'just guessing'. Girls guessed and so allowed their subliminal magnetic sensitivity to guide them to guess correctly. Sadly this research has not been followed up and so remains inconclusive.

The magnetic field affects not only direction finding but also other aspects of behaviour. Stutz[38] considered that the circadian rhythm in gerbils is entrained by the daily magnetic field fluctuations. Rudolph *et al.*[39] found that 60 Hz electric fields as well as artificial magnetic fields affected rat behaviour, there being a decrease in 'emotional reactivity' after exposure. And Walker[40] found that tuna could detect changes in magnetic fields as small as one nanoTesla (nT, less than one 20,000th of the Earth's field). A comparable system in humans could easily be sensitive enough to account for the results of dowsing experiments. This is really quite likely as magnetic fields within the ultra-low and extremely-low frequency range influence humans. Some Indian researchers[41] showed that sudden increases in GMF of about 20 nT are sufficient to lower seizure thresholds and precipitate convulsions in people who suffer from epilepsy, while others[42] found that one hour exposures to 0.1 Hz fields between 10 and 50 nT generate physiological changes in humans. And some people can become so acutely aware of the GMF that facing south makes them ill, while east and west are neutral, and facing north makes them feel better.[43] (Quiet periods in daily GMF variation average about 10 nT. During magnetic storms, e.g. the aurora borealis type of storm, there are huge changes approaching 500 nT. Household wiring and power lines can create magnetic fields ranging between 100 nT and 100,000+ nT.)

This links with my research into the pineal gland in two ways. The first is the strange fact that the pineal starts a process of calcification around puberty. Nobody knows why. But calcification is a bit like the process of our noses becoming more bony as we grow older, bones being primarily calcium, as are our teeth. It has been found that bone, horn, hair and feathers produce their own electrical signals when under stress (pressure), and so are piezoelectric transducers sensitive to electromagnetic fields.[44] For example, in a magnetic field bone heals quicker. Semm *et al.*[45] have discussed a neural basis for the magnetic sense of direction based on their research into the pineal gland, and I propose that this magnetic compass sense could be the calcite crystal part of the pineal gland. It is worth thinking about and researching anyway, and makes even more sense in terms of the pineal as the psychic chakra providing the link point between our material bodies here on this material Earth, and the psychic-spiritual aspect of our beings for which electricity and magnetism seem to be some sort of a link.

This effect is a very important one for us to understand since it seems to be a link between matter and electromagnetic field energy. It makes me wonder about the old divination practices that used bone or tortoise shells for divination and about the Amerindian use of feather headdresses as possibly electromagnetic antennae; the Neolithic gods and goddesses were all horned; the witches definitely had a strong superstition with regard to hair and nails. I think this piezoelectric effect is the clue that links all these magical practices.

## *The earth's magnetic field and psychic ability*

Research in the 1980's[46] by Marsha Adams and Michael Persinger found that spontaneous cases and laboratory experiments of psi are more successful when the GMF is relatively quiet. In all this research the global geomagnetic indices which tell the variations in the GMF for each day were used. Persinger and several others have found that this relationship between quiet geomagnetic activity and increased psychic awareness is true for telepathy and clairvoyance cases but not for precognition.[47]

Persinger also studied reported cases of poltergeist activity and found that poltergeist episodes were correlated with sudden *increases* in geomagnetic activity.[48] He also found that reports of bereavement hallucinations occur when geomagnetic activity is increased.[49] He considers that these correlations hold over decades as well as diurnal time spans, there being some decades that have a noticeably quieter geomagnetic activity than others, e.g. 1870-79 and 1890-1909. He feels that this might explain the historical episodes of times when there has been an upsurge in reported psychic cases. More recent research by Roll and Maher in America, and Braithwaite[50] in England has found that apparitions and other haunting effects occur where there are both natural and artificial heightened electric and magnetic fields. Persinger also noticed that peak spontaneous experiences concerning death and crisis to significant others occurred between 2 and 4 a.m. with a secondary peak at 9 - 11 p.m.[51] This timing is really interesting because it fits in so well with what I have just mentioned concerning the pineal gland, the peak melatonin production occurring six hours after dark, and so, on average over the year, between midnight and 3 a.m.

From the foregoing it is clear that, most often, active psi occurs when the GMF is more intense, and receptive psi when the GMF is quiet. Occasionally, however, the opposite is reported, or there is no relationship. For example, there is a lengthy series of research into remote viewing being done by Robert Jahn and Brenda Dunne at Princeton Engineering Anomalies Research (PEAR) laboratory in America and they have found no correlation. Nor did Chuck Honorton who did research with the Ganzfeld for over fifteen years.[52]

Why this should be so is not yet clear. It could be that their laboratories in some way distorted or shielded the GMF, particularly the PEAR lab which was in the basement of the engineering building with lots of very powerful equipment located in nearby rooms, or that other factors in their laboratories overrode the slight effect of the GMF. As yet we do not know. Also Howard Wilkinson[53] checked some of Persinger's findings and did not come to the same conclusions. That is the fun of research — there is always more that one does not know. More recently, Radin, McAlpine and Cunningham[54] found that although "normal" participants showed no evidence of psi in the Ganzfeld, they did score better when the GMF was low, but "creative" participants, who did succeed in the Ganzfeld sessions, showed stronger psi scoring when GMF was high. Haraldsson and Gissurarson[55] found no relationship between GMF and psi on the day of experiments, but reported that successful experiments occurred one day after high GMF.

Research also suggests that there may be a lunar effect on psi. In a series of telepathy tests, Puharich,[56] found increased psi scores at full and dark moon periods, and Krippner et al,[57] analysing 80 dream sessions also noticed enhanced psi during the full moon period. Radin & Rebman[58] found a lunar correlation in their data of casino jackpot and lottery wins. Over a four year period there were 6 jackpot wins – all of them were during the full moon period. Other games also showed enhanced casino payout during the full-moon period, with peak payout within one day of the full moon. Etzold[59] evaluated 200,000 retro-PK trials covering 8 years and found a significant solar-periodic relationship. He hypothesised that it may be the moon's interaction with the earth's magnetosphere, during the moon's passage through the magnetotail during the full-moon period, that modulates the psi effect. Recently, Sturrock & Spottiswoode's[60] analysis of 3,325 free-response experiments also suggested that there may be a relationship between psi performance and lunar phase. Fraser-Smith, and Stenning et al, have found geophysical evidence that the moon affects GMA (geomagnetic activity) via tidal/gravitational effects.[61]

There is considerable empirical evidence that magnetic fields, whether artificial or natural, can affect a person's experiences. Randall and Randall[62] found a relationship between natural magnetic disturbances caused by the solar wind and hallucinations. Examples of more recent work includes Fuller et al, who evoked epileptiform activity using magnetic fields, and Dobson et al[63] who applied magnetic fields to epileptic patients. Persinger and colleagues[64] have found that artificial electromagnetic fields stimulate visionary type experiences such as people experience during UFO abduction type accounts, evoke memories of significant others, give OBE type accounts, and have apparitional type experiences.

The receptive psi/GMF research has been criticised by Hubbard and May[65] who urge that local geomagnetic measurements are made. This recommendation is a valid one, and most of the research on the pineal gland reported below uses local measurements or artificial magnetic fields. It becomes even more pertinent when we look at the electromagnetic fields surrounding sacred sites, barrows, etc. Because of this we are now involved in doing a long term study looking at the effect of local GMF activity on meditation and psi awareness.

The major basic suggestion from this body of data is that psi ability is affected by the GMF so that when the magnetic field is *different from* normal, different sorts of psi experiences are more likely to occur. This suggests to me that sensitivity to psi is in some way affected by the GMF. We know that certain practices enhance psi effects, e.g. being hypnotised, and it seems as if a shift in magnetic field strength may be a similarly psi-conducive situation.

This link between psychic sensitivity and geophysical variation, is a very clear strong link between the Earth and our psychic spiritual sensitivity. I have noticed in my literature survey of research on the pineal gland that the pineal is also responsive to the GMF. The production of melatonin in the pineal gland, which is almost certainly the precursor for the possibly psi-conducive pinoline (through its effect on DMT production by serotonin), is affected not only by light, but also by the GMF. The fact that the GMF affects the functioning of the pineal gland suggests a neuromechanism by which our psi ability is affected by the GMF.

*The pineal and the earth's magnetic field — the 25 hour clock*

Over the past three decades there has been some interest in the effect of the GMF on the functioning of the pineal gland. Most of the experiments have been performed on animals of various sorts, and we should be wary of generalising to humans. However, the tiny body of research that has been done with humans in no way contradicts the animal research. What follows is a selection of research culled from the neurochemical literature that best illustrates the findings with regard to the effect of the GMF on pineal activity.

In 1980 it was found that the GMF affected the electrical activity of pineal cells.[66] It is well-known now that the brain is electrical, in that we can measure brain waves such as the alpha waves of meditation with EEGs (electroencephalograms); and the magnetic measurements by machines like fMRI[67] have opened fantastic new doors into our understanding of brain functions. But to find that the GMF affects this electrical activity is new, and for it to affect the pineal gland is very interesting. This finding was corroborated in 1983[68] and not only for the cells of the pineal gland but for the autonomic nerves that connect the pineal gland with the retina — its source of information about light, daylength, moonlight, and season of year.

Also in 1983, it was found that artificial magnetic fields, differing only slightly from the GMF, inhibited the melatonin biosynthesis in rat pineals at night by inhibiting the enzyme that makes melatonin from serotonin. It seems as if the *change* in magnetic field is the important factor, because animals habituate to artificial fields.[69] More recently Henshaw and Reiter[70] have found that long-term exposure to high magnetic fields suppressed melatonin in people. Burch et al[71] found that people exposed to 60 Hz magnetic fields at work showed a reduction in melatonin; and Weydahl et al[72], working with people north of the arctic circle, found that a high GMF reduces melatonin production and increases serotonin. However, they found that the circadian rhythm does continue through the polar night, adding evidence for the importance of the GMF as a Zeitgeber (literally time-giver).

So not only does a change in magnetic field affect the electrical activity of the pineal and autonomic cells, but it actually affects the production of the pineal's major neurohormone melatonin, that which is in control of our endocrine-chakra system. Could this be one of the reasons for our mood fluctuations with, say, the full Moon? The pineal is linked with the endocrine system. The endocrine system affects our sexuality, stress levels and emotions, and this is affected by changing magnetic fields. The GMF changes according to the Sun's activity, the rising of the Moon and even perhaps is affected by the other planets though this is not yet clear. Could we have here a physiological link with certain aspects of astrology?

In fact, Cremer-Bartels *et al.*[73] state that the general biological role of melatonin may be interpreted as a translator of environmental conditions. They conclude that the natural GMF variations may be suggested to be the Zeitgeber of the daily cycle in the pineal gland and retina. In other words, the pineal gland is our biological clock involved in regulating everything from our sleep—wake cycle to our menstrual cycle to puberty onset and ageing, and now it has been found that natural variations in the GMF are the Zeitgeber.

This research into the pineal gland's enzyme activity confirms some research done by Wever many years before.[74] He showed that shielding the GMF desynchronised circadian rhythms in humans significantly, even when light perception was not excluded. This suggests that the magnetic component of the regulation of circadian rhythm is just as important as light. When an animal is kept in constant darkness it is found that its circadian rhythm carries on, but on a 25-hour rhythm. Since the tides are caused by the Moon and they change by one hour a day giving a 25-hour cycle, and the Moon affects the Earth's geomagnetic variation, it is possible that this is one aspect of the various factors which govern our circadian rhythm. Thus, magical premises like Moon magic held important by the witches and other pagan peoples are now being researched scientifically and are being found to give us a whole new understanding of ourselves in relation to the Earth and the solar system. It won't be long before this knowledge extends to the stars as well — and at a well-grounded Earth-based level of knowledge: the channellers, mystics and magicians give us our vision, and the scientific method earths that vision so that it can be manifested strongly within each of us on this planet.

Further, it was found that in quails an artificial magnetic field affects not only the melatonin synthesis in the pineal but also melatonin production in the retina is affected[75] and that not only does an artificial magnetic field affect the melatonin and the enzymes that make the melatonin in the retina, but that the pineal response to the GMF is present only when the retina itself is stimulated. There are non-rod and non-cone cells in the retina which have the nerves which are connected with the pineal via the SCN. Exactly how this affects blind people I don't know, because no research has been done with blind people so far. I do know that, on the whole, loss of sight makes one exquisitely sensitive in other ways, and from what little I know most blind people are still sensitive to light *per se,* the blindness being from other causes than a totally non-functioning retina.

This link between pineal gland and retina is interesting in that the retina is a little bit of the brain which has become pushed out in front so to speak. It is this relationship between GMF and light and darkness that is the crucial point here, with the retina sensitive to both. Our eyes are a direct link with the soul of a person, and there is a rich folklore about the 'evil eye' and its opposite, the blessed gaze that heals, the power of someone's gaze to affect you,[76] lovers gazing at one another, the list is endless. This research is suggesting that there may well be a physiological link behind the psychic/spiritual aspects of that lore. Magnetic fields are linked with the phosphene effect in which light flashes in closed eyes. Strong electrical fields also provoke cataract and retinal damage. Light and colour are very important in our lives. How many of us feel incredibly depressed when the grey clouds and mists come over and hang so heavy upon us? A red light in a room gives it a warmth and colour that is very friendly and comfortable and sensual. White fluorescent lights are cold and uncomfortable and give some people headaches and other illnesses.

Researchers have also found that there is what is called a magnetic window and the body does not respond to magnetic fields whose strength is greater than or far lower than this window, which is around Earth strength.[77] Exactly what the implications of this are for those who live in cities surrounded by massive electrical currents from all the wires and cables that increasingly web in those who live in houses I am not too sure. I do know that city culture on the whole is the source of the whole materialistic and anti-spiritual ethos that is destroying our planet, but this is a hopeful bit of research suggesting that although our electrical world is an unhealthy one, it is not affecting our magnetic sensitivity too badly.

But Clark et al[78] have found that radio frequency transmitters and local 60Hz fields suppress melatonin production and oestrogen production in post-menopausal women, and this is linked to an increase in breast cancer. And Reiter *et al.*[79] found that rats exposed to 60 Hz electric fields from conception to 23 days of age exhibited reduced peak night-time pineal melatonin content. Adult rats exposed to electric and magnetic fields show reduced night-time melatonin. We, in our electrified houses, are like these poor rats. Our wiring carries 50 Hz electricity to all our lights, fridges, televisions, stereos, computers, hot water tanks and other electrical equipment with which we surround ourselves. This means that although we may not be affected by the massive electro-magnetic fields generated in some cases, we almost certainly are affected by our house wiring. Graham et al found that stability of melatonin levels were influenced by chronic magnetic field exposure. And Gobba et al[80] measured extremely low frequency magnetic fields in the work place and found that those exposed to more intense magnetic field showed higher melatonin levels on both Monday morning, (post-weekend) and Friday morning (post-working week). Maybe this is one reason why the number of haunted houses and other spontaneous psychic phenomena are not quite so extensive as they used to be before the days of electricity. I do know that in Victorian times the mediums were producing materialisations and having objects floating round in the air and so on. This is to be found nowhere today. Also the magicians of that time, like Aleister Crowley, were reportedly conjuring up all sorts of spirits, and I for one, have not heard of modern-day occultists having such dramatic episodes. Could it be that although we still are psychic our faculty is muted to encompass only small everyday occurrences? Or is this a sign of the ever-increasing materialistic and reductionist ethos in our culture? There's nowhere left for things to go bump in the night in our over-lit night-time world.

Thus, the pineal gland, and particularly its production of melatonin, is definitely affected both by magnetic and electric fields. This could possibly underlie Persinger's apparently contradictory results noted above in which he sometimes finds a correlation with days of low GMF and sometimes with sudden increase in GMF. It appears to be the sudden change in GMF that is the important factor, and the changes in direction of intensity are correlated with different sorts of psi experience. Since melatonin affects a wide range of endocrinal and neuronal functions within the body, anything which affects the pineal gland will have wide-ranging effects. Although there has been no specific research yet on the production of beta-carbolines in the pineal in relation to geomagnetic effects, it is highly probable that its synthesis will be similarly affected. Thus, all this research suggests that through the pineal gland we have a physiological means by which the Earth's magnetic field can change our inherent sensitivity to psi awareness.

*Dowsing and geomagnetism*

Mineral veins and flowing ground-water are associated with geological faults, fractures, old stream channels, cavities in limestone, lava tubes in volcanic rocks and so on. These discontinuities cause small geophysical perturbations, including anomalous local magnetic field strength.

Since all the research mentioned suggests that we are exquisitely sensitive to the GMF, it is entirely reasonable that field dowsing is at least partially a physical response to variations in the local magnetic field. There is definitely a psi component to dowsing as is evidenced by map dowsing, and I suggest that the response to the GMF is in some way connected with the pineal and psychic sensitivity.

There have been speculations for years that field dowsing for water works because some people are highly sensitive to the magnetic field changes that occur when there is underground water. In 1949, Solco Tromp, a Dutch geologist, claimed that dowsers' rod movements often coincided with small magnetic disturbances. In 1962, Yves Rocard, a French physicist, made a similar claim, as did Valery Matveev, a Soviet geophysicist. An article in *New Scientist*[81] expounds these ideas very clearly. The pineal gland could be a brain 'receptor' sensitive to the GMF which enables the dowser in the field to be directly responsive to the fluctuations in the GMF, as well as to attain the correct 'frame of mind' necessary for success.

Dowsers' rod movements do definitely seem to be linked with changes in the intensity of the GMF. Williamson reports that when dowsers look for mineral veins and flowing ground-water, these are both associated with geological features such as faults. Simmons, a geophysicist at the Massachusetts Institute of Technology, has conducted surveys of the GMF around two dowsed wells near Boston. Unlike most other wells drilled into the crystalline rock of the area, these wells yielded huge quantities of water. Both wells were found to be within a narrow magnetic anomaly only a few yards wide. The anomaly resulted from a fracture zone that was channelling the flow of ground-water, hence the exceptionally high yields of the wells.

In 1970, Duane Chadwick of Utah State University conducted a series of experiments that fully met the critics' requirements. He tested 150 people and he carried out control experiments. The results were startling: the dowsers' rod movements showed an apparent link with tiny changes in the intensity of the GMF.[82] More recently, a series of experiments done with a water dowser[83] who was working for Water Aid in Sri Lanka and Africa found that his success rate was more than 80% whereas the success rate of standard geomagnetic equipment used for detecting suitable places for wells, was only 66%. This suggests that a good water dowser is not only picking up the geomagnetic information, but supplementing it with what could well be psychic information.

This sensitivity to the GMF also explains how dowsers find veins of metal ores which are also associated with faults or fracture zones.

Of course, in some cases ore minerals are themselves magnetic.

*UFOs and geomagnetism*

Research into UFOs, ley lines and ancient stone circles suggests that these, like dowsing, also appear to be connected with geomagnetic anomaly and with the magnetic effects of the solar system on our planet.[84] The 'Dragon Project' research mentioned here is very preliminary, but very interesting and definitely warrants further investigation. Sadly in the three decades since it was done nobody has replicated or extended this work. Paul Devereux's book, *Places of Power*, he regards as the most comprehensive material on the work done by the Dragon Project.[85] Since this time he has written lots of other books on related topics, but I shall focus on his earlier book, *Earthlights*, in which he discusses Persinger and Lafreniere's book *Space-Time Transients and Unusual Events*[86] which analyses a range of UFO and anomalous happenings. This book in fact marks Persinger's original interest in the effect of the GMA on anomalous experiences, which then lead to all the research I have already mentioned. Their data suggest that UFO phenomena tend to cluster in specific areas, though there is the confounding effect of population density (where there are more people, more people are likely to see and report UFOs!). When this is accounted for you get both window areas: Warminster, Dyfed, Barmouth and Cumbria — and times: UFOs tend to occur or be reported most often in April, July, August and September.

Persinger and Lafreniere's core theory for explaining UFOs involves seismo-electricity. The areas where most UFOs are reported are primarily areas of seismic-related stresses, that is, areas that are prone to earthquakes, or at least tremors. Not all areas of geological faulting that produce seismic stress will actually move so that we can physically feel it; normally the move is much more subtle than that. In fact, in Britain there are approximately 400 tremors a year, about one a day, but we don't notice them. During seismic strain, pressure on the rock crystals produces electromagnetic fields through a modification of the piezoelectric effect. The fields created by this process then have electrostatic effects such as ball lightning, will-o-the-wisps, and other earthquake and UFO-related light effects, and are also connected with psychological and psychic effects such as poltergeist outbreaks.

Stresses on faults may accumulate over long periods of time. As a result, a localised 'electrical column' could be formed. It is *while the stress is present* that such an effect would be created and phenomena would be likely to occur. The existence of electrical columns produced by accumulating stress would affect all living things as well. When parts of the brain are electrically stimulated, dreamlike states can be induced and vivid imagery unleashed even in waking consciousness. Such stimulating currents are not large and could conceivably be produced by transient currents associated with tectonic stress.[87] This links in with all I have said with regard to the pineal gland and psychic events.

> The Canadians add that while the major source of energy for such a phenomena-producing mechanism would come from fields associated with geological stress, much smaller displays might be triggered by electrical fields generated by meteorological events such as thunderstorms. Other effects could be added to these sources, such as impulses from the sun and moon ... They record that researchers have already shown that magnetic anomalies can be produced at the Earth's surface above appropriate mineral deposits during times of geomagnetic storms.[88]

Thus, earthquakes produce luminous electrical activity: before, during and after the quake electrical discharges to the air take place. However, it is important to remember that mere stress in the rock is sufficient to cause a piezoelectric effect. In general, an overall pattern of UFO sightings is related to earthquake epicentres, but it is the Earth movement and the related electromagnetic field that is the significant factor.

### UFOs and the psyche

Paul Devereux talks at length about the psychic aspect of UFOs. I include it here because it is an excellent example of the conjoining of our psyches and the Earth, and it illustrates the principles of magic so very well. Devereux, Jacques Vallée, Jung[89] and several others have all considered that the reported form of UFOs and their occupants is created more by our conceptions, by archetypal symbolism, by our twentieth-century collective unconscious, than by any inherent form in the energy field that we call UFO. Devereux mentions that, with the UFO Foo Fighters in the Second World War, which were the first well-known UFOs, 'Some airmen reported that the lights *seemed to be responding to their thoughts.*'[90]

What amuses me here is the transparent connection between the little green men of Ufology and the Green Man of the pagan religion. That UFOs are a modern form of fairy folk becomes more and more apparent the more I look into it.[91]

One of the lores of psychology is that we can only perceive that which we can conceive. Something that is totally alien just cannot be seen by us because we have no conception of it. All of our perception

## Where Science and Magic Meet

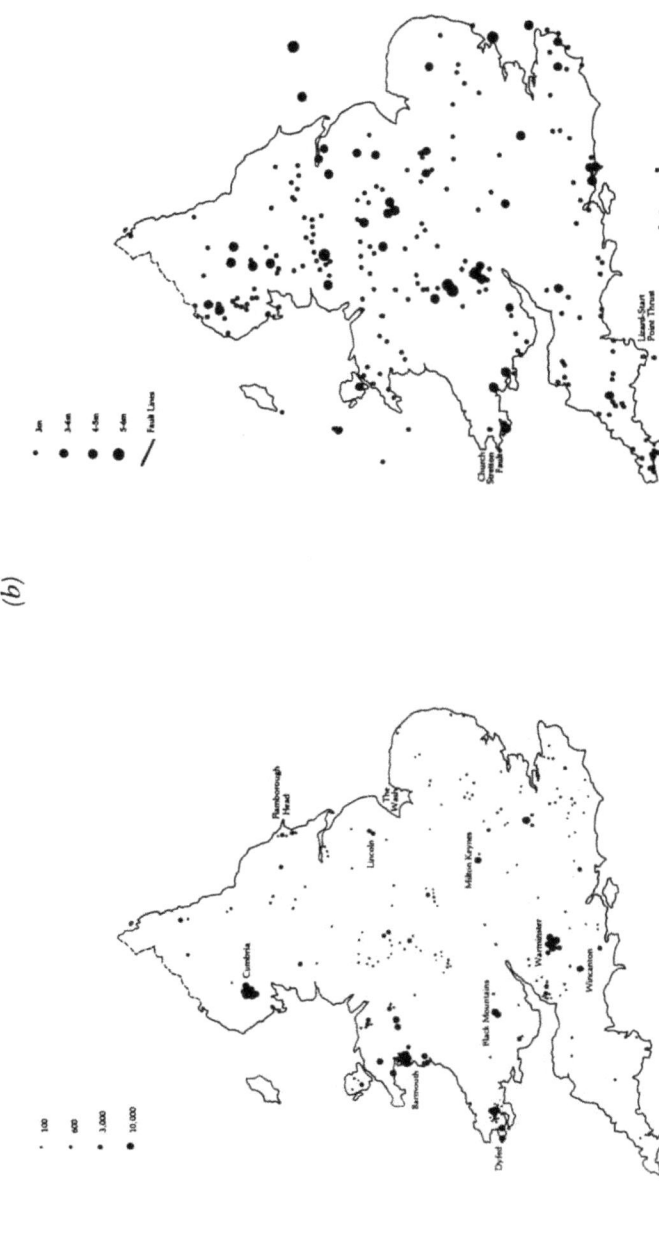

*(a) Population corrected UFO sitings for England and Wales. The 1971 census map provided the detailed information which allowed the corrections due to population bias to be made. The above map is the result, revealing striking 'window' areas. (Illustration reproduced from Earthlights by courtesy of Paul Devereux.)*
*(b) Earthquake epicentres in England and Wales. (Illustration reproduced from Earthlights by courtesy of Paul Devereux.)*

is a matter of matching retinal input with already existing data within our memories. This is what babies are doing when they first explore the world during their first months. Everything they encounter gets picked up, chewed, touched, turned over, chewed again, felt, and so they build up a tactile, visual, auditory image of the object, and in time they can recognise that object without going to all that trouble. Other objects then are matched against that imprint, and so general categories such as cups, toys, food, etc. get built up. But something that is totally outside our world or world-view has nothing to which it can be compared, and if it is at least partly ethereal, then it will be seen according to that image to which it is closest within the person doing the perceiving. Thus we used to have goblins, gnomes, pixies, elves, which varied widely in their specific characteristics from place to place, though there were certain things in common, and their details changed over the millennia: they grew smaller, and they always seemed to wear the clothing of a time just prior to that of the time in which they were seen. Nowadays the collective myth is that of space beings - so we see UFOs — the beings are still human like, but different, alien to us. And they are getting bigger again, some are even larger than the average human, though the 'greys' are small and very like traditional pixies in form.

The hypothesis is that there is some sort of two-way interaction through psychokinesis between our minds and the energy field of the UFO. In Rex Stanford's conformance theory it is proposed that the more subtle and sensitive the material the easier it is for our minds to affect it. It is hypothesised that the energy field that we call UFO is a profoundly sensitive energy form responsive to our subconscious mental cues. This is not to say that some UFOs are not visitors from outer space — some may be: there is absolutely no way at present of telling — just that some UFOs seem to be connected to energy fields related to the Earth, and we perceive it in terms of our cultural mythology. 'The UFO form is "a very wonderful fire" as one witness in the Barmouth wave put it.'[92]

In my discussion of PK and the subliminal mind, I mention that we are able to create thought forms through vivid visualisation, concentration, emotional energy and Zen action. This is exactly what is being suggested here. And this is equivalent to the theories underlying certain hauntings — that a place has an energy which we then put a visual and/or auditory image to.

The geomagnetic anomaly associated with the areas in which UFOs are most commonly found, affects our pineal gland which produces more pinoline, which stimulates production of DMT, lof divinity and all that implies in moral and ethical terms. This is which takes us into a psi-conducive dream state of consciousness where we are both psychic and 'think' at the collective unconscious level of our mind, in dream images, hallucinations, and archetypal primary process thought. It is at this level of our minds that we are most in touch with, or at one with, the world mind, which is manifesting as UFO form. Thus the UFO energy will respond according to the prevailing conditions, such as how close one is, time of day, ambient electromagnetic conditions, psychic ability of the percipient, and so on. UFO researchers have noticed that many witnesses do report other psychic experiences, so they are people who are open to experiences of this nature.

However, with global phenomena like UFOs one is not dealing with a personal, subjective thought form, but an image that is shared by a whole culture — in this case a Westernised culture since UFOs have been reported mostly in Europe, Russia, North and South America and Australia. UFOs are archetypal for our time. The prevailing imagery of the Westernised global culture of the twentieth-century means a technological, and more specifically a galactic, outlook — our aliens no longer come from Mars, they come from Sirius or Betelgeuse, or whichever other star system is stirring us, like the Pleiades! Our science fiction is a similar product of our species-unconscious imagery.[93]

The feeling that I, and this is also mentioned by Devereux, have about such things as UFOs and other anomalous phenomena that appear to behave intelligently is that we are dealing with an aspect of the planet herself as an intelligent being, another aspect of Lovelock's Gaia hypothesis,[94] but at a more mental level. We are part of this planet. Our minds are part of the evolving life of this planet, as are the levels of consciousness of other animals, and so just as we embody one form of consciousness evolved on this planet, so the hypothesis that this planet has a sort of world mind that manifests in anomalous ways is an hypothesis that has a certain appeal to it — at least at a magical level if at no other. It is this aspect of magic that is becoming so popular now with the rebirth of natural magic in Britain and America.

The understanding I have here is that the planet as a whole is an evolving organism, and we as humans are a particular aspect of the planetary evolution of consciousness with our huge neocortex, in that we have built exquisitely beautiful churches, cathedrals, and other buildings, have composed music, have painted pictures, and have a technological standard of living unheard of by other animals. With our language we are both individually and socially self-reflective and aspire to self-transcendence — to our highest idea the conscious aspect of the planet's evolution which we embody. The global mind common to us, and possibly also to certain other animals who also have a developed neocortex, is that mind which is pre-verbal, pre-analytic, pre-logical, the mind form of the limbic system, which is emotional, instinctive. This global mind common to all humans is the primary process mind. This archetypal mind used to express itself in the form of what are universally known as elementals, devas, and, in Europe as fairies. There is a possibility that this form is returning once again.

Devereux goes on to suggest that the UFO phenomenon is a sort of planetary ectoplasm. In the heyday of mediums in Victorian times, some physical mediums used to have a semi-physical substance known as ectoplasm flowing out from their bodies, mouth or navel or vagina, which would then take on the form of a hand, or even a complete body of a person. Devereux is suggesting that UFOs are like this: quasi-physical substance which can take on whatever form our consciousness puts into it. And remember we can understand the alien in terms of inner as well as outer space — there are more things in our mind than we are aware of — and our mind can act as a totally independent being. As Devereux so cynically says:

> On the one hand the mysteries of the universe can be reflected back to the stunned conscious mind of the witness, while on the other — a far more common occurrence — all that is displayed are elements from the common cultural ethos, usually in the comic-strip media imagery that has now become the mental currency for keeping our collective beliefs, fears and dreams alive. If the UFO pageant today resembles nothing more than the incoherent ravings of a fevered mind, then that is the true reflection of our current collective mental situation.[95]

This comment applies as much to the Victorian craze of mental mediumship, with its spirit guides, and the modern craze of channelling, as it does to UFOs. We have to take responsibility for our visions and our voices. Just because we channel something does not necessarily make it 'good' or 'right'. We can channel or experience garbage far more easily than we can channel or experience wisdom, because most of us are very unwise at present. Sometimes I am asked to define channelling and I tend to respond that it is a monologue for which nobody wishes to claim responsibility. Many witnesses of UFOs appear to have had prior psi experiences, which suggests that a person who is psychic is more likely to see a UFO, although I have seen no survey of this sort. It would be very interesting to initiate UFO witness questionnaires in much the same way that there have been OBE (out-of-body experience) questionnaires in order to elicit this sort of information.

The thesis that I am suggesting here is that some UFOs are Earth energies that we react to psychically, hallucinating imagery that reflects our times. The electromagnetic fields associated with UFOs have very strong effects on our psyches, through our pineal gland and the production of pinoline which links us with our collective unconscious, and so the imagery seen is archetypally important for us.

## Sacred Sites — The Wider Aspects of Geomagnetism and Psychic Phenomena

There is a strong folklore linking stone circles, other sacred sites and ley lines with psychic events, people who live in these areas having 'second-sight', apparitions, haunted houses, fairy legends and so on. Psychically they are the most 'charged' area of Britain, and with this hypothesis connecting psi with the pineal gland with the GMF we can at last begin to get a glimmering of understanding into what this 'charge' is.

Persinger and Cameron[96] have noticed a relationship between poltergeist-like episodes and geological fault zones, and this link is considered incontestable fact among certain dowsers who have a wealth of anecdote. A frequently reported psychic phenomenon is that of ghosts. A television programme on ghosts in Britain stated that there are more reported apparitions in Britain and Ireland than any other country in the world, about one in ten people seeing a ghost at least once in their life. That's about 6 million people in Britain who have seen a ghost. One thing that is very apparent in the reports of this type of phenomenon is that the sounds, sights, smells, touch, temperature changes and other disturbances experienced in the case of lightning, the Aurora Borealis and strong static electricity are all similar to those reported in connection with ghosts.

I have already mentioned at the beginning of the book how important emotion is in a magical or psychic event. Hauntings by specific ghosts tend to be connected with traumatic life events, or a traumatic death, of the particular apparition. This means that there is a lot of emotional energy surrounding the life/death circumstances of the apparition. If a place is charged by unusual geomagnetic fields, and the people living in that intense place react intensely, which means emotionally, then it is hypothesised that this emotional energy, in some way affects their energy body, L-field, and that this imprints on to the surrounding Earth field of the place, just as radio waves carrying information can be picked up with crystal receivers. Thus you get the same ghostly action occurring repeatedly as a memory replay imprinted on to that particular energy field, which, like a film, can be re-run for those who are sensitive enough to pick up the atmosphere of the place — atmosphere is an interesting word in this context, in so far as it denotes weather condition, air pressure — and emotional states. Our language again and again points out these underlying connections between the physical and the psychic. Under the right conditions, atmospheric, personal or emotional, an imprint is released. Thus a ghost is an hallucination of an imprint, and certain places can be hypothesised as being inherently more likely to have associated ghosts and hauntings, as well as certain times being peak times for this phenomenon to occur. Also certain people, those we call psychics or sensitives, are more likely to feel this energy imprint and respond to it by seeing or hearing the appropriate form or action. This is obviously only a part of the explanation — a spirit of a place or object is often far more than merely an imprint on a place with associated emotional, electrical, atmospheric charge.

There has been considerable research into the various ancient monuments and sacred sites that are to be found all over our beautiful island[97] and some remarkable correlations occur. These areas are highly correlated with UFO sightings which occur in window areas like Croft Hill in Leicestershire which is composed of granite and syenite and lots of quartz.[98] Devereux found that 50 per cent of the UFO sightings occurred within 10 miles of Croft Hill; 25 per cent within 5 miles. All the UFO sightings were in areas of exceptional geophysical activity, and were at places also associated with abnormal meteorological events — this covers strange thunderstorms, strange aurorae through to Fortean type events.

Comprehensive maps in Devereux's book show the relationship between UFOs and earthquake epicentres, between UFOs and geological faulting, between UFOs and areas with less thunderstorms than the average, between stone circles, ley lines and UFOs and between stone circles, ley lines and geological faulting. This research corroborates all of Tom Graves's ideas;[99] ideas in which he mentions dowsers' findings with regard to ley lines and stone circles. It seems that the energies he and other dowsers were picking up are at least linked to geomagnetic energies that wax and wane according to the cycles of the sun, moon and other planets, though some feel that there might be other energies such as microwave or ELF (extremely low frequency) waves. Of the 286 stone circles extant in Britain today, 235 of them are found on Pre-Armorican rock outcrops. Pre-Armorican rocks are those that are more than 250 million years old (Pre-Cambrian through to Carboniferous) and cover 36 per cent of the land mass of Britain. You don't need statistics to understand the significance of this, but for those who enjoy mathematical figures the statistical chi-square test of this occurring by chance yields the figure of $x^2 = 169.35$, $p < 1 \times 10^{-6}$. In other words, there is less than a one in a million chance that the stone circles were placed on these specific areas of geological faulting at random. The stone circles have been placed on specific rock outcrops. These rocks are extensively faulted. These geological faults are all areas of tectonic stress, leading to piezoelectric effects and geomagnetic anomalies. And this means they are places of unusual psychic power.

> From the work he has currently done on stone circle distribution in England and Wales, McCartney is satisfied that *every stone circle in those two countries is within a mile of a surface fault or lies on an associated intrusion.*[100]

Measurements of unusual physical effects associated with megalithic stones have been made using gaussmeters, which have shown anomalous magnetic readings near the stones compared to the locality; for example, Eduardo Balanowski detected with a gaussmeter several anomalies in the magnetic field around the stone menhir at Crickhowell. Cooper has measured radioactive levels with a geiger counter and certain areas around the site have yielded above-average counts. It appears that an *ionisation effect* is being recorded, which tallies with Don Robins's[101] ideas concerning the condenser effect of stone.

At the Rollright stones, where much of the Dragon Project research took place, audiosonic equipment measured a high-pitched sonic outburst from the stones just at the moment of dawn; and infrared photography showed the same energy burst from the stones at the moment of dawn. Several well-known and well-respected dowsers, for example Bill Lewis, have also measured unusual effects.[102] This effect has also been noticed by ornithologists recording bats near to ancient sites. So there is some interaction between the stone and the energising effect of light first hitting it — a truly '2001' phenomenon. Robins thinks that the stone crystal lattice gives some sort of semiconductor effect. At sunrise only the shorter wavelength visible and infrared radiation is clearing the horizon and it is this that creates the ultrasound effect in the stone monuments. The stones seem to act as some sort of capacitor which is discharged when the dawn light first hits them.

In other words, because many of the British stone circles are built on quartz-bearing granite intrusions, and are themselves built out of ancient quartz-bearing stone, the whole circle is surrounded by unusual geomagnetic fields, and static electric, sonic and infrared discharges occur when such activity as dawn, or hand pressure,

## Where Science and Magic Meet

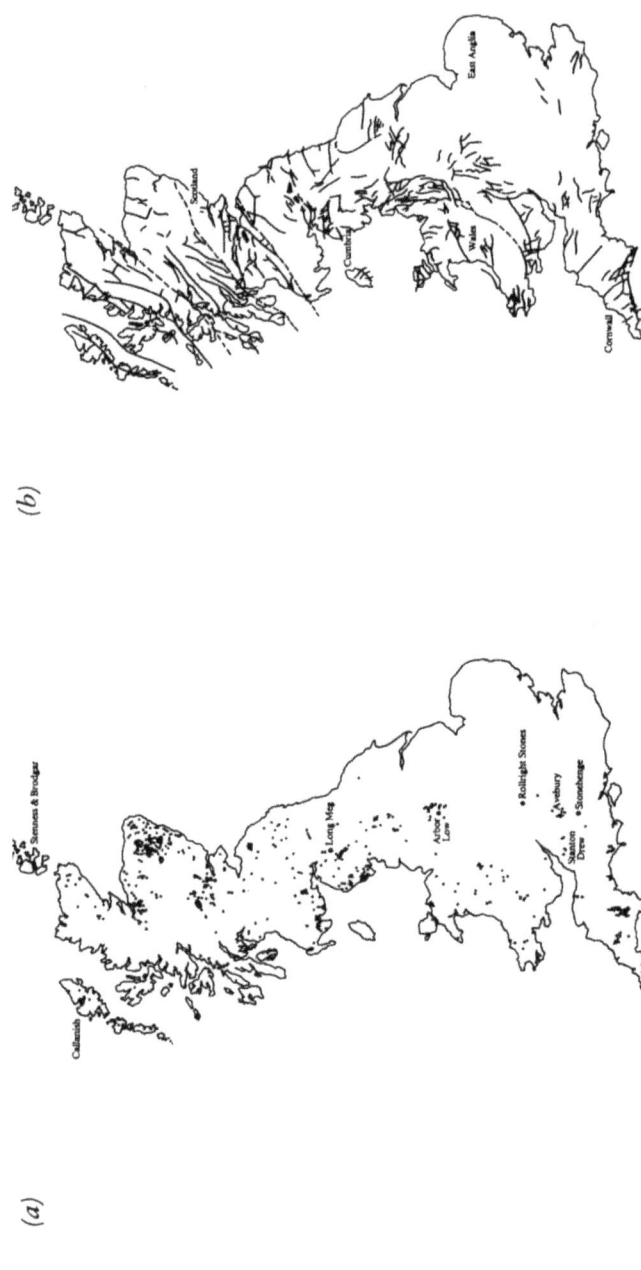

(a) Stone Circles in Great Britain. Notice the high concentration in the north and west of the country, and their total absence in the south east. ((Illustration reproduced from Earthlights by courtesy of Paul Devereux.) (b Geological faults in Great Britain. All the major faulting systems are illustrated. Minor localized faulting, which would unnecessarily complicate the pattern, is omitted. (P. McCartney). (Illustration reproduced from Earthlights by courtesy of Paul Devereux.)

touches the stones.

Because we are electromagnetic beings, we obviously interact with the electrical energies of the stones, and touching them causes pressure on them and an electrical discharge occurs. This is possibly linked with reports of people being thrown back from the stones, the stones dancing and walking to the nearby river or sea at certain times of the year or phases of the moon.

> The suspicion at the moment is that the placing of the megaliths at places of power - at places where an optimum confluence of cosmic factors, subterranean water and geological conditions exists - helps to maximise the energy storage and transduction effects occurring naturally in stone. It is also possible, as Robins suggests, that the groundplan geometry noted by Professor Thorn may also have an enhancing effect. The apparent early prehistoric obsession with astronomical observations also begins to make sense within such a framework.[103]
>
> It is even conceivable that the groundplan geometry of the stone circles was not originally drawn on the ground, but was simply a *result* of the stones being placed in optimum relationships to enhance the effects being sought by the builders. Perhaps it was only after many years of observing how certain patterns kept recurring that groundplan geometry became formalized. In the same way, the marvellous geometry of snowflakes is not prepared on a drawing board; it is the end result of the way certain substances have to be structured under certain conditions. All Nature is a slow motion geometrical event.[104]

This field research is of a preliminary nature and so the findings are only suggestive. It is to be hoped that it will be followed up by further work soon.

The complexity of a magnetic anomaly map is very often indicative of the geological complexity of the area. (It is really not surprising that California is having such a psychic boom considering that the San Andreas fault is in perpetual activity.) In fact Britain is a geologists paradise having almost all the different ages of rock present in one small island. The West coast of Britain - Cornwall, Wales, Cumbria and Scotland - are all areas of intense geological faulting and magnetic anomalies. These are the very areas where you get the majority of the ancient sites, ley lines, UFO sightings and people with 'second sight'. Standing stones, stone circles, long barrows and cursuses; these monuments are all that is left of the Neolithic peoples of Britain and Europe who inhabited these Isles from the end of the Ice Age about 9000 BC till the coming of the Bronze Age

Celts about 1400BC, who continued what the Neolithics had begun until the invasion of Iron Age Celts about 500BC, who absorbed the earlier cultures and formed a patriarchal society with a priestly hierarchy known as Druids. These Iron Age Celts are the people around whom the recent Celtic revival has focused; the Mabinogian, Arthurian legends, Ogham, tree calendars, and so on are relics of the ritual way of life of these people. And now we look even further back to knowledge of a way before the Iron Age.

Silbury Hill is 130 feet high; there are over 200 stone circles; there are thousands of barrows and other earthworks. Work on these sites lasted for thousands of years; the Neolithic people were not crude barbarians. What will our civilisation have still remaining in 5,000 years' time? Attempted renovation at Avebury showed that even with modern machines, erecting these huge stones was a mammoth task. The centuries-long construction of the major monuments tells us that the *place* was the important factor, since at Stonehenge the Bronze Age peoples added to what the Neolithics had started. Bringing in bluestones from Wales — over 200 miles journey — shows that *material* was as important as place.[105] Henges seem to be temples, astronomical observatories and even universities. Scientific and magical functions seem to be inextricably interwoven. And knowing what we do now about the pineal gland and the effect of the GMF on psychic functioning we realise that these places were specially chosen for enhancement of magical working. Tom Graves[106] even speculates that the circles act as some sort of screen from surrounding energies. Almost certainly the barrows do.

*Leylines*

Leylines (or dragon lines)[107] are essentially alignments of sacred sites. The most well-known of these, even called a Dragon Line by the National Trust, is the alignment that runs from St Michael's Mount at Penzance in Cornwall up through Burrow Mump, Glastonbury Tor, Avebury and through to Bury St Edmunds. It is aligned on the Beltane (May day) and Lammas sunrise, and the Imbolc and Samhain sunset.

Naturally if ancient sites are implicated then so too are their associated leys. 37.5 per cent of a random sample of leys showed 'evidence of UFO events occurring on them or in their immediate vicinity'.[108] I have personally witnessed a haunted house situated right on a leyline which has very strange effects on all the people and plants which stay any length of time in the building.

Many people feel that sacred sites, beacon hills, etc. are a sort of Earth acupuncture point, a site of Earth chi energy, and that leylines are the meridian channels for this energy. This is an analogy that I have used for many years, because it seems intuitively to fit. Thus the ancient practice of lighting beacon fires on the beacon hills at Midsummer and other sacred days is a bit like moxapuncture of the Earth's meridians. This analogy fits even better with the recent discovery that acupuncture points can be measured on our bodies using electrical skin-resistance meters, the skin resistance at the acupuncture point differing from the area around. And so with sacred sites, the geomagnetic force differs. Tom Graves discusses this concept in some depths in his book *Needles of Stone Revisited*.[109]

Leylines have a spiritual significance, in as much as there is a feel about them which lifts the spirits, which makes you gasp in wonder, which makes you feel good. Alexandra David-Neel in her book *Magic and Mystery in Tibet*[110] talks about the art of some Tibetans to walk with high speed over great distances, a sort of Eastern equivalent of seven league boots from the fairy tales. These leylines give one the feel that you could walk along them with seven league boots. Because presumably the Neolithic peoples' minds were more in tune with the subtle Earth energies, because they did not have any form of electrical pollution such as we labour under now in the form of electric pylons, power stations, roads, railways, television, radio

*Avebury avenue stone (illustration courtesy of Serena Roney-Dougal).*

## Where Science and Magic Meet

waves, etc., almost ad nauseam, then they would be far more aware of the psychic potential of these places and appear to have had a science using this natural Earth energy. I actually feel that many of the circles are places such as Doris Lessing describes in *Shikasta*,[111] places which are tuned in to stars for psychic communication, harmony of this planet's energies for evolutionary purposes. Stonehenge is a cosmic temple related to more than just movements of Sun and Moon; Callanish generates a strong spiritual atmosphere as well as relating to the whole of the Moon's cycle. Barrows and tumuli seem to be psychic meditation spaces, sweat lodges and initiation chambers, as well as Bronze Age burial places.

We obscure the traces of this vital energy with pipelines, power cables and motorways, polluting the electrical environment at what must be its most finely tuned and vital level; for the connection between the dragon lines, the earth's magnetic field, the electric air, the magnetic sun,

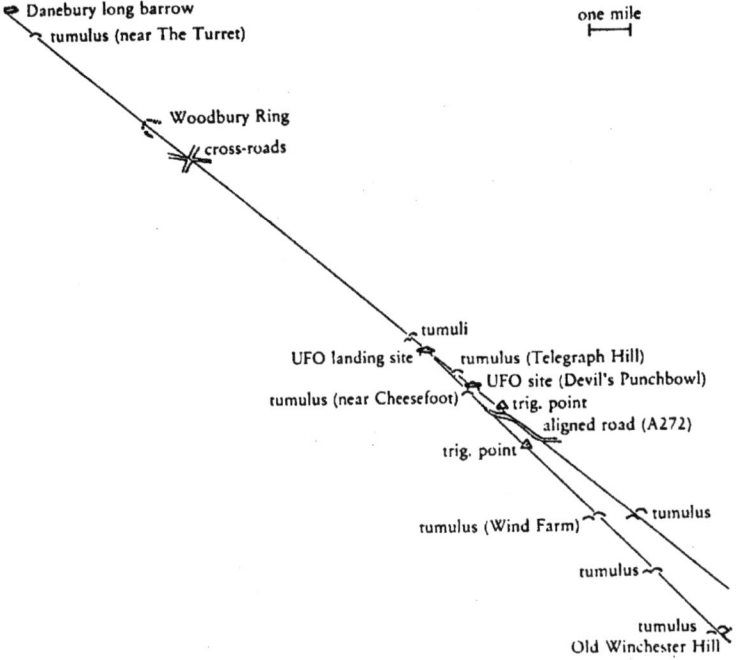

*Plan of the Winchester UFO/Ley alignments. (Note- this alignment was drawn up from a 1:50 scale map many years ago and has not been verified since.) It does, however, show the Devil's Punchbowl and Cheesefoot Head, which are now famous for their annual crop circles. (Illustration reproduced from The Ley Hunter, 75, by courtesy of Paul Devereux.)*

draws us into the overall energy patterns of the whole cosmos, linked electrically and magnetically through the electromagnetic vibrations of light.[112]

*Barrows — ancient Faraday chambers or Reichian orgone energy accumulators?*

Wilhelm Reich, a student of Freud who fled Nazi Germany to America, only to be imprisoned by the Americans in the 1950s for his radical ideas, conceived of a planetary energy which he called orgone energy. He distinguished between the general living energy of the universe and that energy utilised by living things, although the distinction is marginal. One thing I have noticed with regard to the life energy of this planet, and I wonder if I am correct, is that although there is mass extinction occurring at the moment of many thousands of species of plant, insect and animal, the actual amount of life energy of this planet is not decreasing — it is all converting into human form since there is a massive population boom in humans. So are we witnessing an evolutionary trend of the planet in which she evolves into primarily human-type consciousness, for whatever reason or purpose? Worth thinking about.

Reich showed that orgone energy has bio-electric properties rather similar to Burr's L-fields in concept, or Rupert Sheldrake's morphogenetic fields. Orgone energy is supposedly also found in the atmosphere as the blue haze that hangs over the tops of woodland and forest and is the scintillation that can be seen in the sky, although I always thought this was ozone. It is interesting that the blue haze of accumulated orgone (ozone?) energy should be seen in the colour that is traditionally associated with spirit, but it is not surprising. Reich constructed orgone accumulators, boxes made from layers of organic and inorganic materials, which built up surpluses of this etherous energy. He considered that they were helpful in various forms of treatment.

> Prehistoric burial chambers, usually sited at the convergence of ley lines, are constructed in exactly the same way as the orgone accumulator boxes. Layers of alternating organic and inorganic material form both types of chamber. The ancient ones are found all over the world; they all have the same pattern of construction. One can only surmise that these, too, are orgone accumulators, places for concentrating the life force for the benefit of humankind.[113]

Have you ever spent time inside a barrow, such as West Kennet long barrow near Silbury Hill? If you do so, you will find that there is a certain sort of peace inside the barrow. It is now well-known that some Bronze Age tumuli are built of clays that are not local, showing that the *material* used in even these little, exceedingly common structures was of vital importance. I feel that these clays, built sandwiched between an inner stone layer and the outer grass covering, act to shield the inside of the chamber from geomagnetic fields, but I have never seen any research with gauss-meters to corroborate this feeling. I say this because the PhD research I did at City University was done inside a Faraday chamber, and the feeling is identical. A Faraday chamber is made from fine copper mesh and screens out most electromagnetic waves such as radio, television, etc.

There is a suggestion that some tumuli have fire pits in the centre but no wood fire. Instead blackened pieces of stones are found. This suggests a sweat lodge where heated stones are placed in the fire pit. As there was no smoke hole this would make complete sense. Recent connection with American Indian spiritual tradition where sweat lodges are commonly used would suggest that the Neolithic peoples used these tumuli as sweat lodges, which makes me feel that different tumuli and barrows had different functions, as meditation chambers, places for all-night vigil, and so on.

Reich's orgone energy seems to be linked in some way with the GMF, which is linked with electrical energy. And I have been linking this with Burr's L-fields and with chi or psychic energy. But we must remember that electricity is not chi, not the fifth force; it is the vehicle through which we see some of its manifestations — when the fifth force becomes manifest it does so through electro-magnetism. The invisible worlds touch and interpenetrate the physical world and give us some enlightenment about the higher aspects of reality. Not surprising then that we gain this enlightenment through light, the impact of the fifth force, of the ether, on the electromagnetic fields that permeate our universe. Electricity and magnetism are the intermediaries between the material world and the ethereal world, touching us physically but also linking us to other realms of the totality of creation. It is through electro-magnetism that we can perceive the subtle forces that operate in those intangible regions.[114]

Electricity is the nerve that energises matter, life, society. Ball lightning, UFOs, leylines, dowsing, sacred sites, psychic phenomena can all be linked using our knowledge of electrical energy. This is not to reduce these phenomena to the purely physical, but to be a starting point for a fresh way of looking at these phenomena and seeing the common ground between them all. I see this approach as being one facet of the whole, one aspect of our understanding that complements and grounds the mystical, magical, esoteric level of understanding. There is an electrical correspondence to chi energy.

# Earth Magic: Sensitivity to the Earth's Magnetic Field

***As below, so above! (Dod Cartoon from The Ley Hunter Journal reproduced by kind permission of Paul Devereux.)***

> Chi energy, the vital force ... simply manifests itself in electrical terms as one aspect of its complex nature ... and so can be detected through the modulation of those forces.[115]

## Some Implications

Geopsychic magic — is that the reason for the megalithic monuments? The stone circles and the mounds, such as barrows and tumuli, were the sacred sites at which space, time and mind were most effectively linked: the early working temples. They are essentially absolutely practical, as well as sacred, in their function, places where the veils between this world of matter and the elusive world of spirit are thinnest — an elemental technology combining cosmic, atmospheric, geological and human factors.

The megaliths tell us that on this particular spot is something special. The implications of this are enormous. They suggest that the Neolithic peoples who built these monuments were aware of the energy systems involved and manipulated them using critical times such as the solstices or full Moons in order to enhance the geopsychic effects. This is a 'natural science' of which we have no knowledge at present — the interaction of our psyches with that of the planet, and hence with the rest of the solar system, and possibly even with the stars. The degraded relics of this science are present in our myths and folklores, and in some of the traditional occult systems, but they are no more than very degraded relics. The Iron Age Druids whom the Romans held in such high regard, knew perhaps but a fragment of that science of the peoples who built the monuments.

> In studying these occult groups and thinkers, one senses that something more complete, more awesome, so incredible as to be unspeakable, existed

in remote antiquity.[116]

It is noticeable that most megaliths congregate around each end of a geological fault line. For example there is the Wizards (Pixies) Tump at Hinkley Point in Somerset, where two nuclear power stations have been built. Tom Graves suggests that in modern times these places of unusual geomagnetic energy have been used by people for purposes other than the sacred, and that in doing so they are creating negative effects in our land. I feel great concern that the new privatised electricity industry is going to call itself National *Power*. Magically using that name is very bad news because these people are siting their life-destructive generating stations on sacred 'power' points. When is this madness going to cease? We must stop these people from destroying our world by their greed, their avarice and their usury.

To me, the lore about fairies is our oral tradition about this earth magic, the peoples who built the megalithic monuments, identifying and enhancing the Earth's energy at these points, and the ancient religion of these lands. So let us now explore this lore.

# 6. THE FAIRY FAITH[1]

With its remnants of occult learning, magic charms, and the like, folklore seems to be the remains of forgotten psychical facts.[2]
Many poets, and all mystics and occult writers, in all ages and countries, have declared that behind the visible are chains on chains of conscious beings, who are not of heaven but of the earth, *who have no inherent form,* but change according to their whim, or the mind that sees them. You cannot lift your hand without influencing and being influenced by hordes. The visible world is merely their skin. In dreams we go amongst them, and play with them, and combat with them.[3]

The myths and legends of the fair folk are the oldest in Britain and need to be revived. Our culture is like a tree, and as the branches grow higher, so the roots must grow deeper else we are not properly earthed and all our visions and ideals are just head in the clouds.[4]

The ideas associated with fairies strike at the very roots of human belief systems.[5] The Celtic fairy faith is part of a worldwide animism which forms the background of all religions. The Australian Aboriginal spirit race are virtually identical to fairies in that they are always youthful, frequent sacred sites, are only seen by psychics, control human affairs and natural phenomena, are real invisible entities who must be propitiated if people wish to secure their goodwill, and are beneficent, protective beings when not offended and may attach themselves to individuals as guardian spirits. The Phiraces in Thailand are probably the most complete parallel to the Celtic fairy faith, while the Amerindian beliefs are very similar, as are the spirit beings found all over Africa, Asia, China, Japan and India, Greek nereids and nymphs and the Etruscan Italian beliefs. The Arabian djinns and afreets have entered our fairy lore through

Aladdin and his magic lamp.

In the Scottish Highlands fairies are regarded as the counterparts of mankind, outwardly invisible, possessing magical powers, but strangely dependent on humanity — they are of a nature between mortal and spirit best seen in twilight. Magical skill is of the essence of fairy, who inherit it naturally while mankind can only acquire it, although we do naturally inherit psychic ability. Illusion, glamourie, is the supreme fairy art; transforming one thing into another — a rock into a magnificent castle or leaves into gold — the Cinderella tale. Occasionally a fairy will initiate a mortal into their arts, and this appears to be the origin of the Craft of the Wise (witchcraft) in Britain.

Invisibility is a typical fairy art — the fairy mist. Certain elves appear to be permanently invisible unless one has a talisman or spell by which they can be seen. Others have a hat or cloak that makes them invisible, e.g. mantle of Manannan, ring of Luned. In the Scottish highlands, 'fith-fath' is an incantation or word-spell which can render the speaker invisible or transform a person or object into another form, e.g. human into animal. This suggests not only that fairies initiate mortals into their art, but that mortals by their own art can come closer to the world of fairy. There is also a fairy ointment or salve which lets mortals see fairies. Women who nurse fairy children are often given the salve to rub on the child's eyes; if by chance or design they rub it on their own they find that they can then see the fairies at any time. Nearly always the fairies spit on the eye and so blind them when they discover what has been done.

Another of the magical arts typical to the fair folk is the gift of prophecy — the foretelling of events and the future of children, as well as their endowment with certain gifts or qualities, as in the tale of the Sleeping Beauty. In Scotland, what was known as the second sight was originally regarded as a means of beholding the fairies and only later did the term come to be used for precognition. Robert Kirk in his book, *The Secret Commonwealth of Elves, Fauns and Fairies*, revised by Bob Stewart,[6] fully describes the manner by which the second sight could be acquired for seeing the fairies.

There are four distinct groups of fairies:

1. *The coarse and country fairy* (e.g. the brownies of lowland Scotland and north England; hobgoblins; fynodere in the Isle of Man; in Ireland the leprechaun; the English lubberkin; in Scotland the gruagach or hairy one; the urisk, the brollachan or fuath and, of course, Puck and Robin Goodfellow).

There was a strong taboo against calling individual members of the fair folk by their particular name. There were times and seasons when it was unlawful to mention the fairy name. It is the exception outside Ireland to find a definite name and personality assigned to members of the fairy world. Hence many euphemisms came to be applied to them, such as the good folk, the people of peace, Tylwyth Teg (Fair Folk, the fair clan or family), bonnes dames, wee folk, seely folk, good neighbours, pechs, pixies, piskies and pooks. In the Scottish Highlands and in Ireland the fairies were known as the *sidhe* (pronounced 'shee' — the hill folk, the dwellers within the hills or mounds). 'The fairies,' remarks Robert Chambers, 'are said to have been exceedingly sensitive upon the subject of their popular appellations. They considered the term 'fairy' disreputable.'[7] So, from now on I shall call them the fay or fair folk.

The free use of the genuine names of supernaturals was believed to offend them and tempt their vengeance, for if you know the name of a person or spirit you have power over them. The name and the dread power named are so closely associated in the very concrete thought of pagan culture that the one virtually is the other, that is, subjective associations are projected into reality. It is implicit in animistic belief that the name of a man or spirit is a vital part of the individual — to know it and pronounce it presumes power over the person or spirit to whom it belongs. This is the essence of ritual magic. Some fay women even refused to acquaint their mortal husbands with their name, as did some fay husbands. Those fair folk whose names we do know are those who have attributes of divinity, for example, Gwynn, Manannan, Rhiannon. Even speech with the fair folk appears to have been regarded as dangerous in some areas of Britain. On the other hand, if a human learns a fay's name the fay is bound to undertake any task which the human wishes, or free them from any vow they have made to the fay, as in the tale of Rumplestiltskin.

2. *The banshee (bean-sidhe), the 'white ladies' or fairy women.* These are known in Ireland, Scotland, Wales and some parts of England. They are usually attached to ancient families of distinction and so differ from the group above who are associated with the peasant folk. The glaistig water sprite is a form of banshee, the spirit of a deceased woman of the family under fairy enchantment. However, the banshee may be a representative of a Celtic goddess of death. The banshee seems to be the origin of some of the white or 'grey lady' ghosts so common in large country homes. Thus, fairies who are linked with specific houses represent an ancestral spirit and also a deity of the house.

In the Fairy Faith, death is less a change of condition than a journey, a departure for another world. Thus, the dead can show themselves to the living; apparitions, phantom funerals and death warnings abound in Celtic lands. Just as psychics can see fair folk so can they see the dead. After death the soul still exists and travels amongst us. The lore concerning the ancestors is very similar to that of the fair folk: the dead guard hidden treasure; the dead take the living after sunset; the dead can make themselves invisible at will; the dead disappear at cockcrow. Barrows are places of the fair folk and of the ancestors. The dead are also said to frequent crossroads, which, together with barrows, links us back to ley lines and places on the earth where the veil between our material realm and the psychic is very thin.

Another link between fair folk and the ancestors is the taboo against eating either fairy food or food of the dead.

> Human food is what keeps life going in a human body; fairy food is what keeps life going in a fairy body; and since what a person eats makes them what they are physically, so eating the food of Fairy land or of the land of the dead will make the eater partake of the bodily nature of the beings it nourishes. Hence when a man or woman has once entered into such a relation or communion with the otherworld of the dead, or of fairies, by eating their food, his or her physical body by a subtle transformation adjusts itself to the new kind of nourishment, and becomes spiritual like a spirit's or fairy's body, so that the eater cannot re-enter the world of the living.[8]

To eat of food proffered binds one to the giver - a mystic relationship is formed. This compares directly with the Christian communion in which partaking of the body of Christ enables one to enter the kingdom of heaven. The eating of fairy food, making one as of fairy nature, is part of the ancient belief in the communion with supernatural beings.

A Roman writer, Porphyry, mentions an interesting little snippet of information: 'All the various gods, genii or daemons, enjoy as nourishment the odour of burnt offerings.' Burn the toast and give it to the fair folk; which sheds a completely different light on the story of Alfred the Great and the burnt griddle cakes. Also, the fragments which fell to the floor at a meal were believed in the Highlands to belong to the fair folk.

Another aspect of the fair folk which links with their ancestral, or perhaps divine, character is that of time. A night and a day in fairyland is equivalent to a year and a day in our world. (In Celtic

## The Fairy Faith

reckoning the day started at sunset hence the night comes first.) So a fairy year is 365 of our years; therefore if a fairy lives to be 100 that is 36,500 of our years. This is almost, but not quite, immortal! This shift of time and space is an interesting one. To a gnat we are immortal. One of our days is like one of their years. And so to the fay we are as gnats. Never forget this! This lore about time also links us with Einstein's Relativity Theory in which at the speed of light time stands still, i.e. eternity. An immortal is one who is always in the now — living at the speed of light. Is this what we mean by a lightbeing? And at the speed of light all matter becomes pure massless energy. To me this is the essence of that reality we call faery, and also of their craft we call magic. In fairyland one is not at the speed of light because time does pass, but much more slowly than here, so one is in a frequency domain that is part way towards the eternal, timeless realm.

Mortals, if they wish, can live in the world of the sidhe for ever. A magical bodily transmutation occurs if one lives in fairyland so that on returning to this world one must not touch anything, as, for example, in the tale of Bran and Nechtan. In this tale Bran and Nechtan thought they had spent but a year and a day in the Land of the Women, but on returning to Ireland they found that the legends of Bran were very ancient indeed. Nechtan, who left the boat, turned to dust on reaching the shore, so Bran told the people about his travels to the Land of the Women from the safety of his boat. This time shift is the same type of story as in the tale of Rip Van Winkle, though the actual shift in time is different.

3. *The British elf.* Elves dwell in communities whilst the former two types of fairy are typically solitary. Elves are known throughout Britain and tales about them vary only slightly from area to area. They are given to magical practices and possess neighbourly qualities, yet they are resentful of human interference and must be treated with courtesy or else! It has been said that they kidnap mortals, especially babies; exact a tithe of corn and milk; are governed by a king and queen; have a 'sacrifice' tithe every seventh year; marry and bring forth children; enjoy hunting; are jealous guardians of treasure; are good healers; possess great skill in the preparation of herbs for medicinal purposes; borrow goods from humans but always repay in full, often adding some gift or boon; sometimes inflict injury by shooting with a magic dart or arrow, or merely by giving a beating or a spit in the eye; possess the gift of prophecy; call upon mortal midwives to assist in childbirth; if a mortal penetrates to their subterranean dwellings they can be released from their toils only

by the agency of certain spell-breaking rites at certain fixed seasons.
In short, their entire economy constitutes what may be described as a definite culture-complex which marks them out as a class of being or spirit not only distinct from (hu)man, although in many ways closely resembling them, but also from other spiritual beings.[9]

Most of the elven folk, like the Puckish beings, live in wild places in or near barrows, knolls, mounds, castles, duns, brughs, and raths. They also live underground in a fair country as bright as day (Pied Piper of Hamelin), flat and green without hills and dales, in castles with crystal battlements, or a cave brilliantly lit; with apple orchards, crystal streams, beautiful gardens with flowers and birds. The fairyland (Elfane) of Thomas the Rhymer is without Sun or Moon, with streams of blood and water and the continual sound of the sea, and water rushing and clamouring all the time. Woods and forests are also home to the fair folk. The Forest of Dean had tall and dignified beings who married with local chieftains. It is this aspect of fairy lore, together with the deity aspect I mention next, that I link most strongly with the Neolithic people, and their megalithic sacred sites, and with the old pagan religion, often called witchcraft, of these isles.

4. *Pagan deities.* In Wales Gwynn ap Nudd (pronounced Neith) was king, ruler over the goblin tribe and lord of the dead. As a rule, the Hades world (Annwn) of the Welsh pantheon ruled over by Gwynn ap Nudd is, unlike the Irish Manannan's peaceful ocean realm of Tir-na-Nog, often described as a place of much strife; and mortals are usually induced to enter it to aid in settling the troubles of its fairy inhabitants. In the Welsh Mabinogion, and the High History of the Holy Grail, the tales of Arthur are essentially tales of the Brythonic Fair Folk.

The Irish sidhe are a race of beings who are like mortals, but are not mortals; who to the objective world are as though dead, yet to the subjective world are fully living and conscious. Some modern Irish fair folk descend in an unbroken line from the ancient pantheon of the Irish gods, the Tuatha de Danaan, who were the folk of the fairy hillocks. Fairy Ireland was thick with royalty, and a map could be drawn of the various fairy kingdoms which would be very similar to the mortal divisions. These are linked with the 'ancient cult of the sacred queen'.

One of the aspects of divine beings in all the various myths and legends is that they represent for us archetypal ways of being, lessons in life that we all have to learn in our self-development — our aspirations to be better people, to realise our own divine natures and

## The Fairy Faith

our highest ideals as well as our failings that we work to overcome. In other words we create our pantheons from our own subconscious fears, imagination, hopes and longings. Thus, divinities of any particular pantheon are a mirror of the peoples who created them. To rediscover our own Celtic and pre-Celtic, Neolithic, pantheons gives us an insight into the people who created this pantheon, and returns to us our British culture, rather than the Greek, Roman or Norse culture, with whose pantheons we are most familiar at present.

The great Irish goddess is Dana and her people are the Tuatha de Danaan. The modern Irish people, through their veneration for the good St Brigit, render homage to the Divine Mother of the people who bear her name Dana — who are ever-living fairy people of modern Ireland, for Bride or Brigit is one of the aspects of Dana who became Christianised. For when the sons of Mil, the Bronze Age Celtic ancestors of the Irish people, came to Ireland they found the Tuatha de Danaan in possession of the country. The Tuatha are said to have come from the north 'but whether from the earth or the skies or on the wind was not known'.[10] The Tuatha are said to have been intelligent, courteous, magical and excelling in knowledge.

There were three war goddesses; Badb (Bibe), the confounder of armies; Macha, the fury that riots and revels among the slain; and Morrigu, the Great Queen, the Crow of Battle, who infuses superhuman valour. (This name, Morrigu, is very reminiscent of Morgan le Fay of Arthurian fairy, witch and magical legendary powers.) Like the Fates, they spin and weave and have great influence over the lives of humans. As with the Greek myths, the Irish gods and goddesses helped the humans in their wars, as well as having wars themselves, in the eternal mythic battle of good against bad. Thus, the battles of Moytura are a record of a long warfare to determine the spiritual control of Ireland by the Tuatha, representing gods of all that is good, and the Fomorians, representing gods of that which we must strive against. Humans, such as Cuchulain, Conn, Finn and Cormac, fought in this battle and the Tuatha fought with humans against the invading Danes.

The three war goddesses often exercise their magical powers in the form of royston crows, in Scotland as hoody-crows, in Brittany as magpies, in Wales as crows, in Scandinavia and Germany as ravens. This connection of the magical with crows extends even so far as the sacred ravens in the Tower of London, and the crow as a special messenger in Tibetan divinatory lore.

The large circular barrow, New Grange, on the River Boyne in Ireland is strongly associated with the Irish fair folk, the Tuatha de

Danaan. It was built by Neolithic people, and, as the burial place of the High Kings of Tara, was the sacred and religious centre of ancient Ireland. Some of the Tuatha de Danaan live in Erin in barrows, and some live overseas in a domain which is best described as a land of the gods, a paradise of immortals, Tir-na-Nog (the land of youth), Manannan's isle in the West, an overseas paradise. This is a retreat of the gods as well as the land of the ancestors, in which they reside in bliss and to which favoured mortals may be admitted while still alive so that they may share the happiness of the Celtic divinities. Peace and plenty are its chief attributes. Plenty of wines and ales, beautiful scenery, the people noble and friendly beyond all human standards, old age unknown, no sickness, flowers everywhere, mead and fresh milk flow, and everyone is on permanent holiday. The Arthurian equivalent is the Isle of Avalon. In Scotland this paradise is called Sorcha, in Wales Annwn, in Cornwall Lyonesse. There is considerable correspondence in descriptions of the two lands, underground and overseas, in all the different strands of fairy lore.

> Fairyland exists as a supernormal state of consciousness into which men and women may enter temporarily in dreams, trances or various ecstatic conditions; or for an indefinite period at death.
>
> Though it seems to surround it and interpenetrate this planet even as the X-rays interpenetrate matter, it can have no other limits than those of the universe itself.[11]

Manannan Mac Lir, the founder and ruler of the land of immortality, bears a silver apple branch — a wand in the form of a branch, like a little spike or crescent with gently tinkling bells upon it. Irish druids made their wands of divination from the yew tree, which was the Celtic symbol of rebirth. Could this tree lore have come down from the Neolithic peoples via fairy lore? This sacred branch of the apple tree bearing blossoms or fruit, or sometimes just an apple, is the passport of the Otherworld, gift of the queen of Tir-na-Nog, often serving as food, sometimes making sweet music. Their music is heavenly — hence enchantment. I feel that sound energy is a vital clue to their magic and to the magic of the stone circles.

For example, in one tale of Cormac, he undergoes three trials and goes to Manannan's land and is given the cup of Truth, which is a symbol of having gained knowledge of the mystery of life and death, and also the Apple Branch, which is a symbol of the peace and joy which comes to all who are truly initiated:

> For to have passed from the realm of the dead, of the fairy-folk, of the gods,

and back again, with full human consciousness all the while, was equivalent to having gained the philosopher's stone, the elixir of life, the cup of truth and to having bathed in the fountain of eternal youth which confers triumph over death and unending happiness.[12]

Irish heroes were avatars or reincarnations of the early gods, that is, related to the Tuatha de Danaan. Goddess Eatin becomes the mortal wife of the Irish king Conchobar. Dechtire, his sister and mother of Cuchulain, is called a goddess. Cuchulain is an avatar of Lugh, a sun deity. Thus, the Tuatha women married the Irish Kings and so continued to rule Ireland for many generations after the Sons of Mil forced the Tuatha underground. And their descendants ruled after they had left this world for fairy land.

> And so it is that long after the conquest the people of the Goddess Dana ruled their conquerors, for they took upon themselves human bodies, being born as the children of the kings of Mil's sons.[13]

I find this a fascinating aspect of the social hold of the Tuatha de Danaan in Ireland, the lore that the high kings of Tara could only reign by marrying a Tuatha woman — hence the line was matriarchal and although the Bronze Age Celts had nominally won the Isle, the Tuatha continued to govern through the women who were considered to be the spirit of the land. This is the essence of goddess worshipping matriarchy which seems to be the hallmark of the Neolithic peoples. The Picts in Scotland were also matriarchal. Obviously the fay are both male and female, but female fair folk stand out in the folklore, partly to glorify the male Celtic hero by his association with them, and partly because many of the ancient goddesses survived as fair folk.

Hunting and riding in procession are the most commonly reported activities of the fairy court — riding by moonlight with jingling of silver bells along the fairy paths, which can be linked with ley lines. Horses found in the morning all a'lather were considered to have been elf-ridden. This belief originates in the Wild Hunt — the gods, the supernaturals, the providers of food, were the controllers of the game supply and they themselves hunted it down. The Sluagh of the West Highlands were actually thought of as hunting men with bows and arrows during their flight through the air, so that they might capture their souls. The Sluagh are sometimes regarded as fairies, at other times as the dead, this link between fair folk, ancestors and divinity being inextricably intertwined.

As with all things, making distinctions and separate groupings

*Where Science and Magic Meet*

**Puck, the famous shape-shifting Hobgoblin (closely related to the Welsh Pwca and the Irish Phooka. (Illustration reproduced by kind permission of Brian Froud from Fairies published by Pan Books)**

like this is an artificial exercise, for these four different groups shade one into the other. I separate them out so that we can see the various threads more clearly. In the same manner I separate out the various origins behind all our different beliefs about the various sorts of fair folk, though once again there is no clear line between the different aspects.

1. *Elementals — devas, sylvan spirits and deas or 'mothers'*

There are four kinds of elemental or devic fairies according to the four elements:
- Earth: gnomes, goblins, pixies, corrigans, leprechauns, and some elves;
- Air: sylphs, aerials, flower fairies;
- Water: undines, lake fairies;
- Fire: will-o'-the-wisps, salamanders (seldom found in Celtic Faith).

It is through these daemonic beings that we interact with the divine. This is an important point to me. At one level we can understand ley lines, sacred sites, UFOs, haunting as some sort of electromagnetic earth energy that affects us physically. On the other side we can understand all of this in terms of fairy lore; 'conscious beings who are not of heaven but of earth, who have no inherent form, but change according to their whim or the mind that sees them, who direct the magnetic currents of the earth.'[14] These words were written a hundred years ago, and to me they are the link between modern science, myth and magic.

The size of the fair folk varies from thumb size to human stature depending on type and era. In all likelihood fair folk of larger stature were ancient gods while the diminutive forms are the devic aspects of faery. Shape shifting by magic may partially explain the variations in size, but also the different types of fay, and also cultural expectation. The belief in the fluid nature of spirit underlies this. Sea gods particularly are prone to shape-shifting which reflects the ever-changing character of their element, whilst the earth spirits are more stable in their form.

## 2. *Poltergeists and other psychic spiritual aspects*

The following psychical phenomena have been linked with the fair folk: collective hallucinations and veridical hallucinations; objects moving without contact; raps and noises called 'supernatural'; telepathy; seership and visions; dream and trance states manifesting supernormal knowledge; mediumship or spirit possession.

Gervase speaks of 'follets' who pelt the houses with stones etc., annoying the people in much the same manner as the brownies in Berwickshire. This is a typical spirit interpretation of what is now called poltergeist, and associated with chaotic psychic phenomena often linked to an adolescent who is undergoing severe emotional disturbance at the onset of puberty, but occasionally requiring some other form of explanation, of which the fay is as good as anything else that has been thought up.

As M. Camille Flammarion said of mediumistic seances:

> The greater part of the phenomena observed — noises, movement of tables, confusions, disturbances, raps, replies to questions asked, are really childish, puerile, vulgar, often ridiculous, and rather resemble the pranks of mischievous boys than serious bona-fide actions. It is impossible not to notice this. Why should the souls of the dead amuse themselves in this way? The supposition seems almost absurd.
>
> There could be no better description of the pranks which house-haunting fairies like brownies and Robin Goodfellows and elementals enjoy than this.[15]

*Where Science and Magic Meet*

Compare the above quotes with the information I have already given you about haunted houses on ley lines (fairy paths); with modern reports of crisis apparitions which are so very similar to traditions about fairy apparitions giving news of death. There are hundreds of carefully proven cases of phenomena or apparitions precisely like many of those which the Celtic peoples attribute to fair folk. These phenomena are, so to speak, the protoplasmic background of most religions, philosophies, or systems of mystical thought yet evolved on this planet. Fairy phenomena are in one aspect essentially the same as 'spirit' phenomena, so the belief in fair folk ceases to be purely mythical, and fairy visions are to be understood in the same terms as the one I have been propounding with regard to UFOs; that certain types of sightings of the fair folk do arise from a sort of mental suggestion arising from the earth energy or intelligence we call fairy acting on the percipient's subliminal mind.

This mental mystical part of ourselves is always there, merely changing its form every so often. UFOs are fairies in modern guise, representing in outer form the subconscious archetype of our planetary mind at this time — contact with other beings in the Universe.

### 3. *Crop circles and circle dancing*

Corn circles can be seen to be both a recent and old phenomenon that was previously attributed to the fair folk as witnessed in a poem that Geoffrey Ashe mentions in his section on the fair folk in his book *Mythology of the British Isles*[16] It is by Richard Corbet (1582-1635) and entitled: 'Farewell, Rewards and Fairies':

> Witness those rings and roundelays
> Of theirs, which yet remain,
> Were footed in Queen Mary's days
> On many a grassy plain;
>
> But since of late Elizabeth
> And later James came in,
> They never danced on any heath
> As when the time hath been.
>
> By which we note that fairies
> Were of the old profession;
>
> Their songs were Ave-Maries,
> Their dances were procession.
> But now, alas, they all are dead;

Or gone beyond the seas;
Or farther for religion fled,
Or else they take their ease.

I find this poem very interesting, because the time of the passing of the fairies is that of the witch hunts, and the murdering of the pagan peoples, who were closely associated with the fair folk. Thus, the times of Elizabeth I and James VI and I were at the peak of the witch trials; and the curious reference to the old profession makes me suspect that he refers to magical ability, called the craft in humans, and to the worship of the goddess (Ave Marias were their songs), and the pagans had to flee for their lives from the Puritans at this time.

And now with the crop circles it seems as though they are dancing their rings and roundelays yet again. The crop circles phenomenon has been increasing and changing dramatically over the past thirty years. I know that it was not a known phenomenon by the farmers when I was a child, and other than fairy rings, which were commonly interpreted as mushroom circles, there were no folk references to it. The linking of mushroom circles with fairy rings is only a partial link because of the link of fairies with hallucinogenic mushrooms such as psilocybin and amanita muscaria, but does not explain the poem's lament that the fairies are no longer dancing their rings and roundelays, for the mushroom rings did not cease at the time of Elizabeth and James, whereas perhaps the crop-circle equivalent of that time did cease. And with the neo-pagan resurgence of the past few decades, or whatever other social mythos you wish to take, perhaps the time is right. Though there may have been the occasional circle,[17] there was nothing like what we are witnessing now.

I have also noticed that when first reported in the early 1980s the rings were single large rings, and that since then they have increased not only in number but in complexity of pattern. Whatever is causing these rings has a sense of humour, for no sooner have the researchers decided that they have understood how the rings are formed and have defined the limits on their formation, than rings have appeared which defy these limits. This has a very pixieish quality to it.

Another link I make between corn circles and fair folk is the ancient practice of circle dancing within the stone circles. Sir John Rhys identifies Stonehenge with the famous temple of Apollo in the island of the Hyperboreans referred to in the journal of Pythesas travel, and says:

> The citizens gave themselves up to music, harping, and chanting in honour of the Sun-god, who was every 19th year wont himself to

appear about the time of the vernal equinox, and to go on harping and dancing in the sky until the rising of the Pleiades.[18]

This dancing would have been circle dancing as in the national Breton ronde or ridee at, or in, such cromlechs which are, like the dance, circular in form. It is thought that these circle dances are memories of ancient initiation dances; they are definitely highly spiritual dances, related to both Sun and Moon. To dance these dances in the stone circles is strong magic for invoking spirits such as the fair folk represent. Circle dancing is universal and is still practised by, for example, some Amerindian peoples, and is once more becoming increasingly popular. It hits a chord within people at this time and so feels incredibly good to do.

This circular initiation dance has come to be attributed to corrigans in Brittany, to pixies in Cornwall and England, and to the fair folk in these and other Celtic countries. Circle dancing like the Breton ones used to be danced at weddings in the Channel Islands, the revolution being around a person not a stone, and the people bowed to this central person. Evans Wentz notes that if the dance was pictured as a circle with a dot in the centre, we have the astronomical symbol still used by astrologers to represent the sun. He says that in Guernsey the sites of principal dolmens or standing stones 'were visited in sacred procession, and round certain of them the whole body of pilgrims solemnly revolved three times from east to west'.[19]

Evans Wentz also notes that the place for holding a gorsedd (a sort of modern Welsh initiation), under the authority of which the Eisteddfod (an annual Welsh festival of song, poetry and dance) is conducted, was also within a circle of stones, face to face with the Sun as there was no power to hold a gorsedd under cover or at night, but only where and as long as the Sun was visible.

Have you noticed the synchronicity between the new trend of circle dancing and the reappearance of corn circles? As far as I can tell the craze for circle dancing started about 1980, after it was introduced to the Findhorn community in Scotland. The first reported crop circles were also approximately 1980. I know that Colin Harrison started teaching circle dancing here in Glastonbury in the early 1980s, and that it has become more and more popular since that time, so that you can even do it at evening classes now. I wonder if there is some sort of a link, other than the synchronistic one. Circle dancing seems to be the most ancient form of dance that we know of. You even see Paleolithic cave paintings of women dancing in a circle. My suggestion here is that the dancing in circles being done by so many thousands of people now all over the country is a sort of sympathetic magic which is recharging the Earth so that phenomena connected with spirits of the Earth are reappearing.

In the two decades since writing this, crop circles have become art forms and are no longer just simple circles. The discussions now have become as complex as the circles themselves and so I prefer to leave this to people who are more knowledgeable than I.[20]

### 4. Memories, myths and legends of the Neolithic peoples

There are many strands in the lore about the fair folk that suggest we are hearing a memory of people who lived in these lands from the end of the Ice Age, 10,000 years ago, until the coming of the Bronze Age Celts. These are the megalith builders, so the more we can glean from the tales, the more we can understand about these ancient monuments, and the magical reality they represent. The Dravidians in India still worship at ancient dolmens in what is still remaining of their forests, and in the mountains. 'The Dravidians are considered to be the builders of the dolmens as are the ancient peoples of Britain.[21]

The first and most obvious link is that the elven folk and the sidhe are associated with *burial-mounds, stone monuments and circles,* e.g. the spriggans are found only in cairns, cromlechs, barrows or detached stones, with which it is unlucky to meddle. Most stone circles have legends linking them with the fair folk, for example, the Merry Maidens in Cornwall and the Fairy Piper.

Perhaps the most telling aspect of this lore is that the *fair folk hated iron;* this is a dislike shared by ghosts and other spirits in many parts of the world. If a man fixed his dirk (a dagger or knife) in the door of a fairy mound it could not be closed. Infants could be protected from the fairy kidnappers by iron in the crib.

The Neolithic peoples did not use iron, so in time iron came to be religiously regarded as efficacious against spirits and fairies. More though: the philosophy underlying the magical properties attributed to iron 'is based on mystical conceptions of virtues attributed... to various metals and stone and a careful examination of alchemical sciences would probably arrive at an explanation.'[22] This is partly psychological, and partly electromagnetic because iron is very strongly magnetisable and this affects us psychically. In understanding the electromagnetic energies of sacred hills, stone circles, etc., and of our own bodies, so we begin to realise how strongly iron affects us and our environment physically, and so hence presumably psychically. It is not just our electric society in which we are never more than a few miles away from an electricity cable of some sort that is affecting us at every level, but our ubiquitous use of iron. I feel that something went drastically wrong at a very deep level — literally a disaster *Shikasta* style[23] at the end of the Neolithic which brought on the Bronze and Iron Ages, and which led to the degeneration of our cultural history and the present state of our planet — the life-destroying force was able to get a foothold on to this planet at that time.[24] By being aware of the seed of this force, so we can heal that harm done to all those who held fast to the principle of love during the dark times. On the other hand I also see the shifts as part of the evolution of consciousness, but more of that later.

The third aspect of the lore that links us with an earlier race of people is the *connection between the fair folk and agriculture*. They traditionally own all the common and wild land until it is cleft by spade. Is this another instance of antipathy to iron? or Mother Nature being disturbed? It is regarded as a distinct breach of elfin law or privilege to till fairy soil, or to remove stone, timber, or leaf from its precincts, i.e. sacred space, particularly barrows, cromlechs, stone circles, etc. If you cut a thorn tree growing on a spot sacred to the fair folk, or if you violate a fairy preserve of any sort, such as a fairy path (leyline) or by accident interfere with a fairy procession, illness and possibly death will come to your cattle or even yourself. This is a link with Neolithic sacred sites, a method of preserving them from destruction. If we had kept this taboo until now, perhaps we would not be so bereft of trees or wild and open spaces — bring back the taboo! These sacred places were possibly originally the places that the remnants of the Neolithic peoples actually lived.

However, the lore says that they will help on tilled land turning the soil until ready for planting — often with a bargain built in to the deed so that they received the ensuing crop. They were thought of as exercising control over the crops. In Savoy they are said to have taught the people the art of agriculture. Agriculture is absolutely central in a survival culture and so became embued with knowledge that was passed on through the ages and gradually became ritualised, mythologised, etc. The cross-quarter days are important agriculturally rather than being turning points in the solar year, and I find it very interesting that, in the Celtic calendar, these were the important festivals. In Scotland and Ireland, right up until the last few decades the cross-quarter days were public holidays. We are heirs to a marvellously intricate system of agriculture, but developing that system took millennia and in the early days it was vital to hold the

## The Fairy Faith

information as clearly as possible.

'With Christianity this ritual aspect of agriculture became distorted and disorganized — what remained was a vague but ineradicable conviction that so-and-so must be done or the powers would be displeased.'[25] Nowhere do we see more clearly than when examining the roots of our pagan heritage, how anti-mother, anti-Earth the Christian religion is. They worship the sky father (Our Father who art in heaven) the sun god born at the winter solstice, and deny any homage to the mother, to the Earth. And fairy lore and the pagan religion that is associated with it, is primarily a religion of the Earth. It was this that was suppressed so vehemently by the Christian Church.

The Tuatha de Danaan of Ireland were also concerned with agriculture and the institution of festivals and ceremonies connected with it. The country fairy such as the brownie is the guardian of the farming peasant, who assists them to grow food successfully, so relieving them of the hunting-gathering way of life — an essential part in the evolution of humanity, that occurred in the Neolithic times. Thus the fay exact tithes of corn, milk etc. In the Western Isles of Scotland the top grain of corn on every stalk belonged to the good people. In that region if a person had the 'ceaird chomuinn,' the association craft — a special fellowship with the fair folk — he or she could compel the fair folk to come to their assistance for planting or reaping whenever they chose. 'This fact is eloquent of an ancient pagan cultus of which the fairies were the presiding gods or spirits', namely the Old Religion.[26]

The fair folk were also highly *moral and ethical.* There is an extraordinary zeal shown by the fair folk for neatness and decorum in all the circumstances of life. If everything in the house was not absolutely as it should be the house cleaner was severely punished — pelted with stones, pinched and tormented until they cleaned up. This is the fay aspect of poltergeist and haunting phenomena. Vice and greed were invariably punished; the deserving poor took the places of the rich; the plain but virtuous became the beautiful. The elves appear as a moral force; they are invariably on the side of justice, as in Robin Hood and Maid Marian legends in the fairy tradition. One who does a fay a service must never ask more than just payment for it. This idea comes from the belief that fay spirits, in their ancient aspect of gods, personify the sense of justice and right-dealing. This aspect of religion is vital to our human growth: we need a moral and ethical superstructure to guide us, and this the fair folk provided. Their code is akin to all religious spiritual paths, the Perennial Philosophy. The fairy faith does not separate religious ritual practice from everyday life, so every facet of life was imbued with religious significance.

The early Celts recognised an intimate relationship between humans and nature; unseen forces, not dissimilar to what Melanesians call *'mana,'* looked on as animate and intelligent and frequently individual entities, guided every act of human life. As long as people kept themselves in harmony with this unseen fairy-world in the background of nature, all was well. This is the essence of the new pagan 'Green spirituality' which is arising once more today. In this code, as soon as taboo was broken, disharmony set in. The whole background of taboo appears to rest on a supernatural relationship between humans and the other world. Almost all taboos ought to be interpreted psychologically, or even psychically, and not as ordinary social regulations.

In all parts of Britain it was considered essential to keep on good terms with the fair folk. They were good neighbours, but relentless enemies. It was held to be good to keep a bucket of water in the highways overnight so that the fair folk could bathe in it, to leave food for them, etc. If this custom was not respected then their enmity was assured. If they are harmed they wreak woe by firing magical darts or arrows (Neolithic), which bring about partial or entire paralysis which enables the fay to take the person to fairyland, or else the soul was spirited away and the body left. This is called the fairy stroke, which is a supernatural injury inflicted magically, sometimes causing epilepsy and almost invariably fatal in its results, but which could be cured by a 'fairy doctor', that is someone who specialised in curing complaints induced by fairy agency. These ideas sound very similar to our modern conceptions of near-death experiences, or even out-of-body experiences, and also to psychosomatic illness in which the mental state of the person is responsible for their physical dis-ease. This links also with the folklore surrounding hexing, in which a shaman, sorcerer or witch causes harm to another through their magic.

There were many charms and spells by which you could protect yourself against elves or keep them at a distance. I find this fear of fair folk very intriguing because it is so contrary to all the rest of the fairylore. I wonder if what started out as just retribution and an awe that ensured their ways were preserved, then became perverted later by Christianity and the witch trials. This fear, in its origins, seems to be part awe for the magical beings who built the henges, barrows etc. and who held so much wisdom and knowledge, and part the animistic spiritual knowledge that any being of great power is over and above the mortal's own needs and wishes and sometimes the mortal has to submit to the higher forces. Both witches and fair folk were feared for their powers to harm you — with fair folk that harm only came if you infringed one of their lores.

Linked with this lore concerning their moral code is the tradition of the friendly fairy. If one assists a fay, often in the guise of old woman etc., then you will receive aid when you most need it, e.g. Cinderella, the Red Etin, Puss in Boots. In some Scottish variants of Cinderella the fay appears in the form of a red calf or a grey sheep. On the whole however, the helpful fay in animal form is not so usual in Britain as it is on the continent. Fair folk used to gift people with good qualities or material treasures: caps or cloaks that make one invisible; chests that hold an inexhaustible supply of grain; excellence in one's craft (as in the story of the Shoemaker and the Elves); boots that make you swift of foot; magic stones which transport you where you will; or cows with super milking powers.

One of the crafts they taught was that of *healing*. The fay are renowned for their skill in healing wounds and diseases, for their knowledge of herbs. At times they communicated this knowledge to mortals, and it became one of the hallmarks of a wise woman, or a cunning man, that they knew about herbs and healing.

In Scotland and Ireland they are also skilled in *crafts* — smiths, cobblers and brewers. As spinners and dyers the elfin women were unrivalled, thus resembling the Fates of classical mythology. Fair folk wear green; or fawn skins; or graceful linen garb of black, grey, green or white; or red, blue and green; or green with a red or blue cap with a feather. Basically their apparel and speech depends on place and culture.

In fairy lore, green symbolises eternal youth, resurrection or rebirth as in nature during springtime. Thus in the initiation ceremonies, on return to the physical plane after the trance the initiate was dressed in green, for they had penetrated the Mystery of Death, as in the tale of Cormac which I mentioned earlier. Green symbolises the birth of the world and the moral creation or resurrection of the initiate. Many mythic peoples have green especially attributed to them: Melwas had a green cloak; Robin Hood and Maid Marian dressed in green. Gwenhywvar (King Arthur's wife) tells her knights to dress in green, and in one tale had a green horse.[27] This symbolism concerning green is found in other times and places as well. Thus, at the Festival of Al-Khidr in Lower Egypt, Al-Khidr always dressed in green as he is immortal, unchanging. In masonry, green is the symbol of life, immutable nature, of truth and victory. In the evergreen the Master Mason finds the emblem of hope and immortality.

Knowing this about the symbolism of green and the fair folk, makes the Green Movement even more appropriately named, for it is the physical aspect of the neo-pagan spiritual movement to honour and love and cherish and care for the Earth, and the fair folk are the beings of the Earth.

The fair folk are most especially renowned for their *music and dancing* — circle dancing particularly in grassy places, and among the stone circles. Men who were kidnapped tended to be pipers or minstrels who returned with the gift of fairy music in their hands.

There are many tales that concern *marriage* between a fairy and a mortal. Such a union was hedged round by rigid taboos. The fay mistress must be permitted to retire from public gaze at certain times. The mortal lover must not reveal his association with his fay mistress or allude to her fay nature or touch her with iron. If any of these happened she would disappear, sometimes with any children she had borne. In general the fay woman made the first advance to her prospective lover — remnants of the old matriarchal ways here. Sometimes the fairy lady appeared to her lover in the guise of a deer (as in the tale of Finn and Sava), hare or other animal. She often gave him a gift which protected him. Occasionally mortal women had fay men as lovers, often surrounded by problems such as wanting to know his name, the lover disappearing and the woman spending years and many trials searching for him. Often girls were abducted to be wed to the fairy king.

This intermarriage was apparently very necessary for the fairy peoples, which suggests on a physical level that the Neolithic peoples were dying out and needed fresh blood so that there should not be too much inbreeding and weakening of the line, and, on a psychic/spiritual level, the maintaining of the ethereal within humanity for the evolution of our consciousness. Children of such marriages are renowned for the second sight, which could be the origin of the supposed hereditary witch families — those families that seem to have psychic abilities as an hereditary trait.[28] There are several cases in witch trials of the witch claiming marriage to a fairy. How many of us have fairy blood? A classic example is the tale of the McLeods of Dunvegan who were helped by a fay woman.

Most parts of Britain had such marriage practices. The fairy records of Wales abound in legends of elfin sweethearts, e.g. Pwyll and Rhiannon, The Green Knight, the tale of Melusine, the fairy lady who wed Guy, Count of Lusignan. Thomas the Rhymer and

## The Fairy Faith

his marriage to the fay queen meant that he lived for 7 years in fairyland. Most tales though have the elfin lover in this world, rather than the human going to fairyland.

As mentioned earlier, in Irish stories of this class the hero usually obtains 'the sovereignty of Ireland' through his gracious act. Thus the fairy lady is the spirit of the land and by being courteous, kind and caring, he proves himself a worthy man to take on the stewardship of the land, as well as the matriarchal descent from the Neolithic peoples. By the Middle Ages these tales had altered to make the woman shrewish and desiring of dominion over man.[29] I suspect a Christian influence here.

Linked with the lore concerning marriage between mortals and fair folk is that concerning *changelings*. A changeling was known by its wan and wrinkled appearance, its long fingers and bony body, its fractious behaviour and voracious appetite, its large teeth and fondness for music and dancing. A changeling might be tricked into betraying itself by doing an unusual act, e.g. boiling water in eggshells which the changeling then reacted to by saying that he had seen a certain forest grow and die three times but had never seen such a thing before. An elvish cast of eye is witness to the fairy paternity of an infant, also pointed ears like those of Puck — or Spock. Compare here with ideas relating to the evil eye — or its opposite the beneficent eye — the power of the gaze as every lover knows — or the jealous glance. There's more to eyes than we have yet discovered. I came across a fascinating Victorian book called *The Evil Eye*[30] which talks about many superstitions and beliefs closely related to the fairy lore, from goddess lore and horned beings, through to charms and hand gestures.

Kidnapping also occurs. Most frequently they abducted a nursing mother to suckle one of their own, or a midwife, this again linking with the feeling of a race dying out or, as a tradition found around the world, indicating an ancient belief that were supernaturals to partake of mortal milk they would acquire some sort of vigour of life they needed in order to exist here — as though they needed that extra physical dimension, being in themselves too ethereal for this plane of existence. Sometimes children were said to be spirited away whilst they were sleeping, but also stolen by the fays when their mothers were working. In some cases the fay left a piece of wood which was given volition and the semblance of humanity by fairy magic in place of the human child. If you sleep on a fairy mound you are likely to be captured and taken below. A kidnapped person could revisit human places after seven years, but generally found it impossible to readjust and returned to the fairy mound. Often only the spirit of the person is taken leaving the body in a coma. This kidnapping lore finds its modern equivalent in the UFO abduction lore so prevalent at the moment.

There are several theories regarding the lore about changelings: one I have already mentioned — the Neolithic peoples needing new blood to prevent their dying out and/or the enhancement of psi abilities among the new Celtic peoples. Another is that of reincarnation: the ancestors steal the souls of people of all ages, and substitute one of themselves; Spence thinks that originally the belief was not associated with abduction, but rather with the notion that ancestral spirits awaiting rebirth ensouled the bodies of new-born infants of their own kindred. A large number of changelings when discovered appear as old men who know the area from times long past. Such a changeling is obviously the ancestral spirit of a distant generation, i.e. a reincarnation belief, an old soul. This, in time, after the encroaches by Christianity and through a confusion of mental processes, was exchanged for a belief that the spirits of the ancestors actually took an infant's body. In certain cases only the spirit of the children appears to have been exchanged, in others the body; in others both body and soul, in others a magically inspired image of the child is left behind. The ancient belief is that soul and body are as one, when an individual falls sick their soul has been carried off by some spirit: a spirit interpretation of out-of-body or near-death experience.

An interesting tale which recurs again and again concerns the hunchback whose hump was removed by fair folk because he reminded them of the name of a day in the week which they had missed when chanting a rhyme on the week-names. I find this a fascinating snippet of lore which occurs time and again in many variations on the theme. It suggests to me that the fair folk held a different calendar to the Iron Age Celts and is well worth investigating further.

## Connection with the Old Religion: Philosophy of the Fairy Faith

I consider that these legends of the peoples who built the megaliths, and their magical society, are the ancient forerunner of the pagan religion of these lands which was finally ousted by Christianity during the times of the witch trials in the seventeenth century, merely 300-400 years ago, and which is returning now in a new form. This was an Earth Magic religion of a people intimately linked with the

Earth and aware of the energies of the different places and how to use these energies psychically - magically. For our growth into the new we have to let our roots grow deep down. That is the problem with American culture. Because they denied the ancient history of their land, the thousands of years of Indian culture, so they stopped themselves from having any roots that could grow deep. It is as though they have covered the whole land with a layer of concrete and then put a few inches of earth on top of it and they are trying to grow in those few inches of earth. It just doesn't work. And that is why so many Americans are now looking towards Celtic witchcraft or to Amerindian shamanic philosophy for their spirituality.[31]

The most obvious link between the fair folk and the old religion is to be found in the witch trials. Many witches claimed marriage or other alliance with the fair folk. For example, Donald MacMichael who was tried for sorcery in 1677 at Inverary said that he consorted with the Fairy King who was of human stature. Bessie Dunlop, an Ayrshire witch tried in 1576, was visited by a fay queen. Andro Man tried at Aberdeen in 1597 said he had been husband to a fay queen and had children by her. 'The Scottish witch trials of the 16th and 17th century are heavy with reference to fairy royalty.'[32] Most of the points which I have mentioned that link the fairy myths with the Neolithic peoples are also pointers that link the fay folk with the old religion. What I would like to turn to now is the underlying philosophy that links the two.

These are the people whose heritage the modern witches and neo-pagans claim. The findings of psychical research and parapsychology support wholeheartedly the psychic elements of this heritage; that of modern physics the philosophical heritage, funny though that may seem.

## *The science of magic*

The science and art of magic *is* the science and art of the mind, the psyche, together with our emotions and our soul and spirit. Magic obeys psychological laws. The world of spirits, of psychic phenomena, is the world of the mind.

Research into mediumistic seances, OBEs, NDEs, ghosts and poltergeists suggests that active and intelligent disembodied beings (spirits of the dead or crisis apparitions or OBEers) are able to act psychically upon people in much the same way that people are known to act psychically upon one another.

The fairy faith suggests that we can understand certain of the fair folk in such terms, i.e. as disembodied intelligent beings possessed of magical powers and so able to cast spells in accordance with the animistic doctrine of ghosts and spirits with their inherent *mana*, being able to perform magical acts. From this it follows that the old witch and druidical doctrine, working together with the fair folk in their elemental, psychic, divine or even physical aspect, enabled Druids or witches to control the weather and natural phenomena connected with the vegetable and animal processes, and other magical acts.

According to Evans Wentz, the first thing taught to a neophyte was self-control. The magical science taught that, by formulas of invocation — by chants, by sounds, by music, by movements such as dancing in circles, by meditation, by fasting, by going at night at certain phases of the moon to the sacred places that these invisible beings frequented — the fair folk could be joined with to help at a magic level. The original magic was to work together with the elemental beings, with the devas, with the Earth energies, and their link with the rest of the solar system and the universe. All those of us who no longer follow the old patriarchal attempt to control Mother Nature prefer to teach ourselves self-discipline and pit our wills to control our own unruly minds and emotions and personality problems, so that we can best live in harmony with the universe. In fact the patriarchal magic of control over spirit beings, such as in tales of Victorian Golden Dawn Rituals, clearly dates from after the witch trials, for all the witch-trial records talk about marriage, or consort, between the witches and the fair folk. The essential facet is that self-control, self-discipline, is the absolute basic first step in training the mind. I have noticed this again and again with people coming to learn yoga. The very first lesson they have to learn is the self-discipline of actually coming to the class each week. The mind is a tricky thing and will find any and every excuse not to do that which it knows is good for it. Working with the mind is like working with a recalcitrant three year old. Only when you have taught your mind that you *will* do what you want, that you are in charge and not it, can you hope to get anywhere in any form of self-development, spiritual growth or training your psychic abilities, which are the tools of magic.

The particular magical tools for linking with the fay spirits are chanting, sound and music. This is what we are beginning to learn about now. So many groups of people are getting together and chanting the new and old chants that groups like Prana are making known to a wider audience. Tibetan overtone chanting, yogic OM chanting, even Gregorian chanting is being revived by thousands all over the world. And some are even chanting and dancing in the stone circles. The same with magical sounds. Have you ever sat on a moonlit night in a stone circle with the Aboriginal digeridoo playing? It is pure magic. Some people talk about energy being raised. I know what they mean — it's that tingle along the spine, that shiver in the body; and more and more are experiencing such magic. And music; the playing of pipes, of a whole range of drums and other instruments that come from the dawning of humanity's music. These evoke that same quality. And of course circle dancing. None of this is high magic; it's natural magic, peasant magic, pagan magic, the magic of the Earth and which brings the Earth alive - simple, spontaneous, as the mood takes one, coming from inspiration and not from dogma. This is the birth of a new spirituality which is as yet formless, nameless and growing. This is the essence of all religion, this atunement of spirit, this growing of oneself so that one becomes more and more sensitive and more and more able to tune in to the feeling of the place and time - without losing your stability. That's another reason for self-discipline. This is why we have to work on our minds, our emotions, our personality, our being.

*The mysteries*

It is now recognised that most of the mysteries of antiquity, such as those of Eleusis, were psychic or mystical in their nature, having to do with the initiate's entrance into the underworld or the invisible world while out of the physical body, or else with direct communication with gods, spirits and shades of the dead while in the physical body. All these mysteries were performed in darkened chambers from which all light was excluded, the person often staying in such chambers for days. This is interesting in connection with the pineal gland and the effect of the 25-hour body clock when light is excluded - in such a chamber your mind and body run on Moon time.

It is quite probable that New Grange, West Kennet, and the other major barrows were temples in which were celebrated ancient mysteries at the time when people were initiated; and as such they are directly related to a religion of the Tuatha de Danaan or fairy folk, of spirits and of the dead. People made pilgrimages to the places of the ancestors and the fair folk. It appears to have been customary to spend a vigil of three nights and days without food or sleep and during this time one of the fair folk, or spirit of the ancestor, would appear and speak with the person, or grant their prayer.[33] This is directly comparable to what we know of the Greek initiation mysteries and temples at which one could invoke the help of divine beings through dreams.

Having spent many a night's vigil in a stone circle and in long barrows and cairns, and summer solstice on top of Silbury Hill, I can most certainly appreciate that spending a three-day fast inside a tumulus would be a deeply mystical experience, especially with a mind trained to leave the body, or to undergo deep meditation, or some other altered state of consciousness. Fasting in itself makes one lighter, less earthbound, more prone to waking hallucinations and other mystical experiences. It is of interest to note here that all the initiation chambers, like New Grange, are specifically oriented towards a specific star, or sun or moon rise. They are astronomically precise as well as being Faraday chambers in power spots. For example, New Grange is oriented 43°60' SE, as are Gavrinis and Stoney Littleton longbarrows, and so are oriented to the winter solstice sunrise.

The practice of pilgrimage has, of course, been continued by the Christians and in Ireland the Christian pilgrims used to use the stone troughs or beds that are sometimes found in the barrows, like the sarcophagi found in the pyramids. There seems to be an added element in lying or sitting in particular stone that accentuates the mystical effect of the place. These shallow basins are found in the cells at Lough Crew, New Grange and Dowth.[34] So, once again we have this mystic connection with stone; probably granite which is heavily crystallised stone — imagine what that would do to your electromagnetic energies, let alone at the psychological level.

Finn MacCoul's Lake, also called St Patrick's Purgatory, was used right up to the seventeenth century as a place of initiation and pilgrimage; then it was destroyed by the English. It was considered a place of entrance to the Underworld. The person fasted for 15 days prior to an all-night vigil in which a mystic bridge was crossed to a celestial city. These rites lasted from 1 August to 15 August (Lammastide), now Christianised as the Assumption of BVM — another 10-day season of an ancient Celtic Festival, Lughnasadh, similar to the twelve days of Christmas, called Yuletide, celebrating the winter solstice.

Evans Wentz[35] considers that at Dun Aengus, the mystic assemblies and rites conducted in such a sun-temple, so secure and so strongly fortified against intrusion, represented a somewhat different mystical school, one very much older than at New Grange. Like others he considers that all the megalithic structures were places for practical spiritual purpose, and notes that since the ancient Celts never separated civil and religious functions (no split between secular and spiritual), such temples could have been as frequently used for non-religious assemblies as for initiation and other ceremonies. In the Isle of Man the sacred mound is still used for confirming the government's new laws.

Let us note this point and take it to heart. There is to be no separation of our everyday and our religious life. Every breath, every thought, every word and every action is both secular and spiritual — none of this Church-for-one-hour-on-Sundays-and-be-a-bastard-for-the-rest-of-the-week nonsense of so many within the Christian religion. This above all is what the science and art of magic teach.

The chambers, the stone troughs, are very similar in basic structure to certain of the Egyptian, Greek, Peruvian and Indian temples, with their chambers, stone chests, etc. The symbology of facing East to the rising sun and rising moon, to new birth, dawn, life etc., and facing West to the setting sun and setting moon, to night, death, etc., are the two parts of the whole found in every religion, and implicit in all of our spiritual teachings, the Perennial Philosophy.[36]

## The rebirth doctrine of the Celts

Like most beliefs bound up with the fairy faith, a belief in rebirth is re-emerging at this time. Also still surviving and now returning in rather garbled form is a belief in karma: that certain afflictions in this life can be related to one's behaviour in a former life. While it is wise to be wary of most New Age beliefs concerning karma because it can lead to a sort of fascism, it should always be borne in mind as a possibility, because there are certain instances where a person's congenital disabilities or pattern of life experience can be best understood in a rebirth context. And it is an essential aspect of the belief in evolution of the soul — the more we develop our selves in this life the more we evolve our souls in general.

This leads one to related beliefs such as that 'the kind of soul or character which will be reincarnated in the child is determined by the psychic prenatal conditions which a mother consciously or unconsciously may set up.'[37] It was Heathcote Williams in his play *The Immortalist*[38] about longevity, who first led me to consider deeply the manner in which most people are conceived. His use of words is so intense that I shall quote him directly.

Reincarnation is fairly chaotic. It's a kind of cosmic pyramid-selling.

Good fucks lure down good spirits. Bad fucks lure down bad spirits —for reincarnation purposes. And most fucks are pretty paltry: someone comes back from a hateful job thinking *Fuck* the Boss, *Fuck* everything — *Fuck You*. What kind of a conception is that going to lead to?

A lot of low spirits lurking round the basements and cellars of the first astral plane are constantly being attracted by the exchange of stale and vitriolic hormones for their reincarnation. Very few high spirits are being lured down by the exchange of abracadabra orgones.

How we fuck is who we are, and most people fuck like mindless jackhammers on the blink. They haven't the faintest idea of what they're doing — but the results of what they do are transforming the whole nature of the planet.

I have been noticing the children who are being born to those who are awakening these past two decades, and I feel that you can clearly tell who has been conceived in love, who has been conceived deliberately with loving intent, rather than by accident. It literally makes a world of difference and I pray that every woman deliberately conceives every child in pure love for the sake of this planet. For this is the way that high spirits are brought to this Earth plane, and we desperately need high spirits now.

Among the great nations of antiquity — the Egyptians, Indians, Greeks and Celts, this reincarnation doctrine was taught in the Mysteries and Priest schools, and formed the cornerstone of the most important philosophical systems like those of Buddha, Pythagoras, Plato, the neo-Platonists, and the Druids. In fact, of all the major religions the Christian and Moslem seem unusual in not having a reincarnation doctrine. This doctrine is now well researched by Ian Stevenson and I encourage anyone who wishes to explore a scientific evaluation of reincarnation to read his books.[39]

During medieval times the rebirth doctrine lived on among alchemists and mystical philosophers and those groups that survived persecution, and it has come down from that period to this through orders like the Rosicrucians and the traditions of modern Druidism.

The ancient Celtic doctrine of rebirth represented for the people an explanation of the complete cycle of human evolution; it included what we now call Darwinism - which explains only the purely physical evolution of the body - as well as a comprehensive theory of human evolution as a spiritual being, both apart from and in a physical body, on the road to the perfection which comes from knowing completely the Earth and the spirit planes of existence.[40]

## The Fairy-Faith Philosophy

In the fairy faith rebirth doctrine, evolution of the soul is the vital aspect as is shown in the following seven premises:

1. There is a spiritual world of causes which is the implicit background of the phenomenal 'world of effects' in which we live. This is identical in basic idea to Bohm's implicate order (see page 78) and to Buddhist philosophy.
2. The material substance, our body, is merely a means of expression for life, a conductor for a force which exhibits volition and individual consciousness — which we call our soul.
3. The soul is an indestructible unit of invisible power (or force) able to exist independently of the human body.
4. The personality is a temporary combination wholly dependent upon the ego — there is no personal immortality.
5. All evolutionary processes, from the lowest to the highest organisms, illustrate a gradual unfolding in the phenomenal world of a pre-existing soul-monad through an ever-increasing complexity of specialised structures, this complexity being brought about by natural selection. Together with this premise is the one that spontaneous generation of life is impossible on our planet. There must have been a life force before its physical manifestation or its physical evolution began. In this premise life is equivalent to psychical power, to a reservoir of consciousness. It is the 'soul-monad' that survives, which is the immortal principle which gives unity to each temporary personality. It is the soul-monad which is the bearer of any evolutionary gains made in each temporary personality through which it reflects itself. It is the permanent evolving principle.[41]
6. Some races have evolved out of the human plane of existence into the divine plane. Hence the gods are beings which once were human and in time the whole human race will become divine. There exists a need in nature, by virtue of the working of evolutionary laws, for people to strive to reach divinity. 'Magicians are able to produce magical effects because they are able to control this "soul-stuff", and our evidence would regard all spirits and fairies as portions of such universally diffused *mana* "soul-stuff", or, as Fechner might call it the "soul of the world".'[42] (The word psyche is Greek meaning soul, life, breath.)

In conclusion, the essence of the fairy faith is a mystical animism. Mysticism is fundamentally the same in all ages and among all peoples. Modern mysticism (derived mainly from Oriental sources) has affected Celtic mysticism as handed down from the dim past, because the two occupy a common psychical territory. Ironically, with the new physics, an animistic view of man is more in harmony with scientific premises than any other. So now let us explore this new/old mystical religion of our land as it is re-emerging at this time.

# 7. NATURAL MAGIC: THE GODDESS REAWAKENING

## Women and the Psychic — Return of the Subliminal Mind

The second sight is a particular aspect of the psychic. This term derives from Scotland and refers primarily to those who were conversant with the fairy folk, and only later came to be referred to precognition, to divination of the future. It is connected with witches and with fairy lore, and it is particularly connected to Celtic women, the women of the Western Isles of Scotland who kept the old ways and the lore within their hearts and who tried to transmit it on to the next generation.

It is interesting that women are considered to be blessed with the second sight in Scotland more than the men. Most of the famous mediums, channellers as they are called now, in the past century have been women. However, there have always been psychic men, and there seems to be a cultural bias here. I was talking to a Russian healer at a Society of Psychic Research (SPR) conference and she said that yes, there were the wise women in the villages still in Russia, but that in the north you had the male shamanic cult, and in the south you had the women. From anthropology I know that there are some cultures where men are the psychic seers, healers and clairvoyants for the tribe or nation, and in others women are. When men take over this function it is often with the aid of ritual and/or psychotropic plants, or, as with the Tibetans and Yogis, from extensive mental development. In Britain we always seem to have had both, the wise women and the cunning men, the witches and the wizards, but on the whole this aspect of our being is considered to be particularly feminine, in its nature. When, in what follows, I talk here about feminine nature, or patriarchal culture, I am not talking about individual men or women per se, but about cultural aspects which can be portrayed in this way. Both men and women have a feminine side to their nature; both men and women have been deeply affected by patriarchal culture.

As we finally begin to emerge from the dark eons of the Kali Yuga so are women once more realising what being a woman is about - in the West at least. This is not happening for most women in most parts of this planet. In reawakening to our own nature as women, as distinct from people, so we embody the reawakening of the goddess. Women are very different from men, not only in physical terms but also in mental and emotional terms. We seem to be more emotional without having to deliberately work at it. It seems to be something to do with woman's nature as child bearer. Our minds and bodies seem to be more closely linked.[1] Perhaps this is purely due to social and cultural influences, or perhaps it is part of a basic natural polarity such as the Taoist Yin-Yang. I don't know.

For example, it was pointed out in an article in the *New Scientist*[2] that psychology has only just begun to really look at emotion, and that 'to say someone is being emotional is insulting'. This is because up to now most experimental psychology has been almost a purely male-type intellectual province and has been totally divorced from personal experience — the extreme example of which is the Behaviourism of the 1900-50s, in which consciousness itself was considered non-existent because it couldn't be measured. As the feminine side of our being reawakens so experimental psychologists are beginning to realise that all their research has been at a very superficial mechanistic (stimulus—response) level, and they are beginning to realise that they need to look below the surface to factors that underlie our everyday behaviour. This ability to explore and live within the dark, within the underworld of dreams, emotions and intuition, is the feminine side of our being. Many people in our western culture deny this part of themselves, pretend that they don't want or need to be aware of the subsurface levels of themselves.

That some people are now able to study parapsychology at university is an example of the goddess reawakening in the ivory towers of academia. As far as I am concerned parapsychology is about deepening one's awareness at every level of one's being, potential or actual — what Maslow calls actualisation, some traditions call enlightenment, others call divinity, becoming a fully realised human being, fulfilling all our potentials.

Most of my work is concerned with this opening up to the deeper layers of the psyche. First I teach people how to remember their dreams; then how to attain, and work within, the hypnagogic state; then how to connect with the collective unconscious and work with archetypes and symbols; and finally to develop creative imagination. I also teach more traditional meditation and breathing techniques that help develop other aspects of the mind, like focus and concentration, discipline and detachment, that are essential if one is to open fully the channels between the conscious and the subconscious. All of this is related to opening up psychically — to developing intuition and the second sight — but the actual development of psychic abilities is not the prime focus. I feel quite strongly that development of oneself as a whole, healthy, happy person is of prime importance; the other aspects seem to happen of their own accord as one becomes a more sensitive, open person.

## *The fear of psi in our society*

Those people who are born with psychic abilities as a strong and undeniable gift have had it repressed strongly by our society and so need to reconnect with this aspect of being; for others in whom the gift is not so strong, learning to open up psychically is often a very good path for opening up in general to the deeper layers of their being and their spiritual nature. In the process of society's coming to live a psi world-view, a lot of the present fears connected with the 'occult', and with witchcraft and paganism in particular, will have to be faced, as this is part of the white people's history which profoundly affects our present attitudes towards psi.

As a child in a Roman Catholic convent I was taught that I was by definition eve-il. Woman (Eve) was the cause of the downfall of the whole of humanity and just by being female meant that I was inherently sinful. Every aspect of me, but particularly my pubescent body, and any so-called female way of thinking, was suppressed as harshly as they knew how. Repression of the psychic, the intuitive part of me, was part of the whole repression of the female way of being.[3]

When I look at history, I see how women have been persecuted and nowhere is this more clear than in the witch hunts. In the witch trials it was primarily women who were incriminated as wrongdoers. I feel that this was an attempt by the Church to divest women of whatever remaining dignity and self-respect they had, and to make them totally subservient and enslaved to men. For those who do not yet know about the horrors perpetrated during the 400 years of persecution of those who were considered to be witches I suggest you read Starhawk's book *Dreaming the Dark* or Monica Sjöö and Barbara Mor's *The Great Cosmic Mother* or Christina Hole's *Witchcraft in Britain*.[4]

We need to remember the political and social conditions of this time which ended in Descartes and the 'Age of Reason'. There are some estimates that in the 1300s the Black Death killed between one-and two-thirds of the population of Europe, and the remaining people were living under the most awful conditions, feeling that God had sent them this horror because of some evil that they had committed. First the gypsies were the scapegoats, as they always have been, then the Jews, and finally the witches. Then in the 1600s came the Plague and once again the population of Europe was decimated. The witch trials increased dramatically during this period. It is very probable that the people turned against the witches at this point because they were traditionally the healers, the herbalists and midwives. The witches were there to ensure the fertility, peace and prosperity of the land and they had manifestly failed — the people's lives had been turned into a nightmare from which very few survived, for the second time in three hundred years. The people literally lost their faith in the Craft, and it was an easy matter for the Church to impugn not only that the witches could not keep death and disease at bay, but that they were actually the cause of these terrible diseases.[5] The popular conception of witchcraft also changed, and the witch, once a respected and esteemed member of the community, became a near-outcast from whom respectable men and women shuddered away.

The trauma engendered throughout a whole continent by the horrors of the Inquisition and the so-called witch trials is still manifest today, and is probably the single most important factor behind present-day scepticism towards psi — and also in parapsychologists' unwillingness to be associated with occultism. We must not forget that the Roman Catholic Church forbade, under penalty of excommunication, the practice of any psychic arts, even healing, and the Roman Catholic Church ruled over most of Europe for nearly a thousand years right up to the Protestant Reformation. Today we may laugh at the idea of Excommunication, but in those days it was practically the equivalent of being a leper. Since Roman times there had been edicts against the use of magic for murder or other crimes, but during this time psi itself was outlawed.

It is difficult to explain the extent to which the trauma of the Inquisition still affects us today except by analogy, an analogy of which most of you will have heard. In the reign of William of Orange and Mary in Britain (1689—1702), a clan in Scotland called the Campbells went to visit their neighbours the Macdonalds, bearing a note from William of Orange stating that the Macdonalds should be punished, because he thought they had not accepted his authority over them although, albeit reluctantly, they had done so. The Campbells did not act openly and were invited in by the Macdonalds as guests and treated with every hospitality and courtesy. The next morning the Campbells blocked both ends of the pass of Glencoe and murdered every single Macdonald; men, women, children and babes. The horror of this massacre still echoes today, as anyone with the name of Campbell living in Scotland knows!

If such a small localised event can cause such ripples over the centuries, consider the effect of the sustained murders by the Inquisition and the witch trials over the whole of Europe for about *four hundred years*. The Inquisition led directly to the Protestant movement and the intense hatred of Papists in Britain, a hatred so intense that they actually imported a German monarch with only minor blood connections to the British royal line to rule the country rather than have the rightful heir who was a Catholic. And that meant a lot in those days. The witch trials which started in Britain under Elizabeth I, and became worse under James VI and I, both Protestants, led initially to intense fear of those who were psychic, a fear exemplified by the folklore of the wicked witch or evil stepmother such as we have today in our children's stories, and eventually, during the so-called 'Age of Reason', led to a denial of the reality of psi itself. During the witch hunts it would have been necessary for a mother to tell her child to shut up if they showed any psychic ability — for fear of their lives — and so psychic abilities began to be suppressed in childhood, and fear surrounding any sign of psychic ability caused emotional disturbance in the child with regard to psychic ability. After 400 years of persecution (10 generations of people), everyone denied any possible hint of psychic ability literally for fear of their lives. It must have been a bit like living in Stalin's Russia with all the suspicion, informers, mistrust and fear - except far worse and for far longer. And so we get the society we have had for the past 200 years in which some people deny that there is such an ability.

So we see here in the two great tragedies of the Black Death and the Plague, together with the Catholic Church's desire that all the people in Europe were Christian and subject to its rules, a reason for the present-day fear of the psychic, and the denial that psychic powers exist. I have written on several occasions[6] how I consider that the present-day academic and establishment denial of the psychic is in large part due to the fear of psi engendered by those years of persecution. The very fact that the scientific establishment denies the existence of psychic abilities and blocks dissemination of information concerning them is evidence of their fear. For example, it is impossible to get research grants from normal scientific sources to do research in this area. All grants are from private foundations.

There is a psychologist called Charles Tart who considers that much of the reason for the elusiveness of psi in the lab, and for the incredibly small psi effects seen, is that we are all subconsciously afraid of psi.[7] With such a history of centuries of persecution for having psychic abilities, followed by centuries of jeering, is it any wonder that people are afraid of psi?

Even to this day there are those who from ignorance or narrow minds consider that exploring the psychic, or joining a non-Christian religion that they call occult, is immoral. The strength of the feeling against the 'occult' is still very apparent in our society. Just recently there has been a High Court decision upholding some fundamentalist Christians' request that evening classes on research into psychic phenomena be forbidden in their area. Parapsychology, the scientific study of psychic abilities, is seen by these people as evil, as satanic, in no way different from what they loosely term 'the occult', and so they bar it from the classrooms for adult education. This is extreme fear, paranoia and bigotry and it is still very much alive in our society. I laughed the other day when I read in my local newspaper how local people were outraged at a 'black mass' wedding that was said to have occurred recently. Apparently some people had had a fire outside and had been doing circle dancing around it. I cannot see what is harmful in dancing outside around a fire, but it engenders all sorts of fears in some fundamentalist Christians and other ignorant people. The days of the witch hunts are not so far behind us; the attitudes are still prevalent in our society.

Although you can suppress inherent human abilities through fear you cannot get rid of them because they are natural human abilities. It's a bit like trying to deny feelings. The ability to see into the future is as much a part of our nature as is the ability to sing and dance and love one another — and the Puritans tried to stop us from doing that as well. Even in the Age of Reason, in the eighteenth century when psychic phenomena were denied totally as being completely nonexistent and merely a peasant superstition and fairy tale, there were people who displayed psychic abilities. In the nineteenth century this by now sporadic phenomenon, at least as far as the influential academic city people were concerned, once again came into prominence with the Fox sisters and their table tipping and spirit rapping, and so mediumship and spiritualism came into being. Thus psychic research started. Since 1954 and the repeal of the witchcraft laws we have been able to admit openly to religious magic and there are now dozens of different cults and millions of people who are exploring the religious aspect. This is one aspect of the reawakening of the Goddess in terms of the psychic.

> Nowadays, the attitude to the Craft is beginning to change once more, as it is to the metaphysical generally ... we can see that the most terrible persecutions of the witches was an admission of the strength of belief in their powers. With the advance of what some would call materialism and some common-sense, witchcraft was taken less seriously, and as this happened, measures in force against it became less and less stringent ... I am not saying that the only factor producing this altered attitude was disbelief in the powers of the witch . . . The curious thing is that today the disbelief which could be said to be the beginning of the decrease of witch persecutions, is also on the wane. Within the last 30 years, so much supra-physical phenomena, discredited formerly as superstitious nonsense — has been established as scientific fact, that the claims of witch-powers no longer can be laughed out of court.[8]

## Witchcraft and 'Black' Magic[9]

Witchcraft, also often called Wicca, or the Old Religion, or the Craft of the Wise, has recently had a certain revival in popularity, especially among young people in Europe and America. One thing that really bothers me though is the constant undercurrent, particularly from fundamentalist Christians but also from other sectors of society that equate witchcraft with what is commonly called 'black' magic, though 'harmful' or 'bad' would be better words. Therefore I shall use the term 'harmful magic' instead of 'black magic'. When you look at the way the psychic is propagated through our culture it is in terms of the wicked witch who always tries to harm people, especially children; witchery and wickedness (wicca - wicked) became more or less synonymous. Folklorist accounts of witchcraft, or the popular image of witchcraft, is always in terms of harmful magic and evil. For instance Hansel and Gretel or Peter's Russian tales. All of our Germanic fairy tales which are read to very young children tell of wicked psychics in whatever form. Only in the Celtic fairy tales do we have folk stories which tell of beneficial uses to which the psychic is put. And so the people's fear of the psychic is engendered from infancy.

> In all studies of witches and magic, one point must be kept in mind, that when anything regarded as out of the ordinary course of nature is brought about by human means it is called a miracle if the magician belongs to the beholder's own religion, but it is magic - often black magic — if the wizard belongs to another religion... *This is markedly the case in the Christian records of the wonders performed by witches*[10]

So I want to tackle the ever-present aspect of right conduct, or morality, or ethics (three terms which I shall use interchangeably) within the practical application of psi. All forms of power, be they electrical, nuclear, solar, physical, mental, can be used for good or bad. It is very important for everyone to realise this and to ensure that ethical standards are initiated now, in order to prevent psi talents being used for immoral ends by the political, military and technological forces current in the world at present.

In the old days the use of psi abilities for unethical ends was called black magic. *To do, or to be, 'black' is a state of mind or personality. The basic principle behind harmful magic has been, is, and always will be the same; the use of the power of mind to gain illicitly something one has no right to possess, and which causes harm to others.*[11] In the 'black' witch, the 'black' magician, or any other person who uses this power, there is a fantastic egotism, a form of paranoia (the grandiose sort), which stops at nothing in order to satisfy itself. There is the desire to control others, to have power over others; there is the desire to appear bigger and better than others; there is the desire to appear very important.

The basic idea of harmful magic is to destroy any object or principle rated as 'good' or 'holy'. We shall never get rid of harmful magic and its 'sins of the mindless' until the sickness within the human psyche, which is its source, is healed. Basically if someone is an uncaring, inconsiderate, selfish, greedy person and they use psi in order to further their avaricious ends then this is harmful magic, because they are harming others on this planet.

Quite possibly the nastiest magicians of our time are those scientists and intellectuals who deny that there is 'godness' in the world, that there is spirit in a tree, a brook, in a person; those psychologists who were so mindless that they tried to say we were 'stimulus-response' machines feeling and thinking nothing — without a soul, without any spirit. In proposing this machine view of humans and the world they were causing great harm, because in denying the human soul they were denying the higher aspects of our being, our moral code that enables us to overcome those sicknesses within our psyche that keep us from being loving, caring, beautiful people. To the extent that they realised the harm they were doing they were profaning the 'good' within humanity.

This is harmful magic and it has created a materialist age that is literally murdering the planet. To deny the God of the Jews and the Christians has probably been a blessing for he was a patriarchal, authoritarian, jealous and vengeful god threatening destruction and agony to those who did not do as he said. However, also to throw out the visionary mystical god, which is our highest ideal, has led us to the darkest hour of our planet. *For we need a highest ideal, we need a conception of the highest good possible, so that we have something to aim for,* because what good is an ideal if one does not constantly aim for it? And those who deny any goodness in this world, or any spirit, or any soul, are thereby speaking out for the forces of evil and darkness and the material wilderness in which we are lost.

Anyone can be harmful. There are more 'black' scientists, politicians, military personnel, police, industrialists, farmers, etc., than there are harmful occultists. Witness the crazy escalation of armaments, the billions of dollars, pounds and roubles poured into the manufacture of weapons of increasing destructiveness and horror. Witness the increasing filth being poured into our once sweet waters by avaricious industrialists who care not what harm they do as long as they get rich; witness the poisoning of our once good Earth by avaricious farmers who have lost their love of the land and are now agribusinesses; witness the mindless violence on our city streets, violence that is worst in America, the richest, most materialistic, country on the planet, witness the desecration of the oceans with the continent sized islands of floating plastic and the overfishing. This is harmful magic of the most disgusting and awful form — and we are surrounded by it everywhere we turn.

The equating of witchcraft with harmful magic and evil has been kept alive in the public image by films, horror books and so on. The equating of witchcraft with orgies, and 'black' magicians as having sexual powers, is similarly exploited. The old ideas that harmful magic was the use of magic by uneducated village people does not seem to be consonant with these feelings. It seems to me more that at the turn of the ages into Bronze, and particularly Iron Ages, that men for some reason became far more warlike, aggressive, power-seeking, and all the other bad traits that we now associate with harmful magic, part of that control-over-the-spirits syndrome that was so prevalent in patriarchal ritual magic. When you are seeking to have control over others, so you get the most awful ego problems, and become puffed up with pride, egotism, and all the other bad traits to which people are so prone. Those who have done harmful magic are mostly city people seeking power and control over all things, Mother Nature included. Therefore, training the mind so that it is effective at a magical level requires an ethical and moral training at some level, otherwise you will become self-destructive more than anything else, and will end up in mental hospital.

### *The magical power of love*

> And so the crescendo rose until it seems that the Church and people were giving more attention to the Devil than to God ... According to the Church, it was evil which was omnipresent. Beauty was evil; happiness and pleasure were sins ... sex was virtually synonymous with wickedness; it was the bonne bouche used by the Devil to lure men into sin. But the Devil had an active partner in his work of corruption — women. From her high position as the giver and keeper of life and good, woman was degraded to the level of a social evil; an impure creature with whom man had to associate in order to perpetuate the race. The whole business of procreation was regarded as distasteful, disgusting even ... [And still is by certain Jewish sects who wear a garment in bed with a special hole in it for the purpose of permitting the act that will conceive children!] ... A fear of women was bred of this unhealthy attitude to sex ... [There was a] preoccupation with all manifestations of evil, with cruelty and dirt ... deification of denial and pain, when mortification of the flesh was the highest service one could offer one's Creator. In what unlit corner of the human subconscious can have spawned the idea that the chief pleasure of the Universal Principle of Love lies in the sight and sound of suffering?[12] (words in brackets added)

Anyone who has read any history of medieval times and later the Puritan period will recognise the truth in the above quotation, unreal though it may seem to us now. By contrast, pagans have always considered sex to be a natural function; neither unclean nor shameful; not to be used promiscuously though. Perhaps because of the witches' recognition of the great power — the tantric potential of sexual energies — it was regarded as a sacred act. There are many women and some witches who feel this way about sex to this day, though not all pagans have this reverence for this most beautiful act of union between two people. When two people 'make love' they are literally doing just that — creating love between them which then flows out to those around — we all love to be with young lovers!

— and the more love we can create in this world the better for everybody. Such an act is the ultimate in good and so is sacred and holy in every sense of the word. To degrade such an act and to 'fuck' someone, with all its overtones of violence and aggression, is not only degrading for oneself and the other person but is a tragedy for the world, because the opportunity for something blissfully beautiful has been lost and perverted into lust which is mere gratification of the senses, which causes the heart to harden, and the potential to love deeply and truly to be lost.[13]

There is a saying attributed to Wicca which goes 'Death and the devil have no power over love and laughter.' I find this saying slightly strange in that to a witch death is an integral part of life and there is no such concept as the devil, that being a Christian manifestation, so I change the saying to the conception that love and laughter can override all the bad that can possibly happen to us. This, to me, is the principle code of witchcraft. The witches consider that happiness and the joy of being alive is essential to our well-being as is evidenced by their saying: 'Be happy and ye *shall* be wise.' How many are there among us who can look deep into their hearts and know profound happiness? And how blessed a religion that looks first to true joy and happiness as the means for spiritual growth — as do the Yogis with their aim for nirvana or spiritual bliss, or the Buddhists with their aim for emptiness.

## Archaeology — A search for the Origins of the Craft

Witchcraft, as a part of the universal animistic paganism, claims to be the oldest religion in the world, as old as the human race itself; the cradle of all sects, cults, and theologies of humanity. The Anglo-Saxon word 'wicca', or wise one, is considered by many to be the origin of the word witch, and so traditionally witchcraft is called 'the Craft of the Wise'.

The story of witchcraft, according to modern popular conception,[14] goes back to the dawning of consciousness when humans began to personalise the manifestations of Nature, the religion called animism by anthropologists. Its twin deities are the two main concepts of male and female deities which are found in the pantheons of nearly all religion; which Carl Jung calls the Archetypes of the Collective Unconscious. They are personifications of the pairs of opposites, which manifest in the world of form; of light and dark, summer and winter, life and death, seed time and harvest.[15]

The cults of the Great Mother spanned the world, and her names are as innumerable as are the aspects under which she was

*Where Science and Magic Meet*

*The now famous cast of the Wise Woman of Willendorf from the Gravettian—Aurignacian site of the Upper Paleolithic Age (Illustration courtesy of Natural History Museum, Vienna)*

adored — Hecate, Artemis, Isis, Brigid, Diana, Ashara, Ishtar, Cybele, Cerridwen, Hera, Mary, are but a few. If you look closely at the statue of the "Venus" of Willendorf you will see that it is not a representation of a fertile pregnant young woman, but of a post-menopausal woman, the sagging belly and breasts are typical of the female body post-menopause. And on her head is a woven basket. Could she have been a tribal elder/priestess/shamaness in trance? When a woman stopped bleeding and her body shifted to a perpetual "pregnant" shape was this considered particularly special? – she was the "grand" mother in every sense of the word.

The Great Mother appears to have been the dominant figure right up until the Bronze Age, when myths suggest that the male came to dominate the female, as in the story of Zeus. Until then the male was the consort of the female, and this appears to have been enacted also in political terms with the queen being the ruler and the king her consort. By the end of the Neolithic and the beginning of the Bronze it is now thought that the people worshipped two deities, the Mother Goddess and the Horned God (who is possibly a descendant of the totem animal),

George Leland's book, *Aradia: the Gospel of the Witches*[16] is said to be culled from existing witch communities in Italy and to form part of their ancient heritage. This book emphasises that witchcraft was the nature religion of the matriarchal society that existed in Europe prior to Christianity.

It is becoming increasingly clear that the earliest civilisations were matriarchal, lasted for millennia, were peaceful, and worshipped the goddess. Even the Hebrew word for God in the Bible, Elohim, is plural — and is *not* masculine! In the twentieth century BC in Egypt, Isis was revered; in the fifteenth century BC in Turkey, Arinna; from the eighteenth to the seventh century BC in Babylon we find her by the name of Ishtar. Minoan Crete is claimed by some to be the last of the ancient goddess-worshipping civilisations. Some authorities would extend goddess worship as far into the past as the Upper Paleolithic Age of about 50,000 BC, and from our knowledge of the witches, we have also to extend this worship of the goddess right into the present day.

In one of her aspects the Great Mother was revered as the Moon goddess. The Moon was thought to govern fertility; her light was the creative agent; at her festivals, fires and torches flared, candles were lit in a form of sympathetic magic to reinforce her light and persuade the fertilising force to do its work. This ancient belief is beginning now to find scientific confirmation with bio-dynamic agriculturists studying the effects of planting seeds at different phases of the Moon, and of harvesting vegetables and herbs at different phases. For example Koliski spent fifteen years studying the effects of the Moon on plant growth, and Maria Toon devised a Moon calendar for planting.[17] Similarly Dr Jonas in Czechoslovakia has been having success with birth control and fertility problems by utilising astrological charts and phases of the Moon.[18] In a book called *Lunaception* Louise Lacey[19] describes how one can regularise one's menstrual cycle by keeping the curtains open while the Moon is full, and that during this time ovulation will occur, so this is another form of natural contraception most probably linked with the pineal gland. I find it interesting that whilst all other mammals ovulate according to season, the sun's cycle, humans ovulate according to the moon's cycle.

Just as the Hindus have a triple god symbolised by Brahma the Creator, Vishnu the Preserver and Siva the Destroyer, so does witchcraft preserve the image of a triple goddess: Maiden, Mother and Wise Old Woman. *The Water Babies,* a Victorian children's story, has a beautiful version of the triple goddess with Mrs Do-as-you-would-be-done-by, Mrs Be-done-by-as-you-did, and Mother Cary (Cerridwen?), who has learnt the ultimate trick in creation which is to get living things to create themselves![20]

The god of the witches and of modern pagans is the ancient horned god who appears in many ways, principally as the great god Pan in Greek mythology, or Cernunnos in Celtic mythology (pronounced with the C as the Gaelic Ch, as in loch), whose name is also found anglicised as Herne, the Hunter, and is drawn into the landscape at Cerne Abbas in Dorset. The horned god of the witches is the male escort of the goddess, with certain symbolic functions which are anything but diabolic. There is evidence of the horned god from Paleolithic times; in the Caverne de Trois Freres in Ariege, which dates to late Paleolithic, there is a figure of a man clothed in the skin of a stag and wearing antlers. Another painting shows a similar figure surrounded by a circle of women. Pan was the protector of all animals, as is so beautifully described in *The Wind in the Willows* by Kenneth Grahame;[21] nothing could be further from the Christian devil.

Probably the most striking thing about archaeological findings from the Neolithic period are the enormous number of goddess figures from all areas. Horned figures start to be found during the Bronze Age. In the Near East the figures were male or female, and the horns are those of cattle, sheep or goats (e.g. Isis, Hathor). The horns were a sign of divinity. European examples of the horned god are to be found in the Minotaur, Dionysus, Cernunnos and Pan, who is first encountered in the fifth century BC and is the closest to the totem figure of Paleolithic times. A most interesting survival into modern times of the horned god is the Puck Fair of Killorglin, Co. Kerry, Eire, which takes place at Lammas (Lughnasad, 1 August).

In no case does the horned god have any link with the Christian Satan, the 'embodiment of evil'. He was the god of hunting; the woods and the fields were his. He ruled the horned beats and thus was the god of farmers and shepherds. In some cases he was the god of death, as in Herne the Hunter. It is possible that Judaism and then the Christian Church, in attempting to eradicate the former pagan religion, created their devil in the guise of the former god, but one must not see in every horned idol a representation of the devil, the Principle of Evil. The Christians absorbed the earlier Goddess image of Isis as Mary, Mother of God.[22] Let us note here that we are talking now about relics of a religion that seems to have its origin millennia ago. The Goddess era is the time when the megalithic stone circles, long barrows, tumuli, henges and other ancient, sacred structures which survive to this day, were being built in Britain. Witchcraft seems to be a relic of those pre-Celtic Neolithic people. The round barrows were built in the Celtic Bronze Age, and the later phases of

*Natural Magic: The Goddess Reawakening*

*Cernunnos as depicted on the Gundestrup bowl (Illustration courtesy of the National Museum of Denmark)*

Stonehenge were in this era too. The Druids come from this Bronze Age period.

The early churches, too, were usually built upon sites that had been used by the Old Religion. Pope Gregory made an edict that when places or concepts were associated with beliefs too strong to be uprooted, that they should be given a veneer of Christianity. Thus many of the Christian festivals such as Candlemas (Purification of the Virgin) together with St Valentine's Day, All Souls' Day (Hallowe'en), Christmas, May Day, Easter, are relics of former pagan festivals. This process has also opened the gates of Christian sainthood to many aspects of the Great Mother and her consort of whom St Bridget is a classic example. There are now several books available which deal with these correspondences in detail.[23]

Further, if one looks at churches all over the country, 'witch-marks' can be seen in the carvings; pentacles and pentagrams, carved figures of the Great Mother and masks of the Horned God, often called the 'Green Man' because he is represented surrounded with scrolls of oak leaves or acorns, or with leaves seemingly growing from mouth, ears and hair. The Great Mother is sometimes symbolised by what is called a 'Sheila-na-gig', a word of uncertain origin, though found more often in Ireland than in England, and representing the female principle of fertility; though they are crude by today's standards they

## Where Science and Magic Meet

express veneration for this principle.[24]

As we take this view of witchcraft being the ancient shamanic religion of the West and the Middle East, we can perceive that the idea promulgated by the Catholic Church that the witches worshipped the devil was very much a perversion of the truth. The only reason that this stigma of evil was given to our local shamans and their god is a direct result of the Catholic Church's attitude in the Middle Ages which openly declared war on the native religion of Europe and by the end of the seventeenth century had virtually wiped it out, just as all the Jews had been banished from Spain in the fifteenth century.[25]

In our present day, equating witchcraft with harmful magic is a perpetuation of the evil done towards these people by Church and state from medieval through to Georgian times. Do we really wish to uphold the 'black' deeds of the Catholic Inquisition by continuing to label witchcraft as 'black magic'? Or can we now openly, and without prejudice, look at what the faith of the witches consists, and let them practice their craft in their own way in peace?

## Paganism and its Relationship with Magic

Paganism and witchcraft are different from most of the occult groups in that popularly they are now considered to be intrinsically a part of the original animistic/shamanistic spirituality endogenous to tribal people in all parts of the world. Most occult groups that I have come across follow a symbolism, form and code that can be traced back to Egypt, or to the Hebrew Qabbalah, or even to Greece or Rome. Thus they are imported religious systems, either during the Middle Ages by the Crusaders, or by the Romany gipsies, or during Victorian times when the great treasures of Egypt and Greece were being plundered in the name of archaeology. Part of the rebirthing of paganism and witchcraft is the incorporation of teachings and principles outlined by Native American Indian, South American Ayahuasceros or Australian Aboriginal traditions, such as medicine wheel or dreamtime or ancestor teachings. These are all animistic religions and as such are part of the same heritage as the indigenous paganism of Britain.

The derivation of the word 'magic' is from the Chaldean 'Magdhim', the meaning of which is wisdom (the Three Magi or wise men), allied with the qualities summed up in the term philosophy. From this root the words image and imagination are also derived. I make this brief allusion to root words because I feel we can learn so much about a topic from its word derivation. Magic is linked with wisdom — so is the imagination. Science means knowledge, and so this book is aiming to be a synthesis of wisdom on a base of knowledge.

While 'magic' is important to animistic religions it is not all-important. The psi apparent in the parapsychologists' laboratories is the purest form of magic extant, for all the occultists that I have met follow a religious system or philosophy of some form or other. The only philosophy that parapsychologists openly follow is that of scientism — pure curiosity, the seeking after knowledge — although in private many will admit to an underlying religious philosophy. Unfortunately, at this time, scientism has no accepted or defined moral code, so its findings are forever being put to immoral use. Through the history of religion we can see a development from the great goddess as mother of all, to many gods and goddesses representing different principles or aspects, through to male Trinities, to the One God, and now as we slowly come into Aquarian spirituality, to a more abstract mystical ideal. All religions have acknowledged the power of magic within their structure, but to concentrate on magic and the development of magical powers as an end in itself would be to put the cart before the horse. This applies to most pagans and witches as well, although in this era paganism is such a loose structure that the practice of it depends more on the individual than upon any formal lores.

## The psychic arts

From this idea of the origins of witchcraft we can see that the Craft started as a Nature religion which worshipped the Universe in all Her beauty. However we must not forget that the witches were also renowned for having developed and used techniques designed to assist a person to utilise their psychic abilities.

> Witches were feared because they could do things that the majority of other people could not do ... when the exaggeration, the dramatisation and distortion are discounted, the fact remains that if witches did not get results they and their Craft would soon have been forgotten.[26]

While we may consider a matriarchal religion to be a thing of the past, the skills of the witches are once more in evidence today, and it is for this reason that it is a good idea to learn how to utilise and tap these latent abilities within humanity, and so maybe help to heal the Earth and its peoples of the terrible wounds inflicted on it by so many industrialists, politicians, scientists and warmongers of the twentieth century.

Theology is ultimately political. The way human communities deify the transcendent and determine the categories of good and evil have more to do with the power dynamics of the social systems which create the theologies than with the spontaneous revelation of truth from another quarter.[27]

Pagan magic is a discipline relating to the study and use of psychic powers for beneficial purposes as part of learning to "Know Oneself," as part of one's whole spiritual, emotional and psychological self-development, and therefore can be seen to be a Western equivalent of the Yogic and Buddhist disciplines, interest in which is becoming more and more popular. For the past century, many people have been eager to learn from Eastern teachers, and readily see the spiritual and good aspects of their techniques and arts. One ideal which they hold in common with the pagans is to pursue the spiritual path of self-knowledge and self-development first and foremost, the psychic arts (or siddhis) being considered a red-herring or sideline which you gain almost automatically when you reach a certain level, and are *not* to be pursued as an end in themselves. I urge all those who are interested in the development and use of psi to take note of this tenet.

So, why do so many people abhor Western magic and brand it as bad when shamanism, and religions that incorporate psychic abilities from other lands, are at least tolerated? Perhaps it is the claimed extent of magic power (psi abilities) that leads to this fear of it. Perhaps it is a fear that people are more interested in gaining power than in spiritual development. Or that psychic abilities will be put to bad uses, such as spying by the military as in the American "Stargate" project.[28] For our culture is very frightened of the psychic, and magic and witchcraft are more linked with the psychic than any accepted religion. I came across a saying once which I consider very apt here. It goes like this: 'Strangeness begets fear; fear begets hatred; hatred begets anger; and anger seeks a victim'.[29] Perhaps now that magic, in the form of parapsychology, is becoming better understood and more acceptable, this fear and hatred of witchcraft and magic can begin to diminish, so that their techniques can be examined objectively without all the emotional overtones.

For we do need to examine their claims and see what is real and what is fictitious. There are so many cults, and people are so lost, that an awful lot of damage can be done by people who are ignorant or unscrupulous. If a leader's heart is 'in the right place' and they are ignorant there will be only minimal damage, but if someone has set themselves up as a leader just for ego reasons of glamour, power and glory, then their followers can be severely hurt, as in the Jonestown massacre, or some of the followers of Alex and Maxine Saunders who set themselves up in the 1960s as 'King of the Witches' — what a ridiculous title. I personally know of one casualty of their glamourising and fooling around with young people in those heady days, and I have heard that his case is not unique. Glamour is something of which we have to be incredibly careful, which is why I stress the *self-development aspect*. We all know what glamorous means; we all know the effect that a glamorous person has on people; how much power over them they gain merely from the charisma and outward show. We all know the glamour and the glitter of the stage and the circus; there is one aspect of witchcraft and magic which is very susceptible to this glamour and of which we must be very wary, for our egos are only too ready to trap us; humility is an essential - as long as it isn't false humility! A glamorous person is literally 'fascinating', 'entrancing'.

It is really interesting to see how we use our language. What do fascinate and entrance mean? It literally means to put another person into a trance. And we all know the power of a hypnotist over another person. It is the same thing as in a ritual. The young and gullible, those who trust others, those who are ignorant, are the most susceptible. And it is not just power-seeking magicians who use this power of glamour over others. We see it all around us, especially on the ubiquitous television. Television is glamorous; it holds people hypnotised; I have been in homes when it is always on; even though the people are not consciously watching it, they are subliminally absorbing everything that comes out. And most of what comes out is propaganda for the establishment viewpoint, advertising stuff that you don't really need (because if you did need it there'd be no necessity to advertise!), or violence, cops and robbers, soap operas, etc. One day take a show and analyse exactly what went in to the making of it and you will be horrified at the cultural programming that is going on so insidiously in every home in the country. The subtle glamour that makes wealth the only thing to be strived for; that makes big business and big industry the ultimate in wealth and glamour; that jeers at simple people with their faith in spiritual matters; the high priests of scientism discoursing learnedly on materialist topics. Listen also to some of the lyrics in pop music which is aimed and impressionable pre-teen and teenagers – some of it ain't so nice.

## Traditional techniques

Astral travelling (or out-of-body experiences) is an aspect of magic that is traditionally specifically linked with witchcraft — we all have this image of a black-hatted woman on a broomstick flying through the air. There is a popular belief, and a suggestion by some anthropologists, that witches used various herbs, called the Fairy Ointment, to make a flying ointment that gives the subjective impression of astral travelling. This was highly hallucinogenic and so whether or not the witches actually physically flew is questionable. It has been suggested that they only resorted to these drugs during the centuries of persecution when they could not meet together for fear of indictment. There are certainly plenty of books, both magical and parapsychological, giving a variety of techniques for learning how to go out-of-body. The use of hallucinogenic drugs is another instance of the 'discovery' by science of something which the Craft has supposedly known for centuries. This use of local herbs for their psychoactive properties is another link between witchcraft with the shamanism of other continents, especially the Americas where the shamans use a vast variety of different plants for their spirit healing, psychic and psychoactive properties. Michael Harner, in his book *Hallucinogens and Shamanism*,[30] gives a detailed account of the witches' use of datura, mandrake, henbane and belladonna as a 'flying ointment'. Belladonna in combination with aconite is also known to produce falling dreams in sleeping people. The same mixture apparently could also be used to change oneself into an animal, the differing effects being achieved according to expectations and desires. This corresponds to the shamanic idea of one's power animal.

> Unlike classical shamans, the sorcerer in Europe had his trance encounters with the spirit world on occasions distinguished from his *manipulation* of that spirit world. I believe the reason for this major distinguishing feature of European witchcraft lies in the nature of the drugs that they were using. Specifically, the solanaceous hallucinogens are so powerful that it is essentially impossible for the user to control his mind and body sufficiently to perform ritual activity at the same time. In addition, the state of extended sleep following the period of initial excitation, sleep which can

extend for 3 or 4 days, together with the typical amnesia, made this hardly a convenient method for daily practice of witchcraft. Furthermore, there is some ethnographic evidence that too frequent use of the solanaceous drugs can permanently derange the mind ... Thus, the fact that traditional European witchcraft involves the separation of trance states from ritual operations may be largely due to the problems of coping with the particular hallucinogens they used. This ... also raises the question of whether shamans have to be in a trance state at the same time that they are engaged in their manipulative activities.[31]

A more gentle way of entering a visionary state is by using fly agaric or psilocybin mushrooms. The mushroom cult is widespread today among young people and many of those who consider themselves to be pagans of whatever sort, but not among some of those who consider themselves to be serious practitioners of the Craft who consider that mental training and discipline are the appropriate route to follow for spiritual and psychic development, and that opportunities for gaining mental, emotional and spiritual strength and knowledge are sacrificed by using drugs. This belief is also found amongst Yogis and Buddhists. However, the great value of psychoactive substances lies in their giving the person an experience of another state of consciousness, one which can be a mystical experience of the highest order, a spiritual understanding of this universe we inhabit, and frequently a firsthand experience of the clairvoyant reality. This experience profoundly moves the person, and it is this which is of such value. Once experienced, never forgotten. The world is never the same again, and this can be a necessary spur to undertaking the arduous training and self-discipline that self-development requires.

Another of witchcraft's traditional powers is mesmerism, or hypnotism as it is now called — is there a difference between the two other than in name? This I have already discussed in Chapter 1, and I return to it again because the essence of pagan ritual, as described by Starhawk,[32] is creating the right atmosphere, and the techniques of mesmerism and hypnotism are integral to this. It is well-known in parapsychology that the right attitude is essential. In witchcraft one initially creates a safe space and in doing so one sets the scene by banishing all inhibiting thoughts, feelings, attitudes, etc. and then calling in those elements, aspects, abilities and powers that one wishes to use. This is excellent psychology and I should imagine that every successful experimenter does this to some extent in their experiments. But most parapsychologists tend to work alone, whereas a lot of those who practice psi within paganism do so in a group. And when you have a whole group focusing upon a topic, say healing a friend, then the charge is that much greater, and the possibility for effective psi is greatly increased. Recently there have been large scale healing studies in hospitals in which groups of people have sent healing to patients with significant results.[33] And so slowly science is confirming the traditional practices.

Also good teachers of magic stress the importance of self-discipline and training the mind. No one can use psi effectively if their minds are as wishy-washy as most people's minds are, running about all over the place totally unable even to stop their thoughts, let alone control them. To use psi effectively one has not only to train the mind to go blank, to focus, to concentrate but also to visualise strongly so that an image can become almost tangible. It is this which is trained in magical disciplines. Pure psychology!

As I discussed in great detail in Chapter 1, symbolism is the language of the unconscious, through which the deep levels of the 'collective unconscious' can be contacted. In witchcraft, as in other magical systems, there is the belief that symbolic representation of a power, or abstract aspect, or energy, forms a channel between a person's unconscious and the power behind the symbol, and that their awakened imagination enables them to make contact with it.

> The core of the witches' faith, of which all else is trappings, is this belief in the Unseen Invisible, the Supreme Cosmic Power . . .. the world of reality which lies beyond the reach of human sense ... we know that the evidence of our senses is evidence only of illusion. The world in which they operate melts away in the light of the scientific discovery in modern quantum physics that matter is not solid, that it is a form of energy — as is mind — that the two interact, that nothing is *in reality* as it seems to be. The sky is not blue, roses are not red; they only appear to have these qualities. The witches have never acknowledged the limitations of the sense world; they have always lived, moved and had their being in the domain where the subtle forces now called ESP operate.[34]

Illusion, its meaning and the part it plays in the development of understanding is an important part of British paganism and fairy teachings; the function of illusion on these levels is to create an atmosphere. The purpose of this is to condition the mind by the seemingly miraculous to the reality of phenomena which also appear miraculous, because not subject to physical laws. But one must never forget that in ritual, illusion is symbolic. Under the pretence and behind the mask, reality is there. Parapsychological research has recently come to endorse this view, especially in the work of

Batcheldor.[35] It's all in the mind!

Illusion feeds imagination, and disciplined imagination is necessary to the opening up of new stages and degrees of experience. The operative word here, as the Craft and all the old teachings emphasise, is *discipline*. Imagination and illusion are tools, not ends in themselves. Imagination must be controlled, illusion must be recognised. Perhaps the reason why so many parapsychologists have had difficulty in obtaining clear and strong results in their experiments is that they have not used participants who have had the necessary training in this disciplined use of the imagination. The basic idea behind these methods is that out-picturing sets in motion a force which brings what is imagined, or imagined intensely, into manifestation — perhaps not immediately but, if you keep on working at it, in time you will create that which you wish. One theory is that the working principle of the operation is resonance.

Powerful aids to the building of illusion, or inducing change in consciousness, are rhythm, vibration as used in drumming, but also as used in poetry and song. Perhaps this is the reason poetry, chanting and song have always been associated with rituals and magic. Poetry has been called 'the language of the subconscious', possibly because the subconscious appears to more easily accept suggestion given to it in rhythmic form. Poetry is the natural language of the Craft, which has always used it, not only in spell-making but also to convey its wisdom and mysteries.

Basically all paranormal phenomena have the same source whatever label is put on them - the use of a subtle force to modify matter or conditions. According to witchcraft, the spoken word too has power, so long as the power is not dissipated by idle chatter or lying, so long as it is spoken with meaning, and that an intention once uttered is carried out and the promise fulfilled. Therefore we should all learn to watch not only our words, but also our thoughts, as the Buddhist scriptures counsel. There is a popular superstition that your ears burn when someone is talking about you and research by Braud & Schlitz found that when there is a CCTV camera focused on you and someone in another room is looking at you via the camera, that your autonomic nervous system responds[36]! So if just watching another person affects them – think how strongly an angry thought would hurt! Like any other power magic must be treated with respect and care.

Another of the traditional arts of the witch is that of healing, both psychic healing and with herbs. This, possibly more than any other aspect has become increasingly popular with a wide range of spiritual healers, holistic healers, acupuncturists, radionics, masseurs, herbalists and others all plying their craft. Very few indeed of these healers would consider themselves to be pagans or witches, but the holistic philosophy behind their healing practices is in this age-old tradition. These healing practices are being explored by several parapsychologists from many different angles. William Braud has done years of excellent research on distant healing (called DMILS – direct mental influence on living systems); Douglas Dean examined the physical changes in water from holy wells, or water held by healers while Lawrence LeShan and others have been training nurses and others in the medical professions in Therapeutic Touch.[37]

And finally, the arts of scrying, tarot reading, clairvoyance, etc. are incredibly popular these days, and advertisements for a clairvoyant or tarot reader can be found in most local papers these days. Many but not all of these people would consider their world view to be essentially an animistic one, even if they do not actually consider themselves to be witches or pagans. And these are the very arts which parapsychology has explored in such depth.

## Modern Conception of Paganism and Witchcraft

The popular modern conception of witchcraft in its origins is that it was basically concerned with ensuring fertility, with ensuring survival, which is a natural and understandable concern of hunter gatherers and early agriculture. The link with the fairy faith is of paramount importance here, as is knowledge of healing, herbs, midwifery etc. The village witch, therefore, was the one who grew the medicines, herbs, for the people of the village and was basically the village doctor, the wise woman of the community, a priestess, and counsellor. These people were the European equivalent of the American Indian shaman, to whom no such stigma of 'evil' has been, or probably ever will be, attached. They would also have played a leading role in the ceremonies and rituals surrounding the various holy days.

Where though does the modern conception of the witch as an old hag with a pointed hat riding on a broomstick come from? Some archaeologists say that the broom was a sacred symbol to many early peoples.[38] The broom can be seen as a potent symbol of cleansing, also as a sort of wand in times when ritual objects were forbidden. The pointed hat could well link with the cone of power, another fabled aspect of the Craft and their psychic arts. This would make sense if there is some aspect of electrostatics that links with psychic ability and the pineal gland. Just as a church steeple attracts electrostatic energy, which is the reason for lightning conductors that thereby earth the energy, so any pointed object becomes intensely charged with electrostatic energy. This is another aspect to horns as well, they too are pointed and so would become highly charged. I feel that there is more to it than this, and suspect that we have a lot to learn about shape and its focusing of energy, in the same way as the use of horns becomes clearer when we link them with electromagnetism, earth energies and psi-conducive situations.

The cone of power is said to be an auric emanation resulting from a circle of people chanting, dancing, meditating and generally raising psychic energy together. Lethbridge studied what he considered to be this energy. His experiments show it to be, as he says in *Ghosts and Divining Rods*[39] an invisible but demonstrable cone of force, surrounding all form, animate and inanimate, which is how witches have described it.[40] Even more possible is that Kirlian photography, and the corona discharge measured by this means, is in some way linked with this invisible but measurable energy. We have yet to explore and understand fully the electrical and magnetic properties of our bodies and how these interact with the electrical and magnetic properties of our planet and the weather.

According to Starhawk a cone of power is created by the group of people sitting in a circle within their safe space that they have formed. They then do a group meditation and breathing exercise, the culmination being the visualisation of a cone of energy forming around the group. This energy is then seen as flying off to do whatever work the group wishes to be done. The group then earths the energy. Again this sounds like excellent psychology from a parapsychological viewpoint, and I can imagine that, with a group of people who know each other well, who have worked together well, and who have trained and disciplined their minds to do the meditation and breathing and visualisation exercises effectively, quite a psi force could be raised. It all sounds to be perfectly in line with Bacheldor's principles mentioned in Chapter 2.

Since the earliest days of human history, the circle has been considered the most efficacious means of raising and conserving power, witness the cave paintings in France. It has frequently been used in the construction of places of worship such as Stonehenge and other ancient circles, or in the mosques, or in certain Templar churches, e.g. Cambridge. The power of the 'magic' circle as a means of generating forces of what we now call psychic or parapsychological abilities is well-illustrated in the oft-quoted claim that the witches called up the storm that defeated the Spanish Armada in the sixteenth century, and also the fog that prevented Hitler's invasion in the twentieth century. Witchcraft in the 1950s and early 1960s was often connected with the name of the late Dr Gerald Gardner, and he states that this fog was achieved by a large conclave of witches creating such a circle and raising a cone of power.

*Covens*

Traditionally witches would either be solitary or, far more likely, join with a group of others for celebrating the festivals, learning from each other, or for purely practical reasons as with any group working together. This group was called a coven. Remember that before we had books all knowledge was entrusted to memory and, if the knowledge was not to be lost, it had to be passed on to others who were in the next generation. Therefore it was essential that those with the knowledge trained successors. This, to me, is the most vital (life-giving) aspect of covens.

I feel sure that our modern conception of a coven is something far removed from the pre-witch trial days since they must of necessity have then become underground secret institutions, and so the modern idea of initiations with binding by cords and vows of secrecy, and all the other paraphernalia feels to me to be a distortion. All we can do now is to trust our own intuitions and, as well as trying to revive lost knowledge and lost ways of being, we also have to build our own new way of being. And in a society in which there are no witchcraft laws, we do not need to bind people with vows of secrecy and silence, so that perversion can be laid aside, and a new understanding of a coven as a group of people working together to grow their own psychic spiritual aspect of being in perfect love, trust and harmony is an ideal that can now come into its rightful place.

Members of a coven meet weekly, to learn from the most experienced member, to share information, as a healing circle to send healing to whoever needs it, and any other business that needs to be done, in much the same manner that prayer circles, healing groups, women's and men's groups meet together. Group psychotherapy could even be a possibility — the imagination can run riot.

*Festivals*

The most important pagan festivals are held quarterly, on the Solstices and Equinoxes, and cross-quarterly.[41] The cross-quarter days are the turning points of the seasons, when the shift in the changing day length becomes really noticeable. The first cross-quarter day is Hallowe'en which still retains its pagan memory to this day, and is represented in the Christian calendar as All Souls Day - or All Hallows. In Celtic this day is called Samhain, and is the ending of autumn and the turning point into the beginning of winter. This is the time when all the harvest is in, the animals slaughtered and the preparations for winter complete. It is the beginning of the dark, the end of the old year and the beginning of the new, for in the Celtic system (as in the Hindu) dark always comes before light and so the day begins at sunset, the lunar month at the last quarter of the moon, and the year at Samhain. By lunar reckoning the fire should be lit at sunset when the Moon is in her last quarter, and this tradition is still partially held by lighting bonfires on Bonfire Night, 5 November.

The next festival is Imbolc on 1 February (St. Brigid's Day, the Christian tradition celebrate this as Candlemas on 2$^{nd}$ February, the day after which one no longer needs to use those precious candles, as the sun is now strong enough and the days long enough). St Valentine's Day is the popular folk memory of this festival which came into being after the calendar change when 10 days were 'removed' in order to bring the calendar back into line with the solar year. (We get the same with Winter solstice and New Year's Day being 10 days apart, and Spring Equinox and April Fool's Day.) This is the darkest part of the year, when winter is at its hardest; yet it is

*Imbolc ceremony on Bride's Mound (Illustration courtesy of Lesley Delamont)*

also the time when the first signs of spring may be seen, snowdrops, the days are noticeably beginning to get longer, and the folk customs which are the relics of the Old Religion mirror this feeling. If the lunar aspects are to be respected the feast is held at midnight, when the moon is dark, the fire being lit then, to represent this point marking the turning, out of winter and into spring.

The next cross-quarter day is May Day in the Christian calendar, or Beltane in the Celtic. This is the ending of spring, and the turning point into the beginning of summer, when the hawthorn blossoms, the richness of Nature rushing forth in all the trees flowering, grass growing, baby birds hatching. This feast is held at the first quarter of the Moon, the fire being lit at sunrise, these all being the beginning of the light half — of the day, the moon, and the year.

The third feast is held at Lammas, or Lughnasad in the Celtic, 1 August. This is the ending of summer, and the turning point into autumn. The hay is in, the grain harvest is about to begin, and so the fire is lit and the feast is held at full Moon, in the middle of the day, when the light is at its height and is about to decrease.

The solstices and equinoxes mark the turning points of the Sun; Winter Solstice, Christmas when the night is longest; Spring Equinox, Easter; Summer Solstice, when the day is longest; and Autumn Equinox, like the Spring when day and night are equal. These are the calendrical points, by which we keep in touch with the turning of the Sun and the turning of the Earth on her path around the Sun; they are the mid points of the season, Christmas is traditionally considered to be mid-winter, the summer solstice, mid-summer. These are the feasts of the Sun; the cross-quarter days are the feasts of the Earth.

The Moon is very important in most religions, and many people will meet at dark of the Moon or when the new crescent Moon is first seen in the sky (Moslems still using this for start of Ramaddan, and all Hindu festivals are based on a lunar calendar, as are the Chinese and Tibetan), or when the Moon is full. In fact Easter is held on the first Sunday after the full Moon after the Equinox, the only one of the Christian festivals still linked to the moon, though this means that Lent and Whitsun are also moon related since these are timed by Easter, and are a Sunday next to dark of the moon six weeks either side of Easter.

The holydays are marked by meditation, circle dancing, music, chanting and feasting. Circle dancing and processional dancing were especially important and relics of these are to be found in Morris Dancing, folk or country dancing, especially those with their hobbyhorses or jesters. At Glastonbury morris dancers go on to the Tor at dawn on May Day and dance the sun up. On Burrow Mump, another prominent hill, on the dragon line not far from the Tor, morris dancers dance the sun up at summer solstice.

*Tree, Stone, Water*

An essential part of natural magic is the love that is felt for trees, holy wells, springs, and the stone circles and other sacred places.

Wells, fountains, lakes and springs are visited and offerings made to them — to the spirit of the waters — for harmony in marriage, for rain, to the fair folk for protection from bad spirits, to cure sickness, to raise a wind and for divination. Pins, coins, buttons, pebbles were all cast into water. Dishonour to the divinities of wells and springs has brought destruction on the land. There is still a custom in Derbyshire of well dressing in which pictures are made using flower petals and this is placed around the well on the well's Saint Day. In fairy lore there is a belief that one can reach the Sidhe by passing under the waters of a well.

With regard to trees, many people have been strongly influenced by Robert Graves's book *The White Goddess*[42] in which he mentions a tree calendar and the sacredness of trees and their relation to different times of the year. Most people have heard of the Druids and their special love for oaks, the fair folk are linked with Hawthorn and apple, and there is a saying that as long as there are the oak, ash and thorn in Britain then all will be well. I know of people who are planting tree circles according to Graves' calendar, and I know of people who are linking trees to other calendars according to their own research and beliefs. As yet this tree lore is fluid and growing anew as it should at this time.

In France there are several chapels dedicated to Our Lady of the Oak; some with whole oak trunks enshrined in the wall beside the altar, which suggests that the feelings connected with oak are linked with the ancient goddess religion.

And of course the stone circles and megaliths are being loved and visited by pagans all over the country, particularly to celebrate the turning points of the seasons and the moon. The quantity of research into stone circles, their lore and alignments over the past few decades, is quite phenomenal.

*Paganism today*

In the Craft and the new pagans today there are many different ideas and opinions. For instance, the emphasis on witchcraft as a religion is seen as such by most of the modern Craft movement, but some covens and some of the traditional witch families see it mainly as a Craft, a body of knowledge or techniques for self-development.[43] The Old Religion is so anarchic at present that no one could truly be said to be representative of the pagan traditions as a whole — and that is how it should be. The Craft is long overdue for an overhaul now that natural magic is the popular religion of so many of the young, rites and ceremonies are occurring spontaneously rather than from set dogma or formula, and the practices of the Craft are being examined objectively through the auspices of parapsychology and the scientific appraisal of the magical.

Paganism has survived through millennia because its philosophy is natural, and hence evolves to adapt itself to changing conditions, and is relevant to life whatever the context, because it is primarily a Nature religion. Thus as we eve-olve, as does all of Nature, so our religious and spiritual needs change. And, as the psychologists are fond of telling us, our needs and motivations are the prime movers in our lives. So people today, without reading a word of learned tomes on ancient religions, are manifesting ancient religious rites at the ancient sacred places, because that is what they need to do, and so they are doing it. If you ask them why, for example, they wish to congregate at Stonehenge for the summer solstice, only a very few articulate ones will be able to tell you why, and even they will be fuzzy in their logic and rationale, because the driving force is not reason or logic, but an unstoppable need to be there then. We are all children of our planet, and for some reason our needs and motivations include the magic of sunrise at a sacred place. It leaves a feeling, it quenches a thirst which cannot be logically explained, but which is none the less real. Others feel a need to see the sunset at the winter solstice; the last setting of the Sun for that year, for the next morning the first sunrise of the next year's Sun will rise, and the solstice marks that turning point and there is something truly magical about acknowledging that point and being there and seeing it. It makes you high; it makes you want to sing and dance; it makes you want to shout and run and skip. Hurrah for a new Sun (son) is born today! It is not hollow and empty ritual; it is alive and tangible and gives nourishment to the soul and the spirit. And just as our body needs food and water, so does our soul. And this is why so many people are becoming pagans. Our planet needs us to be in tune with her for she is being murdered by those who have cut themselves off from her, who worship only the father in heaven, or worship money and material things alone, and have forgotten to love, revere and take care of their Mother the Earth.

## Natural Magic: The Goddess Reawakening

Women literally physically embody the goddess. The goddess isn't some abstract impersonal ideal out there somewhere. The goddess is the essence of woman as we should be in our true glory. So, as we women are awakening, becoming stronger, more loving, caring, so the Goddess is becoming stronger in our minds, hearts, beings, and thus in our society. Any man who says he worships the goddess, must learn first to love women, to respect women, to care for women, to listen to women, to hear women, to consider women, and so shall he help to renew the Earth for women are closely in tune with the Earth, in harmony with Nature. Recognise that we are all an integral part of the Earth and Nature, that humans do not stand apart from this world but are the conscious mind aspect of this planet. When a man hurts a woman in whatever way, whether physically, emotionally or mentally, it is the same action in essence as the destruction of the forests, the poisoning of the waters and the air and all the other ways in which men are damaging our beautiful planet, by attempting to have power over, to control, to use and to abuse. This is the essence of man's nature with regard to woman and to the planet at this moment in time, and those men who are seeking the pagan spiritual way must learn through their relationships with women how to change their behaviour. This sounds hard, and in the 20 years since it was written is not so obviously relevant in Western culture, but having spent much time over the past six years in India, the patriarchal ways of living and behaving and thinking are still obviously prevalent over much of the planet. In fact the abuse of girls and women over much of the planet is quite sickening, and us humans really do need to make a major shift in our awareness in this regard.

Paganism is an all-embracing way of life, requiring self-discipline in order to live it every day in every way in every facet of our lives; pagans are children of the Aquarian Age and one of the characteristics of this era is the enjoyment of life, and joy and laughter have always been a key-note of the Craft.

Wicca and the psychic have important complementary roles to play as the trends and pressures of the next cosmic phase grow. The discoveries of modern physics and parapsychology and the old witch beliefs are so similar that no Inquisitor would notice the difference. Precognition, telepathy, clairvoyance, projecting and activating thought forms, blessing plants or crops are all part of ancient witch teachings. I find it very useful to notice the similarities and differences between the scientific approach and the magical approach that paganism exemplifies so well. To me, there is no essential difference in the phenomena, only in the words and in the underlying philosophy behind the difference in terminology.

Once it would have seemed impossible that witchcraft and science could have anything in common; they have come together because of the growing realisation that there is more to life than the five senses tell us of. The belief in the paraphysical world has always existed in a deep layer of the human mind, from which the Victorian philosophies of rationalism and materialism have never actually dislodged it. The approach developing today looks on phenomena which cannot be explained by known laws as a challenge rather than something to be shunned. They are there to be rediscovered — most of today's world-shaking 'finds' seem to have been known centuries ago. The teachings of philosophers, metaphysicians, mystics and seers in past generations are echoed in scientific pronouncements today.[44] The discoveries of modern science relating to the invisible world and its forces can be a means of presenting great spiritual truths in a form in which they can be recognised and used by the majority, instead of the enlightened few, in overcoming human problems and in achieving a fuller, more satisfactory way of life. The discoveries of science relating to the psychic world make possible a more rational attitude towards its phenomena. Science is opening the door, which once it slammed shut, to a world in which anything can happen, because it is a world in which mind affects matter, and mind has no discoverable limitations.

In writing this chapter about witchcraft and paganism, I do not intend to suggest that the ancient matriarchal religion and associated political and social systems are either better or worse than the later patriarchal religions with their associated socio-political systems under which we live today. Nor do I in any way advocate a *return* to them. Rather I advocate a progression whereby we transcend both of these in order to incorporate the 'Divine' within ourselves. The symbol of Aquarius is the water-carrier, a noticeably androgynous figure. The human race has evolved from the childhood of Mother worship, through the adolescence of Father worship, and now we are at the stage whereby we must grow up to adulthood, take personal responsibility for our thoughts, words and actions, and become Father and Mother ourselves. We need to learn how to harmonise the subconscious with the conscious so as to create the supraconscious within our own selves. Some see this more in terms of the equalisation of the right with the left as seen in terms of cerebral hemispheres.[45] In its outer form this can be seen as the harmonisation of the matriarchal and patriarchal religions — a joining of East and West, of the old and the new. Where science and magic meet there do we truly have the crack in the cosmic egg.

# 8. THE EMERGING RELATIONSHIP BETWEEN SCIENCE AND MYSTICISM

The emerging interface between science and mysticism is a topic I have been cogitating over for several years now. This is a tricky area to talk about because people hold such strong views; their emotions are closely linked with their belief systems. Having said that, now that the Christian Church has lost its power over people, so are people redefining for themselves what spirituality, religion and God mean to them. It is in this spirit that I write these thoughts, realising that these are thoughts in process, so to speak, and only half-way there.

## Definition of the Term 'God'

Ultimately, God is invoked as the Prime Cause of the Universe we inhabit; the ground of all being; in and of everything; that which is outside of space and time - eternal, ineffable and infinite. This term is neither male nor female, since it is prior to this world of duality. Those aspects we invoke as male god and female goddess are part of this 'ground of all being', which is sometimes referred to as Great Spirit, or Universal Mind. Because the term God is now correlated in my mind with the Christian conception of a Father in Heaven, I shall avoid using that word and use other words and phrases as they feel suitable.

Wilber[1] notes that there is a second aspect to God, or Spirit as he calls this concept; namely, the Highest Principle - that which we strive after in our highest ideals, the source of our morals, values and ethics. And he warns us not to confuse the two conceptions of God. Obviously, as the Prime Cause, Godspirit is intimately related to psi and magic and to every other thing, event and process in the whole Universe, since without the Creation there would be nothing. This definition does not actually help us at all because everything is Spirit. Thus we must turn our attention to the second aspect of Spirit as conceived by humanity — Good as the Highest Ideal. It is with regard to this aspect of Godspirit that all the religions of the world have spoken throughout the ages.

*The relationship between Godspirit and psi-magic*

Theoretically, psi-magic makes living beings, in particular human beings with their heightened mental awareness, potentially omniscient and omnipotent. These are abilities normally attributed only to divine beings so the natural psychic has, throughout historical times and probably before, always been accorded a special place in society as shaman, prophet, seer, oracle, etc., since as psychic they manifest a link between humanity and divinity.

Psi-magic, therefore, plays a vital role in the relationship between God, Goddess, or Spirit, or Mind, or whichever term you prefer, and humanity. Good as the Highest Spirit has been the guiding ideal for every culture and civilisation that we know about, and has therefore appeared in a different form in virtually every age and every religion, since this aspect is our human ideal and the form taken reflects the society, the culture and the individuals who make up that society. We create our own gods and goddesses according to our needs.

By way of illustration of this point, let me briefly discuss the Christian God — the all-male Trinity, with Our Lady in a rather ill-defined role as Mother of God. Traditionally, in Christianity psi-magic has been seen as the *means* by which God communicates with human beings, since whenever He has manifested in this world, it has always been via the 'miraculous': apparitions of angels, the divine qualities of the Son of God, whose ministry began with a miracle and was sustained throughout by a wide variety of miracles.

Already it can be seen that the Christian God is intimately linked with psi-magic. This is particularly the case when we look at the Roman Catholic section of Christianity, where psi is considered to be such a particularly important attribute of divinity, that manifestations of psi, such as stigmata (a peculiarly Christian 'miracle'), levitations, healing powers, are a prerequisite for a person to be 'beatified' after their death, i.e. called a 'saint', which means they are considered to be very holy, to have lived an especially good life, to have lived up to the principles of the Highest Ideal. The psi-magic is seen in these cases as being evidence of the 'Hand of God'. In other words, the psi events are seen as God manifesting through the saint, so that the psi becomes God's special 'tool', or 'means' by which He shows Himself on Earth.

Another way by which the Christian God manifests through psi-magic, is that of anomalies such as visitations of the Blessed Virgin Mary, the Shroud of Turin, miraculous pictures on the walls and windows of churches, etc. These events all run counter to the so-called 'Laws of Nature' and so are seen to be miraculous and the work of God. They are all examples of psi-magic. In fact, I think it is the Christian conception of God and His relationship with psi that originally brought about the idea that psi is 'super'-natural and 'para'-normal. This in turn meant that any person who manifested psi-magic and was not specially dedicated to the Christian God became labelled a 'witch', or a 'tool of the devil'. These concepts are changing now, but I feel that in order to understand present relationships it is useful to know our history.

If we travel to the far East, Hinduism, Buddhism and Taoism are the major religions. The Hindu religion also has an all-male Trinity who manifest in this world using psi-magic as a tool in a manner reminiscent of the Christian and Jewish Gods. There are also a plethora of minor gods and goddesses similar to Egyptian, Greek, Roman, Celtic and other ancient religions. These too are held to be divine partially on account of their possessing prodigious psi-magical talents. In these cases, as with most earlier religions, psi-magic can almost be considered a necessary attribute for divinity.

However, there is in the East a 'higher' philosophy, as there is in the West also, but for historical reasons it is most developed in the East, which transcends all these anthropomorphic representations of God. This is the 'Higher Ideal' of the Buddha, of the Sufis, and of mystics, like Meister Eckhart[2] and William Blake; the Perennial Philosophy as Huxley[3] calls it. This is the philosophy with which the 'New Physics' is in agreement[4] of which David Bohm[5] and Wilber[6] speak when presenting their ideas of a holographic universe, or of an implicate order behind our everyday see-touch reality. This is the philosophy of magic and the new pagans.

In this philosophy, God is spoken of as the One, the Void, That Which Is, as Spirit, Cosmic Consciousness, Universal Awareness. The conception is not far removed from the Ground of All Being, yet there is a subtle difference because this Godspirit is still the highest to which we can attain, and so we can have a relationship with and a striving towards this concept of Godhead. In this mystic conception of Godspirit, divinity and psi-magic are not so strongly related. Indeed, the Buddhist and Yogic teachings warn us that psi-magic (termed the clairvoyances or siddhis) is a red herring lest we be led astray by its glamour from the true path to Oneness with the Whole, though both the Yogic and Buddhist teachings say that to attain enlightenment one must first have attained conscious control of psi abilities. This is not to say that those who attain to the Ideal — Enlightenment, call it what you will — do not manifest psi personally, it is just that psi is not any more special than any of the other ways in which one links in with, becomes one with, the Divine Whole within and without. And the Godspirit of the mystics, the Buddhists and the Yogis, does not manifest in the world through psi-magic in particular, but rather is ever-present for those who have eyes to behold. And, above all, psychic, magical abilities are ever-present for all, are perfectly natural, normal talents, which can be developed for particular purposes if so required. The development of these talents is a normal part of the whole spiritual development.

## The Great Chain (or Nest) of Being

Psi-magic, or magic as I shall refer to it now suggests, in a very concrete form, that there is a realm apart from the physical, a purely mental realm. Let us for convenience, take Wilber's diagram as representative of a philosophy common to most religions concerning the realms of being (see page 236). In Wilber's later writings[7] he suggests seeing this Chain (or Nest) more as ripples in a pond going out and out, rather than hierarchical levels. This I like. Some will have more realms, some less, but the basic idea will be the same. Magic, as a phenomenon of the mental realm stands midway between the physical and the spiritual. Religion is concerned with the domains both of spirit and of soul which incorporate into them and supersede all the phenomena of the mental realm.

In this Great Chain of Being, for humans the special quality is self-transcendence — the ability to be aware of and respond to the spiritual world. For animals the distinguishing characteristic is consciousness. For plants it is life.

I would like to pause here awhile to explore briefly the wider concern of the relationship between science and religion, which I see exemplified in the Great Chain of Being. If you define science simply as knowledge, then contemplative religion can be science, e.g. the science of yoga. If you define science as 'empirical-sensory knowledge, instrumentally validated', then religion is non-scientific, since how can one instrumentally validate faith, morality or ethics? Let us listen for a few moments to the words of some of the world's greatest modern physicists:

> Religion proper speaks not of norms, however, but of guiding ideals ... These ideals do not spring from inspection of the immediately visible world but from the region of structure lying behind it ... The natural scientist must recognise this comprehensive significance of religion in human society if he wants to try to think about the relation of religious and scientific truths ...
>
> The care to be taken in keeping the two languages, religious and scientific, apart from one another, should also include an avoidance of any weakening of their contents by blending them. The correctness of tested scientific results cannot rationally be cast in doubt by religious thinking, and conversely, the ethical demands stemming from the heart of religious thinking ought not to be weakened by all too rational arguments from the field of science.[8]

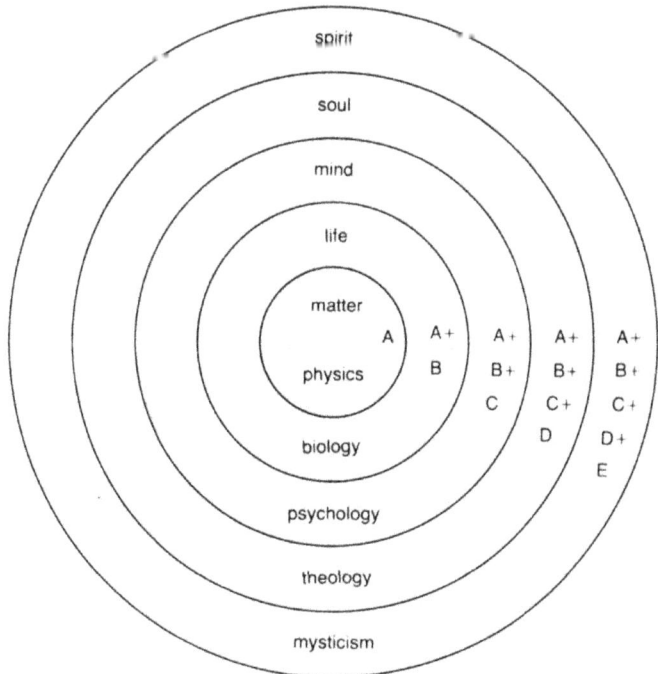

The Great Chain of Being (adapted from Wilber, 1984).

**Wilber Great Nest of Being (Illustration courtesy of Shambhala Publications)**

While being well aware of this essential difference between science and religion, we can take heart that the two are not so antagonistic towards one another as they were. For example:

> It is probably true that the recent changes of scientific thought remove some of the obstacles to a reconciliation of religion with science, but this must be carefully distinguished from any proposal to *base* religion on scientific discovery ... [The new physics] gives strong grounds for an idealistic philosophy which ... is hospitable towards a spiritual religion.[9]

Thus, in our interface between science and religion we can certainly clearly delineate areas of difference. Psi-magic has no bearing whatsoever on the moral and ethical aspect of religion. Psi-magic in itself is a talent, a force, a power, that can be used for good or evil, as is so well exemplified in our mythical and fairy-tale literature. It is the role of religion to turn our minds, hearts and souls towards the good.

If we re-turn our attention to the diagram of the Great Chain of Being, we can see magic as being some sort of 'bridge' across the various levels of being. The mental realm is, of course, the realm of magic, and is the realm I have been concerned with throughout most of this book. Inasmuch as the higher levels incorporate and supersede the basic laws of the lower levels, we should not be surprised that, at a theoretical scientific level, the workings of magic can be understood in quantum mechanical terms, since this is a baseline for everything in the Universe. The various theories relating the mechanism of magic with quantum mechanics are not yet finalised nor commonly agreed upon, but this does not mean that there is not an intrinsic validity in seeking to understand magic within these terms as well as at the biological and psychological level. Further, through psychokinesis our minds act directly on matter, from the decay of atoms through to large-scale poltergeist effects. In fact, as stated by Sir James Jeans:

> Mind no longer appears as an accidental intruder into the realm of matter; we are beginning to suspect that we ought to hail it as the creator and governor of the realm of matter - not, of course, our individual minds, but *the mind* in which the atoms out of which our individual minds have grown, exist as thought.[10]

In a much clearer sense, psi can be found interfacing with life at all levels as the research by Grad, Braud, Le Shan and others on psychic healing of plants, animals and humans has shown.[11] There are some indications that animals possess a certain level of psi ability, though most of this is at the anecdotal level of dogs barking at ghosts, or finding their way home over incredible distances. However, the recent research by Rupert Sheldrake, on the dog who knew when his owner was returning home, suggests that this ability can be tested experimentally.[12] The philosophy of the fairy faith that as life evolves so does consciousness, and with it magical/psychic ability, should not be discounted at our present state of ignorance.[13]

The region of the soul, according to the Great Chain of Being is the region of theology, i.e. of religion, and so it is in this region that phenomena such as spiritualism, theosophy, the accounts of Christian saints, the attainments of fakirs, shamans and so on, can be interfaced with the scientific findings of parapsychology. There is a growing body of literature covering this area such as that of Rhea White, Scott Rogo and Hilary Evans.[14] The most advanced 'science' of this realm is to be found in the East with their 'science' of yoga and meditation, and we can find here techniques for specifically raising magical ability to conscious awareness and control.

The fifth area demarcated by the Great Chain of Being is that of spirit and, as has been mentioned before, this area has really only a theoretical interface with psi, such as in the Holographic Paradigm and Bohm's Implicate Order, which uses scientific concepts to arrive at the mystical and Buddhist viewpoint that every part of the Universe is contained within every other part, i.e. each part contains the whole. Of course, psi is present within the spiritual realm, as is everything else in the Great Chain of Being, but it assumes no greater or lesser importance than any other item in the Universe.

These realms interpenetrate each other, from the greatest good at the angelic level to demons reflecting the harmful aspects of the spiritual worlds. In such a picture each realm is a poor and imperfect model of the next realm, but containing the essence of that quality. When we talk about well-behaved children being angelic, we are unconsciously recognising that angelic quality as reflected in the realm of human beings. We seek to be better, to be purer, to do good: that is the direction that we all ultimately realise is what the human condition is about. That urge is itself a recognition of the evolution of creation, as stated in the Fairy Faith philosophy.

*The interface between pscience and religion*

> I maintain that the cosmic religious feeling is the strongest and noblest motive for scientific research ... it is the most important function of art and science to awaken this feeling and keep it alive in those who are receptive to it ... science without religion is lame; religion without science is blind

... true religion has been ennobled and made more profound by scientific knowledge ... science not only purifies the religious impulse of the dross of its anthropomorphism, but also contributes to a religious spiritualism of our understanding of life.[15]

The above sentiments are as true for psychical research, parapsychology, pscience, whatever you wish to call this branch of study, as for any other science. But above all, through psi one can link with spirits, though not necessarily with Spirit. This strength of link with the outer reaches of the Great Chain should more than anything help to bring our minds and hearts much closer to a realisation of the spiritual within the universe, and to a realisation of the interconnectedness of everything.

The scientific study of psi helps to ground — to earth — psi into the material realm of objectivity. If one can make objective measurements then, in this twenty-first century of ours, it helps us to understand its nature better, because that is the way most Western people's minds work at present. The language of the mystics, yogis or shamans is not sufficiently concrete for many modern Western people to fully relate to, in a way that actually changes their behaviour and mental thought patterns. Even when studying magic at the scientific, objective, concrete level we can never forget that it is a link with the higher reaches of the soul. And, in taking magic out of its old Christian conceptualisation of 'super'-natural and 'para'-normal, and showing that it is a natural and normal talent of living beings, so we radically change our concepts of God and divinity — a change that brings us in line with the Perennial Philosophy, the esoteric religion of all peoples, what Einstein calls the 'cosmic religious feeling'. By realising that psychic experiences are normal, so, by paradox, we realise that each person contains within them the seed of divinity and each person can realise that potential.

## The Social Implications

*The emerging philosophy — a potential change in world view*

Essentially this chapter is looking towards an era in which the whole of society accepts psychic phenomena as part and parcel of life; in which the spiritual regains its place alongside the material. There is one main assumption in this book: that magic is a valid phenomenon. As far as acceptance of magic goes, it seems as if the establishment view rejecting the existence of magic is totally out-of-synch with a large proportion of contemporary society which is, at present, enthusiastically espousing all matters psychic and spiritual;

for example, there are a multitude of new spiritual groups, a new pagan resurgence, diverse occult societies, dozens of personal growth therapies, healing groups, rebirthing, Reiki and all the other cults and movements of the past two decades, too numerous to mention.

In accepting magic, it is not only attitude towards magic that changes. A whole lifestyle and the philosophy that has engendered it are changing as well — and if they don't then the planet will be in very serious trouble.

What is a society like that accepts magic as part of its being? In order to understand this for ourselves moving into the future we have to at least partially disregard all the vast array of prior cultures as possible prototypes, because we now live in a global village as a result of our technological achievements, and so there is an ever-increasing blending together of all the cultures within our planet. Magic within society has a very different meaning when you can pick up the telephone and speak with a friend the other side of the world. Thus certain aspects of the 'Hippy', 'Green', 'New Age' or 'Aquarian' culture, which has grown up within our technologically sophisticated world, are probably the best indicators we have of a future psi-oriented society. The dissidents of one generation are the establishment of the next.

The first thing that strikes me about this sub-culture is that all the technology is used, but the materialistic philosophy and value

*Lighting up Glastonbury Tor for the Millenium (Photograph courtesy of Kevin Redpath)*

systems of the society that created the technology are firmly rejected. Everyone is highly spiritual in a non-religious sense; that is, there is a recognition and a respect for the spiritual aspect of life, but there is as yet no clear *form,* each individual finding their own way through the plethora of spiritual teachings. At present people seem to be trying out various old systems, such as Buddhism, Paganism, Hinduism, Sufism, the Celtic Revival and blending them together into a mystical framework very akin to the Perennial Philosophy. It seems that people are no longer looking for 'a God' to tell them what to do, or to help them when they are in trouble, or to punish them when they are naughty, or to forgive them, but are actually looking within themselves for *all* these things - they are finding the Divinity within their own selves and are becoming self-responsible.

Let me repeat what I said at the conclusion of Chapter 6: we are accepting total responsibility for the power of our very thoughts to create our own reality — which includes our gods and goddesses. We are accepting total responsibility for the care of our planet and all upon it as did the Mother in matriarchal times, and we are disciplining and guiding, as did the Father in patriarchal times, both within ourselves and within society. *We can no longer do wrong and look to a Father God or Mother Goddess to forgive us and put it all right: we must shoulder the responsibility ourselves.* This is the essence of the new/old holographic, mystical philosophy emerging now in our society.

We must be very careful what symbols we choose to represent our Highest Ideal — do we still need concrete images or can we move into the abstract? I have statues of a Chinese immortal, a Thai buddha, a horned god, a neolithic goddess and a modern goddess; all of these represent different aspects — but how necessary are they for our emerging spirituality in which we recognise that we are potentially divine and that Divinity is all of these images and far more?

This philosophy is, of necessity, more abstract than religions of the past, because an adult can work with a greater degree of abstraction than can children, who need concrete examples (myths and parables) in order to understand what they are being told. Neither do adults need the bribery and blackmail implicit in the Christian teachings of fear of hellfire or the rewards of heaven, as did childhood humanity, but rather we can now appreciate and work for mutual good and benefit for its own sake and for the benefit of the whole planet. This is no longer a selfish philosophy emphasising the enlightenment of heavenly reward of the individual; it is a collective philosophy concerning the evolution of consciousness of humanity as a whole.

This emerging philosophy of a magically-oriented society, a society for which the psychic is part and parcel of everyday life, sees the Universe and all life as an interconnected whole, every action and every thought affecting every part of the Universe. This philosophy sees the spirit in every aspect of life. The fairy faith of the Celtic peoples is in some way being renewed and, as Evans Wentz[16] pointed out, belief in magic is an integral part of that old faith, a faith that was very close to Nature. This new philosophy sees our species as going through an evolutionary shift that is greater than each of us individually — a shift inspired and in some way generated by Mother Nature (Planet Earth, or Gaia, inspired by James Lovelock's ideas).[17]

At a practical level, this philosophy says that, as we are wholly part of the Universe,[18] and especially so of this planet, we *must* learn to live in harmony with the planet. The realisation that all is interwoven, interlinked, must surely create a change in practical life attitudes, changes that are apparent everywhere now, after forty years of pushing for them: recycle everything, no more trash cans, no more garbage dumps; electricity from renewable sources together with energy conservation leading to less acid rain, less nuclear waste and less radio-active discharge, less global warming from over production of carbon dioxide; conservation of energy manifesting in such simple ways as washing lines rather than tumble driers; solar-powered cars reducing the noxious fumes which kill our trees; an end to the present debt economy which is slowly murdering the planet as all countries in debt have to over-produce in order to pay back the interest charges. At the individual level this debt economy forces farmers to use pesticides and fertilisers on their land in order to make sufficient profit to pay the banks, thus polluting the Earth yet more. In other ways everyone who has some form of debt or loan is contributing to the over-use of the planet's resources, which is leading to desertification, destruction of the forests, and so on. These, and many more examples of attitude change inspired by the question, 'How is the way in which I live my life affecting the rest of the planet?' are having a radical effect on the way in which we actually live our lives on this planet already.[19] This change is accelerating as more and more within society acknowledge the psychic-spiritual realm as being the *equal* corollary of the material. The USA is seen by most people in the world as being the worst offender in this respect, since it is the affluence of the Americans' life style that is crippling the rest of the world.

Under the emerging holographic philosophy, where the whole is present in each part, and no part can do anything without affecting the whole, it is up to each individual to change their lifestyle in accord with the planet's needs. There is great emphasis on each individual living their beliefs every day at every moment, living a spiritual awareness of the inter-connectedness of everything at every level. Psi is a logical, necessary aspect of this holistic, mystical philosophy.

Many of the aspects of the change in world-view growing among certain parts of society have been triggered by the 'New Physics'. If Newton had not made his discoveries, the philosophy of the Clockwork Universe would never have been promulgated, nor accepted so enthusiastically, though doubtless sooner or later it would all have happened, since that was the spirit of the time.

We have a similar evolution occurring in our time with quantum physics which has taken one hundred years to emerge from the laboratory into a position of influence in the way that people view the world. Books like *The Tao of Physics*[20] have deeply affected contemporary society, whatever the present-day scientific establishment might like to think about them. *The Holographic Paradigm*[21] and *Wholeness and the Implicate Order*[22] all offer a new world-view that is closely linked with the Perennial Philosophy of mystics of all ages, at a time when there is a very deep dissatisfaction with the old materialist reductionist world-view. The journal *Nature* might have considered Sheldrake's book *A New Science of Life*[23] a candidate for burning, but the people loved it. All these new ideas from various scientists are being enthusiastically espoused by a wide range of people. If there had been no Einstein or no quantum physics then this new/old philosophy could not have inspired, and taken root in, contemporary society in the same way. And there is obviously an enormous *need* for this new/old philosophy for it to have been so eagerly and rapidly absorbed and embraced by so many. I feel that *only* when the 'New Physics'/Holographic world-view philosophy is well grounded will the psi world-view come into its own.

The new paganism that is emerging today, whose philosophy incorporates a magical world-view, emphasises the spiritual, the good, the caring aspect of magic because it is very important that this feeling that magic is inherently evil of itself is eradicated. To this end it is very important that we adopt the Yogic and Buddhist philosophy — that it is the spiritual development of the person that is the most important, the psychic abilities, or siddhis, being merely milestone markers along the path of personal self-development, and not things to be sought after for their own sake, or else you become the worst sort of fakir, a faker, rather than a yogi, one in union with the divine.

A central aspect of the new paganism that seems to be an integral part of the new philosophy, is the growing discipline of Earth mysteries, which is linked with the fairy faith and the Craft. There is a need for mystery and magic, and the stronger that need is stamped upon, the more it will emerge in disturbed, psychotic ways. Jung[24] equates spirit with the collective unconscious, and the symbols and phenomena of the occult are typical archetypes. By denying these archetypes so we deny a very deep part of our selves, which then emerges in various distorted forms such as Nazism, fascism, and the hatred of Nature that some people seem to have, which is so rapidly destroying the ecosystem necessary to sustain life. For, having destroyed the Spirit by denying its existence, so now the destructive force is turning to the material aspect of life and is destroying the planet on which we live, and of which we are part. Once one accepts that there is a fundamental spiritual *need* in a person's life, then it is given a fair chance of being expressed in a positive manner that will help the person to develop mentally and spiritually, rather than repressing it so that it harms them by making them neurotic, unfulfilled, dissatisfied, greedy and all the other symptoms of our materialistic world. This is an integral part of the change that is occurring where psi is accepted as a normal everyday part of life.

*Some thoughts regarding Spirit*

Consider for a moment the strange aberration of some of the academic elite of our society in their refusal to accept the existence of Spirit or spirits. The evidence of archaeology and anthropology shows that all cultures from the beginning of recorded history and probably before, and in every part of the globe, have believed in spirits. It is only in the last three centuries, since the Inquisition and witch trials, that there has been a growing lack of belief. But, even in the last one hundred years, which has probably been the most materialistic epoch which this globe has ever known, many people have still retained a belief in Spirit or spirits in some form. Even in our avowedly anti-spiritual culture, the materialistic philosophy is actually confined primarily to the academic establishment and to city dwellers.

*So, the really intriguing question is why materialists should hold their belief that there are no spirits. They have never PROVEN that spirits don't exist; they have just reasoned them away!* Is it not strange? Is it not very strange that the intensely destructive Western worldview, technology and life-style of the past century should coincide with the only period in the whole of human history when certain influential thinkers have doubted the existence of spirit? Why, I ask myself, should there have been such a universal belief if there were no reason for it? And I do mean reason — not proof! And would not the quality of life be improved if we did recognise and honour the existence of spirits — the Spirit of the Earth, the bubbling brook, the old oak tree, the Spirit of the Age. By recognising and respecting the spirit of a thing, we more readily care for it and love it and nurture it; and no society who recognises the spirit aspect of the material world can mistreat it, as our spiritless society has mistreated planet Earth and all upon her.

I am not talking about ghosts and things that go bump in the night, but about the animistic, pagan conception of spirit that is more in line with Jung, or with the ideas expressed colloquially in such phrases as 'Spirit of the Age'. Jung[25] defined the soul as being the personal unconscious, and the spirit as being the Collective Unconscious. He also considered that the soul of a living person becomes their spirit when they die, that is, becomes part of the Collective Unconscious which is the archetypal numinous aspect of all humanity. One of the latest ideas from other cultures which has resonated a chord within the world of the new paganism is the concept of 'the ancestors'. This concept is being pieced together from Aboriginal, Amerindian and Chinese sources, linked with the Fairy Faith, and can be understood as linking in a very strong way with the Collective Unconscious. I often feel that the Collective Unconscious is the mind of the planet, or the 'World Mind' as Julian May[26] calls it. This is the root source of all that is going on in our own little human minds and the various 'ages' that our consciousness is growing through.

### The Energy-Matter Equation

It occurred to me some years ago, when I was trying to explain some aspects of quantum physics to a lay audience, that Einstein's famous equation, $E = mc^2$, can be seen as a symbol of the shift that is occurring from a materialistic world view to a spiritual one. Since Newton the Western world has focused almost exclusively on the matter side of the equation, on the material side of life, and now the shift towards focusing on the energy side of the equation is beginning, as is reflected in language — for example, 'good vibes'. Thus, one can conceive of the spirit of a thing as being the energy aspect of that thing. This is directly related to its matter aspect, but has totally different laws governing its behaviour, as is found in particle physics, for example, the wave-particle aspect of light itself. David Bohm adds a third aspect to this equation – meaning, or information, and proposes the centrality of consciousness in the universe being how it is.[27]

The energy engendered by emotion is probably the strongest energy of our body/minds, with the emotional energy engendered by religious beliefs possibly the most destructive, as seen in the various religious wars around the world during the past 4,000 years. When we concentrate our attention on the energy side of humanity, so we start to concentrate on the spiritual aspect of life, an aspect which is so furiously denied by the materialists who concentrate their attention solely on the matter side of the equation. Thus, society's response to psi or magic is primarily an emotional one, though academics might dress it up with logic and analysis, and people reject or accept magic for exactly the same reason that they reject or accept a spiritual aspect to life. Those who advocate magic must recognise also that there is an emotional component to their behaviour, for if they work without heart then they are soulless machines, and no good will come of their work. We are dealing here with issues that encompass more than the rational — more than reason or the purely intellectual. There is always an emotional component, however much it may be dressed up in logical rationalism. By asserting that there are psychic events occurring in our own and others' lives we open up a Pandora's box in which all the spiritual aspects of life have been shut up and locked away during the 'Age of Reason'. The results of this locking up of our spiritual awareness and the spiritual aspect of life has led directly to the material problems confronting our generation.

The harmful spiritual aspect of the mind has often been conceptualised as demonic, and we disregard demons at our peril for they are reappearing as mass psychosis in our prisons and mental hospitals, which are overflowing with violent, aggressive, crazy people — mainly men it is interesting to note, men having sadly been the active perpetrators of the worst horrors in the name of materialism. And this is not to blame men, they are as much the victims of patriarchal culture as are women. The people who are polluting the earth, the air and the seas, and threatening the whole planet with destruction *must* find their souls, their hearts, their spiritual being. At present they are soulless automatons, heartless creatures, to so destroy our beautiful planet.

I have always seen parapsychology as the 'earthing' of the spiritual. In experiments the psychic is explored in a very logical, rational, exoteric manner; clearly demonstrable proof-ratings are assigned to the different variables. In such a manner many spiritual teachings have been confirmed, for example, that one's attitude or belief about something strongly affects the occurrence of that particular matter. Faith, it used to be called, although now it is 'the sheep-goat effect', and it was said that faith could move mountains. Jesus spoke quite extensively on the incredible effect of faith, which has now become transmuted to attitude, and the Hindus have a whole spiritual path centred around faith, called Bhakti Yoga. Our modern terms are more applicable to our present society, but underneath the change in terminology the concept lives on.

Another example of the 'earthing of a religious concept' occurs when doing a Ganzfeld, or free-response, experiment. The first thing that the participant is taught to do is to become aware of the contents of their mind. This action is what the Christians call contemplation, and the Buddhists call mindfulness, and it is the first step in meditation, the first step in learning how to develop one's mind. The state of consciousness that the Ganzfeld induces is to be found in quite a number of different religions as well, albeit induced by radically different methods such as getting up and chanting at 3 a.m.! The point in common with all of the methods is the aim to create a state of consciousness whereby the conscious mind is stopped, thus allowing one to access material from the Collective Unconscious.

Any person who embarks on a training for the enhancement of their magical faculties *must* be very well developed spiritually. They must have their emotions well controlled. It is well known how powerful human emotions are, and if these find expression psychically there can be disastrous consequences. Poltergeist cases, especially where the focus is an adolescent, are very good examples of the havoc wreaked by undisciplined psychic forces issuing from an emotionally unstable person. The person must therefore develop spiritually to the point where they rid themselves of anger, hate, fear, jealousy, greed, lust, spite, envy, malice, aggression, violence, and all the other harmful traits to which so many human beings fall prey. This is why the Eastern disciplines stress so strongly the importance of the spiritual aim — to become one with the Whole. If a person is to train in magic they must be true to their own self-development first and foremost. I found when I was doing practical experimental work for my postgraduate degree in parapsychology that the participants and I were all undergoing intense personal development — if we were open to that possibility. The actual psi score was merely a milestone marker along this path.

We must not forget that we are working with something of terrific potential power — we must hold the magical arts in great respect and awe. The old Western schools of magic were very aware of the dangers of this power and were *far* more careful who they instructed than we are today with our deadly nuclear power. It was not only the centuries of persecution that forced the practitioners into secrecy, but the realisation that humans in general are a mess and incapable of using magic wisely.

As I have mentioned when talking about the Fairy Faith, it seems as if there was a time when religion and magic were conceptually and practically one. It is therefore vital, with the renaissance of interest in magic today, that the moral code inherent in religious systems be incorporated into any system that explores or develops psychic ability. The great advance brought in by the religious systems in the Piscean age has been the emphasis placed on love, altruism and compassion. Christianity is defined by "turning the other cheek," as opposed to the prior "eye for an eye." Buddhism is distinguished by its great emphasis on compassion together with wisdom. We must be very careful and must not divorce magic from religion — religion in the sense of spiritual discipline and moral code. The witches emphasise that magic must never be used to harm, and should only be used for what is really necessary. Let me repeat again the Wiccan Rede (the counsel or advice of the Wise Ones) which is: '*An ye harm no one,* do what ye will.' They further state that if you do harm, that it will be returned to you threefold, which cannot be a bad belief to have, whether or not it is true.

To acknowledge that energy is interchangeable with matter, and that there is an energy aspect to be considered as well as a material one, is to re-evaluate the *whole* of one's life. When we add magic to this philosophy we have to be aware of the potential effect of our thoughts within the world, for as mentioned earlier parapsychology research suggests that every thought one has has the potential to affect everything else. And my negative emotions can play havoc with those around me, let alone the effect they have on the general world atmosphere — in potential. If, in potential, I can astral travel in the true sense of the words (star travel), and thus can mentally link with any place in the Universe I choose, then I begin to realise the awesome potential of my mind and how careful I must be with my thoughts, since my thoughts not only link with all of creation but can also materialise into physical reality, as is shown by the four decades of research by the Princeton Engineering Research (PEAR) lab.[28] The Buddhist creed of 'Right Thought' strikes home in a very dramatic and immediate manner. Full activation of our magical potential brings considerable responsibility and a dreadful need for a truly spiritual state of being: a realisation of the divinity within each of us.

As a result of our troubled history here in the West, we are only just beginning to make any real progress in bringing spirituality into magic. Fortunately, in the past two decades, this shadow has started to lift, and we are now seeing a re-emergence of magic within religion, as churches admit healers (laying on of hands, the traditional Christian psychic healing method) within their walls, and of religion within parapsychology (the science of magic), as Rhea White's[29] paper indicates, and Giesler's[30] research in Brazil exemplifies. Research on near death experiences, reincarnation, out of body experiences, hypnotic regression, and such like, all help us to real-ise the spiritual aspect of the human being in a very concrete way. They are fresh ways of looking at very old beliefs, so that the twentyfirst-century person who has rejected religious dogmas can at least have some firm foundation for those spiritual feelings which are now emerging.

*Science with a Conscience*

It is no wonder, then, that so many academic scientists are so sceptical of, and derisive about, magic. Whether we like it or not we are advocating a return to a spiritual way of life, one that puts spirit energy back into the materialist equation — and this threatens the whole power base of materialist science which is at present the ruling dogma. They also find it threatening because it means that they have to look into their consciences once again. (I even had to look up the spelling of 'conscience' it is so long since I used the word; con-

***Walking the Mizmaze (Illustration courtesy of Lesley Delamont)***

science, with knowledge.) Inglis[31] called today's religion 'scientism', and when I see advertisements on television no longer using sex or affluence to sell their products but rather a 'scientist' in a white coat, then I know that scientists are verily the most influential symbol of the day — the high priests of the latest religion. This makes it even more important for scientists to regain their consciences, their morality and their judgement of what is right or wrong. It makes it even more important for scientists to admit publicly that matter has spirit as its counterpart, just as light has dark as its counterpart, for then the whole of our city-based society will do likewise. If an advertisement can sell washing powder through scientific approval, then scientists can sell anything! It is important to note here that it is society's view of establishment science that is selling the washing powder. With psi we have the strange situation of phenomena that the majority of people, according to the polls and personal experience, believe in, but which are rejected by establishment science. Should the establishment accept it then the belief system of the whole society would change accordingly.

We all must learn to follow our hearts as well as our minds, since psi is a very emotional method of communication, a feeling more often than a thought. This aspect of bringing heart back into science and society is possibly one of the more important influences that must occur with the acceptance of magic by the whole of society. Just as a healer can affect the growth rate of a seedling or the healing of a wound, so can we affect the healing of planet Earth. The method of science is a sure way to a clear understanding of the matter in question, and so makes a firm foundation for a belief. Then we will have a science with a conscience, a science that not only examines the *tools* of magic, but also the philosophy of magic, and the purposes for which the tools are intended to be used once they are ready for the market place.

# EPILOGUE
# EXPLORING CONSCIOUSNESS: SIX YEARS WITH YOGIS AND TIBETAN BUDDHISTS

## Introduction

The data from parapsychological research is slowly becoming more accepted by the scientific community. The main sticking point is still a good theory into which psi fits. This is essentially the same sticking point for what has been called the "hard problem" in consciousness research – how can the non-physical mind stuff interact with the physical brain?

For the past six years I have been working in India, initially at the world's first Yoga University (Bihar Yoga Bharati) in Swami Satyananda's ashram in Bihar, and then with Tibetan Buddhists at various monasteries in India. It has been an amazing experience, not least because the Indian ashram students showed me very clearly that my knowledge about mind and consciousness was severely limited by the Western approach. I therefore started to make a study of the Yogic and Buddhist philosophies of mind and consciousness and what follows is where I've got to so far, aware that I am still in primary school as far as these ideas are concerned, but feeling that my simple understanding of these complex concepts just might be helpful to others in the West who may be interested in learning about this viewpoint. The Yogic and Buddhist conception of consciousness is a top-down approach similar in some ways with the Neoplatonic philosophy found in the Western Mystery tradition, and also having links with traditional Western animist philosophy. They also are profoundly similar to the Holographic Universe philosophy that I have mentioned as being the best understanding for how psi works that I have come across so far.

## The Vedic and Buddhist Concept of Ground Consciousness (Alaya Vijnana)
## In the beginning…..

According to Vedic and Buddhist teachings, consciousness is sort of the equivalent of the Big Bang or God. This is not at all like the God of the West, but rather, as Swami Satyananda[1] says: "By God (Ishwara) we mean . . . a superior spiritual consciousness." I call this Big C to distinguish it from our personal consciousness. This Consciousness is infinite, eternal — that is without beginning or end — "Consciousness is." Eternity is a concept we are still having trouble with in the West. I recently learned that it was the Buddhists who first conceptualised eternity and it was only in the Middle Ages that it came to the West — which is an astounding thought. So for us it is really a relatively recent concept and may be that is why we have such trouble grasping it. The West needs to come to grips with the concept of eternity and stop thinking that the universe begins and ends – either taking on Fred Hoyle's steady state universe ideas, or that this particular universe has arisen out of the ending of a previous one and will itself eventually end, and another arises in the eternal cosmic dance. I like to visualise the Western concept of the Universe, with a beginning and end, as a straight line. Eternity is a circle – everywhere is the beginning, everywhere is the middle, everywhere is the end; in other words there is no beginning, middle or end.

The TibetanBuddhist conception, as stated by the Third Karmapa, is: "Both faculties and objects arise from the mind. The manifestation of sensory objects and faculties is dependent upon an element that has been present throughout beginingless time."[2] In other words, everything, the whole manifest universe, arises from mind. There is here a problem of translation, in this quote the word 'mind' is being used in the same way as Big C from the Yogis. I shall use the word Consciousness with a Big C to denote the eternal ground of all being. This is the same teaching as Advaita Vedanta, and turns our Western story of how the world began on its head from: "In the beginning there was a Bang," the Bang became light which became matter which formed galaxies, stars, planets and ultimately life, and consciousness is just emerging out of matter (the brain) now – to: "In the infinite eternity Consciousness is" and out of Consciousness matter is formed; Consciousness is the ultimate ground of all being.

Prof. Harishankar Singh, a Vedic philosopher from Varanasi, when discussing these ideas with me, made an interesting remark that the purpose of Consciousness is to provide us with our ethics, our morality for life, our knowledge of good and evil, our highest purpose. Thus Consciousness is both the ground of all being and the highest spiritual aspiration or, as Wilber[3] describes it, the ground out of which the Great Nest of Being manifests. Morality and ethics of humans are very, very different from those of animals – and this is one mark of the difference in quality of consciousness. As Tibetan Buddhists put it – we have a precious human life which makes it possible for us to attain enlightenment.

## Mind, Awareness and Consciousness

Vedic philosophy states that the qualities, or functions, of consciousness are knowledge, will and activity, of which thinking is one activity. In the Vedic and Buddhist frameworks *there is a clear distinction between mind and consciousness*. Separating thinking out is something that I have found to be of central importance in grasping this other view of consciousness. Thought is related to mind and mind is connected to the senses. Consciousness is something much bigger! *Consciousness itself is not a quality, it is reality in all its different forms.* Swami Satyananda[4] says: "The mind cannot be the source of consciousness because it too can be perceived as an object. The mind does not illuminate itself." When you practice some forms of meditation such as Buddhist mindfulness, or Mahamudra, or the Yogic Antar Mouna, you watch the mind, watch the thoughts as they appear and disappear, you 'rest in the awareness.' Yogis and Buddhists conceive of mind as an organ which processes the senses, and is the means by which thought is created. "Consciousness when measured, limited, in space and time, then form and qualities appear — then it becomes chitta (mind)." Swami Satyananda defines yoga as a method, "by which consciousness is disconnected from the entanglement with mind and the manifested world."[5]. This is true too of meditation.

In the Baghavad Gita this conception is pictured as a chariot driven by five horses. The horses are the senses and the mind is the rein leading from the horses to a driver, who is awareness or the intellect (buddhi). The overall direction to the chariot is however given by the passenger (personal atman, the soul) who instructs the driver. We can see here that in this conception mind is very limited; and I find it very helpful to use the word mind with this definition as it is helping me to get greater clarity. The problem in the West is that these words are used in so many different ways and we never quite know in what way the person is using them! This problem is compounded when translations are made of Sanskrit or Buddhist texts. The Vedic conception is very similar to the Buddhist, which talks of 8 consciousnesses: the five sense consciousnesses, the 6th is the mind sense consciousness, the 7th the immediate (or afflicted) consciousness, and the 8th the ground consciousness.[6] The sense of self or ego, continuity of mind, is considered to be linked with the 7th consciousness, which is called buddhi (awareness or intellect) by the Yogis. This also is where actual perception is located because it is the link with our memory and conceptualisation, both of which are needed for us to be able to perceive something for what it is, rather than just a meaningless shape, taste, etc. And what in yogic terms is called the soul seems to be conceptually linked with the 8th consciousness, though Buddhists do not use the term soul. Beyond this, out of which everything arises is Big C, naked awareness, Ultimate Mind – there are so many terms used for this. The Buddhists also have the philosophy that these 8 consciousnesses transform into wisdom as we reach enlightenment. This I like! The ultimate magic!

## Do you have soul?

Vedic philosophy states that there are five primary levels which manifest out of the eternal Consciousness at different evolved states, which they call Soul, or universal Atman. I am just beginning to grapple with this concept of soul and what it means. I always saw soul as the essence of the person, connected with but different from the personality; something which is recognizable, unique in the new born babe and which is still there when the person dies, albeit changed by their life experience. In the Vedic view:

The soul of matter is the unconscious state of pure Consciousness;
the soul of plants is the subconscious state of pure Consciousness;
the soul of animals is the conscious state of pure Consciousness;
the soul of humans is the self-conscious state of pure Consciousness;

this self-consciousness means we can choose good and bad, leading to ethics, morality and the possibility of enlightenment.

In the Buddhist view there are other levels beyond the human, the various levels of different devas and gods, some of which relate to the Western concept of the fair folk.

Thus there is Consciousness, with a capital C, which is the whole universe, and this manifests in us as our soul. From the Holographic Universe perspective, as I'm a part of the whole (from my perspective) and the whole is in every part, this is what is being described here in terms of soul and consciousness. This big C in the Buddhist terminology is known as sunyatta (emptiness). According to Tai Situ-pa Rinpoche,[7] sunyatta is not nothing – it is the whole universe including the formless whole behind this manifest reality since none of it has ultimate entity, everything is impermanent coming and going like a river where the water is always different but the river stays – which I think is related to Bohm's idea of the implicate order, out of which this see-touch reality unfolds. Tai Situpa says that emptiness is where everything has limitless possibilities and potential; there is nothing which is not the manifestation of everything, that is more than manifestation of everything, that is less than manifestation of everything; which once again takes us back to the Holographic Universe ideas.

What I really like about these Yogic and Buddhist concepts is that thinking-mind, awareness and consciousness are clearly separate faculties. Mind is the tool by which we become aware of the senses and is the creator of thoughts. And every meditator knows the difference between the thinking process and awareness, the witness which watches the mind. Consciousness is still not a totally clear concept because it is so multifaceted, but separating thinking-mind and awareness out as two distinct processes, and having a top-down approach to the Universe with consciousness present at all levels makes good sense to me. I am, however, unsure of the definition of the use of words such as sub- and un-conscious in the Vedic usage, particularly with regard to their concept of Atman, i.e., exactly what is meant by matter being the unconscious aspect of Consciousness? However, I am aware of the tangible presence of the stones at Stonehenge and Callanish and Avebury – these stones definitely have a consciousness of sorts – perhaps this is what is meant. And see the power of stone as used in religions, eg. The Kabbah in Mecca, and king-making, e.g. the Stone of Destiny in Westminster. Further discussion with Vedic philosophers is required here!

## Parapsychology and the Vedas

Theoretically this philosophy gives a solid underpinning for an understanding of psi phenomena. Psi is the direct transfer of information without the medium of the senses, more connected with awareness (buddhi) rather than thought (chitta, mind). In fact psi research suggests that thoughts get in the way of psi awareness. With the Vedic philosophy, that consciousness underpins all reality, I am beginning to understand that the active psychic processes, such as psychokinesis or psychic healing, are the motor organs of the Self-consciousness (soul). The receptive psychic processes such as telepathy, clairvoyance and precognition are the sensory organs of soul. Thus psi can be understood as the active and receptive aspects of the soul level of our being. Interestingly, the dictionary definition of the word 'psyche' has 'soul' as one of its meanings. At the psychic level we experience consciousness, at the very least awareness, rather than thinking mind, potential omniscience and omnipotence considered as attributes of the divine, and called the siddhis in yogic philosophy. It then manifests via the normal mental modes, some people becoming aware of psychic impressions through imagery, some through feelings, some intuition, etc.

As I discuss in the Holographic Universe chapter, in parapsychology recent theorising[8] has related the functioning of psi to quantum reality. The Vedic and Buddhist philosophy of consciousness is totally in line with this conception of consciousness being integrated with matter, as seen in such quantum paradoxes as Schrödinger's Cat and non-locality (quantum entanglement). Quantum entanglement (also known Bell's theorem, or the EPR paradox) says that information exists and passes between connected quantum particles instantaneously, i.e. outside of time and space, as does psi. Schrödinger's Cat paradox gives rise to the observer effect, which says that consciousness is central for material reality to take the particular form it does, as we get in psychokinesis, e.g., psychic healing occurring in accordance with the wishes of the healer. Both of these quantum principles, which have been experimentally verified, are in accord with parapsychological data. As Swami Satyananda[9] puts it: matter is the "gross form and manifestation of mind. . . the material world that we see around us is really an expression of the more subtle mental aspects of existence." I could copy this quote several times over with sayings from various quantum physicists, as Wilber does in his book "Quantum Questions"[10]. (Yet again there is a confusion of terms. I think that the words "mind" and "mental" in the quote should really be consciousness.) At the quantum level, matter is localized energy; matter takes both wave (energy) and particle (matter) forms: $E = mc^2$

## The Shaivite Tantric Concept.

Tantric philosophy recognises 4 levels of consciousness (seechart below 1), which are both the manifestation and evolution of the Universe and the individual. Each of these levels are subdivided into 4 making a total of 16. Tantra states that our purpose is to become aware at each of these levels, so that we realise the ultimate state (supra-consciousness), which is one with eternal Consciousness.

Swami Satyananda[11] has written a commentary on Patanjali's Yoga sutras called "Four Chapters on Freedom," in which he describes the four primary states of consciousness as follows:

1. Conscious mind: sthula (gross dimension [of the Universe]); jagrat (waking state [of the mind]) - surface thought and perception of the outside world.
2. Subconscious mind: sukshma (subtle dimension [of the Universe]); swapna (dream state [of the mind]) - individual memory and samskaras (mental tendencies).
3. Unconscious mind: karana (causal dimension [of the Universe]); sushupna [also known as nidra][1] (deep sleep state [of the mind]); collective samskaras and memory.
These realms contain the instinctive, intellectual, psychic and intuitive aspects [of the mind].[12]
And 4: <u>Turiya</u> which is where consciousness goes beyond mind. Turiya means "simultaneous awareness of all three states" which takes us closer to the state of enlightenment.[13] These can be pictured as follows:[14]

|  | JAGRAT | SWAPNA | NIDRA | TURIYA |
|---|---|---|---|---|
| Dimension of the Universe | Gross | subtle | causal | transcendent |
| State of human consciousness | Waking; conscious | dream; subconscious | deep sleep; unconscious | Cosmic consciousness; collective unconscious |
| | <--------manifestation----------------------- ---------evolution--------------------------> | | | |

*The 4 Major Tantric States of Consciousness (Illustration courtesy of Swami Yogakanti Saraswati)*

---

1  Comments within [] are my insertions

Lakshman Jee[15] describes these states as follows:

**Jagrat** is "when the individual . . . . loses consciousness of one's subjectivity and becomes one with the objective world." This is our normal state of consciousness. Most of us are totally unaware most of the time. We are totally caught up with living: working, reading a book, chatting with friends, doing the washing up, etc. Most of us don't watch ourselves, don't watch what we are saying, feeling, being, doing, thinking. In terms of the Universe manifestation this is gross matter. And matter has evolved in this universe from Big Bang through to all the different elements, planets, rocks, etc. In fact, through our agency, it is still evolving in the form of plastics, computers, technology, pharmaceuticals, etc.

**Swapna** is our dreaming state of consciousness, which includes day dreaming, "lost in thought" as we say. Once again we lose awareness and get totally caught up in our thoughts, dreams, daydreams, etc. The aim of meditation is to enable us to remain aware even when dreaming. In Tibetan Buddhism one of the teachings is dream Yoga, which is one of the six Yogas of Naropa, and is where you learn lucid dreaming. The ultimate aim is that when you die you can go through the bardos (the intermediate state between this life and the next) with full awareness. This is considered to be a dream-like state of consciousness. In terms of Universal manifestation this is the life-energy level.

**Sushupna, or nidra,** is the causal dimension, the all-knowing. In our normal state, the common understanding of nidra is of deep sleep, an unconscious state. With increased awareness one can actually become aware in this state of consciousness. Thus it is said that the night of the layman is the day of the yogi. This state of consciousness is the absence of senses and thinking mind,[16] termed the 7th Consciousness in Buddhism, equivalent to the soul level of the Vedic system. In the Universe manifestation it is the Akashic, mental level.

**Turiya** is the state of total equilibrium between individual manifest consciousness and cosmic Consciousness. It is not an interactive state; though full of wisdom there is absence of dualistic knowledge. There is total disassociation from the seeds of gross, subtle or causal dimensions.[17] In Buddhism this state is called 'naked awareness.'

My understanding of these states is that enlightenment is becoming aware in states of consciousness of which we are normally unaware, including the dream state and the 'beyond thought' state.

*Exploring Consciousness: Six Years with Yogis and Tibetan Buddhists*

Awareness is the key concept.

## Tantra and Parapsychology

According to Patanjali's sutras and Buddhist teachings, as we develop our awareness at these different levels of consciousness so we become aware at a psychic level. In the 1970s a theoretical framework for parapsychology, known as the psi-conducive model, was developed from Patanjali's yoga sutras.[18] This led to an ongoing programme of states of consciousness research which has borne rich fruit. I have been using this model as a theoretical basis for research in India, working initially with swamis in an ashram and later with Tibetan Buddhist monks in monasteries, who have done up to 40 years meditation. The findings are still very preliminary but are suggestive that meditation does enhance psi awareness.

As can be seen in the graph below, those monks with the most years of meditation practice show far higher psi scores. For the first 10 – 15 years the scattering of psi scores is like you get with people who have not had any mental training, then after about 10-15 years it all starts to go more positive and get stronger. The magnitude of the correlation between psi score and years of meditation, found in Tibetan study 1 (0.524), was less (0.28) in study 2, and is similar to that between Yoga and psi (0.57). Combining the monks' scores from both the Tibetan studies gives a correlation of (0.73), which is

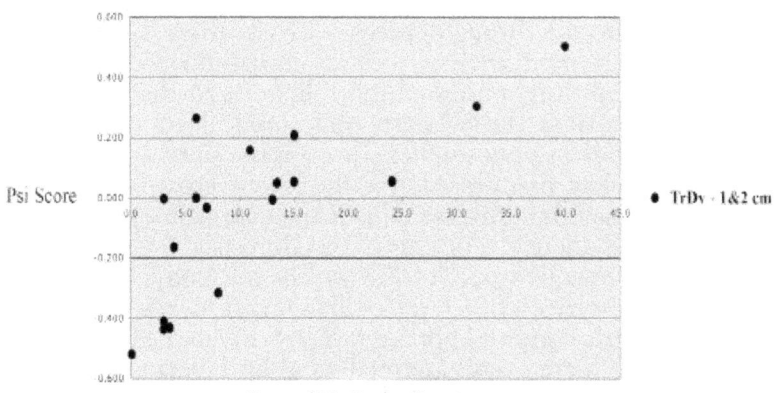

*Correlation of Psi Score with Years of Meditation Practice (Illustration courtesy of Serena Roney-Dougal and Jerry Solfvin)*

shown in the graph here.[19] This research is now being continued at a Tibetan Buddhist monastery in Britain, and it will be interesting to see whether Westerners show a similar pattern. In meditation it seems that we are making the shift from unconscious competence (the natural psychic) through conscious incompetence, which is where all the psi-missing occurs showing our blocks and defences through to conscious competence as we become more and more aware. Tibetans recognize two sorts of clairvoyance: the spontaneous, natural psychic which they consider to be unreliable, and the highly trained lama, who they consider to be 100% reliable. We seem to be beginning to see this here. They also state that years of practice are needed, and this is showing up too.

For me, the central message from the Yogic Tantric teachings is that increasing awareness of those aspects of our consciousness of which we are normally unconscious, dream and pure awareness states, are in fact those states of consciousness which are related both to psi functioning and the samadhi states of meditation. This tallies with the Buddhist teachings about meditation, and with the Jungian concept of the collective unconscious as that aspect of consciousness outside of space and time which is the domain of the psychic.

## 3. The Union of Energy and Consciousness

In the West we normally think of energy, such as a light or electricity, as a non-conscious force that interacts with matter. In the Vedic philosophy, energy has its own form of consciousness, e.g., the consciousness of light is illumination. In the Tantric philosophy the manifestation of consciousness into the different dimensions of the Universe occurs through energy – or vibration (spanda). Once manifested this then evolves towards transcendence. This could possibly be a spiritual philosophical link with M-theory,[20] whereas Quantum theory is the Western philosophical equivalent of the Vedic and Buddhist philosophies. Turiya is the subtlest level and the end dimension of manifestation is the lowest gross level of jagrat, which is the material universe. This is the complete opposite of the Western beliefs about the origins of the Universe.

The subtle energy aspect of life within the body, which I have researched at the theoretical level for many years now, is the Yogic and Buddhist Tantric concept of prana, which flows in the body through the nadis, which are energy (termed wind by Tibetans) channels in the human body. There are 72,000 nadis of which the three primary nadis, sushumna, ida and pingala, run up the centre of, and on either side of, the spinal column. Where these energy channels intersect, the chakras are located. I have discussed the chakras, and particularly ajna chakra, the third eye, from a physiological viewpoint earlier because that has helped me to get a handle on them, but the basic concept of the chakras is as the energy aspect of the body linking body with consciousness – the body consciousness energy. I think that the Tantric concept of the way the Universe works as the mingling of Consciousness with Energy, manifesting from subtler levels to the gross level of matter, which then evolves back to the subtler levels of consciousness is as good an understanding as I've come across yet. The subtle energy aspect of the body seems to be the halfway house between consciousness and the material world. In trying to get some understanding of how mind connects with matter this seems to go some way to a reasonable understanding, particularly with quantum philosophies bringing the two together so clearly now.

Both Yogic and Buddhist tantric teachings use prana, the nadis and chakras for meditation practices and they are considered to be the most powerful way of shifting our awareness out of this see-touch reality into Big C.

## Compassion – the bodhisattva principle

What the Buddhist teachings have, that for me is their most important contribution, is the Bodhisattva principle: that we are aiming for enlightenment for the good of all beings; the altruistic approach. This is of course part and parcel of the Holographic universe philosophy. If the whole is in every part, then every thought word and action of every part affects the whole. So everything I think or do affects everything else which automatically affects me. This gives an understanding of karma which goes beyond the rather mechanistic view that most people have. We are all interconnected, the Universe is an indivisible whole, and compassion, love, is the juice that fuels the evolving principle. The more compassion and love I generate the better for all including myself. Thus, ultimately the psychic level of being is not the overt clairvoyance of a medium or fortune teller, but that sensitivity which is involved in making the choice which takes you in the best direction for the benefit of all people, that oils the wheels of life, that is part of wisdom, the wise decision, the best thing to do for all concerned. It is also that sensitivity, that awareness, that is linked with empathy, where you feel from the other person's perspective, say the right thing, do the best for them, which is part of compassion. So we are looking at the subtle level of the development of wisdom and compassion. Can't be bad really!!

This is as far as I have reached  - who knows where I go to next!

# Notes

## Foreword to Third Edition

1. Articles on my web site are at: www.psi-researchcentre.co.uk/articles. S.M. Roney-Dougal, *The Faery Faith*, Green Magic, 2003.

## Introduction

1. Carlos Castaneda (1968-1974).
2. Graves (1961/1977).
3. Michell (1969/1975).

## Chapter 1

1. There have been a number of surveys of reported experiences starting with the SPR Census of Hallucinations in 1894 (Sidgwick, 1894). In 1946-9, D. J. West (West, 1946-9) did a survey in Britain and in America L. E. Rhine collected over 17,000 cases between the 1930s and 1970s (Rhine, 1977). There have also been polls and surveys of belief in psi: Dale, White and Murphy did a survey in a weekend supplement called *This Week* (Dale, White and Murphy, 1962); Palmer and Dennis a survey in Charlottesville (Palmer and Dennis, 1975); *New Scientist* polled people in Britain (Evans, 1973) and Brian Inglis in *The Times* (Inglis, 1980); there was a Gallup pole in America in 1978 and a survey in Hong Kong in 1980 (Emmons, 1982)
2. For an excellent history see Inglis (1978, 1984 and 1986).
3. Standard parapsychology textbooks which will give you all the details are Wolman (1977/1986); Edge *et al.* (1986); Krippner (1977—97; 8 vols.); Broughton (1992); Irwin (2004); Radin (1997, 2002). From these you should find references to almost any topic you wish to explore. Books that are easier to read are those by Scott Rogo (1982-6); of spontaneous cases, Heywood (1964,1971); L. E. Rhine (1981); of Zener card experiments, Pratt (1940); and of psychokinesis, K. Pedler (1981) wrote a book to go with his TV programme *Mind over Matter*, and Danah Zohar (1982).
4. Subliminal perception happens all the time. Because it is subliminal we don't notice but we are affected by everything which we perceive. These percepts tend to emerge in dreams primarily, but also in our moods, slips of the tongue, things we say without knowing why, and so on. They use it in advertising to affect you without you being aware that you are being affected. Basically everything in your physical presence is heard, seen, smelled, tasted and touched *all* the time. But 99 per cent is screened out 99 per cent of the time. Only the parts that are important and relevant to you ever get through to normal consciousness. The mind is a wonderful tool and normally protects us from overload — when it doesn't we break down. LSD and other hallucinogens take off some of the filters and rearrange things a little so that we perceive — which is different from seeing — things in a totally new light and the world is never the same again. That's why they are so effective at breaking down conditioning. For a recent review of research into subliminal perception see Shevrin et al (1996) & Snodgrass, Bernat & Shevrin (2004).

5. Schmeidler (1976).
6. Parker (1975).
7. Targ and Puthoff (1977).
8. Hasted (1981); Hasted and Robinson (1981).
9. Roney-Dougal (1986b).
10. Capra (1975).
11. Koestler (1972); Le Shan (1974); Watson (1974, 1979); Wilson (1973).
12. Gardner (1978).
13. Watson (1979).
14. Dixon (1971, 1981).
15. Johnson and Heraldsson (1984).
16. Eddington; Schrodinger; Wigner (in Wilber, 1984).
17. Llinás & Paré (1991); Llinás & Ribary (1993). For a full article on this, entitled "Walking between the Worlds," go to my web site: www.psi-researchcentre.co.uk/articles.
18. Jung (ed.) (1964).
19. Gregory (1920/1979); Lang (e.g. 1950/1970); Campbell (1890/1940); Jones and Jones (1949/1982).
20. I could quote long passages from Norman Dixon's books on subliminal perception (Dixon, 1971, 1981), for example: 'The normal location of unconscious processes appears to be in the subcortical brain and in the autonomic nervous system' (Dixon, 1971); or from various parapsychological researchers (Irwin, 1979) who are obviously very interested in possible physiological understanding of the psychological processing of psychic information, but an absolutely typical quote comes from biofeedback research: 'The neurological location of conscious processes seems to be in the cerebral cortex and craniospinal apparatus. The normal location of unconscious processes appears to be in the subcortical brain and in the autonomic nervous system. Information from limbic processes can reach consciousness and vice versa. The limbic system is an important link between emotions and the body. Emotional states are correlated with electrophysiological activity in the limbic system. It is connected to the hypothalamus — the central control panel of the autonomic system including the pituitary' (Green and Green, 1977).
21. Dean (1966); Davis and Brand (1980).
22. Braud and Schlitz (1983, 1989);Schlitz & Braud (1997).
23. Schmidt, Schneider, Utts & Walach (2004); Radin (2005). Schlitz & Radin are also doing staring experiments on the internet (www.marilynschlitz.com).
24. Mavromatis (1987).
25. Llinás & Paré (1998); Don *et al* (1996); Don & Moura (1997).
26. Green and Green (1977). Biofeedback has shown researchers that there is a whole area of our being that shares an essential similarity. As Green and Green from the Meninger Clinic in America who have studied yogis and psychics in their biofeedback research state: 'It is worth repeating that whether we are talking about prayer, meditation, hypnosis, autogenic training or biofeedback training, the issue is the training of the subconscious ... speaking metaphorically, and perhaps literally, biofeedback seems to provide a means for bridging the functional schism between the left and right hemispheres of the brain.' And between the conscious and the subconscious, the scientific and the magical/mystical.
27. Mavromatis (1987).
28. White (1964). Rhea White was an American parapsychologist who edited one of the scientific journals, and was advocating much the same ethical awareness. Most parapsychologists who use these altered states techniques are very aware, loving people. She created an enormous parapsychology bibliography and also a journal called Exceptional Human Experiences.
29. Gardner (1978).
30. Hurley (1985).
31. Watson (1974).
32. Rao (1978).
33. Watson (1974).
34. Stanford (1977b).
35. Luke, Delanoy & Sherwood (2008).
36. Hurley (1985, p.135).
37. Ibid., p. 152.
38. Ibid., p. 164. For an in-depth discussion of these ideas see Finley Hurley (1985). Richard Gardner discusses the power of the subliminal mind in affecting others in *Evolution through the Tarot* (1978): 'Control at the subconscious level is literally "hitting below the belt" as very few can escape from control of their subconscious without very real help. It is from this level that we influence all life forms.'

# CHAPTER 2

1. Roney-Dougal (1979a).

2. Roney-Dougal (1979b).
3. Ibid.
4. Roney-Dougal (1979c).
5. Stanford (1974).
6. Roney-Dougal (1980).
7. Mishlove (1983).
8. Ibid.
9. Drury (1982).
10. Mishlove (1983).
11. Targ & Harary (1984).
12. Palmer (1980); Gauquelin (1974); Beloff (1979).
13. Krippner (1980).
14. Winkelman (1982b).
15. Winkelman (1983, 1992, 2000)
16. Winkelman (1983.
17. Ibid.
18. Drury (1982).
19. Ibid.
20. Roney-Dougal (2006).
21. Stanford (1977b, p. 837); White (1964); Braud (1975).
22. Honorton (1977).
23. Radin (1997, 2006).
24.. Honorton (1985); Hyman (1985).
25. Palmer (2003); Milton & Wiseman (1999); Bem, Palmer, & Broughton (2001).
26. McMoneagle (1997, 2002).
27. Palmer (1978).
28. Roney-Dougal (2006); Roney-Dougal & Solfvin (2006, 2008); Roney-Dougal, Solfvin & Fox (2008)
29. Chilton Pearce (1973).
30. Gardner (1978).
31. Schmeidler and Murphy (1946).
32. Jung and Pauli (1955).
33. Stanford (1977c, 1978).
34. Malinowski (1925/1954); Mauss, (1906/1972).
35. Mauss, (1906/1972); Winkelman (1983).
36. Batcheldor (1968, 1983).
37. Hansen (2000) has written a book on trickery appropriate to this argument.
38. Reichbart (1978).
39. Elkin (1977).
40. Malinowski (1954).
41. White (1977).
42. Winkelman (1983).
43. Morris (1980).
44. Irwin and Cook (1983).
45. Child and Levi (1979).
46. Morris *et al.* (1979).
47. Morris and Harnaday (1981).
48. David-Neel (1965/1971).
49. Roney-Dougal & Solfvin, (2008)
50. Schlitz and Gruber (1981).
51. Schlitz (1982).
52. Mishlove (1983).
53. Batcheldor (1984).
54. Ibid.
55. Ibid.
56. Schlitz (1982).
57. Honorton and Barksdale (1972).
58. Stanford (1977a, p. 334).
59. Ibid., p. 334.
60. Ibid., p. 341.
61. Evans Pritchard (1937/1958).
62. Stanford (1977c).
63. Ibid.
64. Kennedy (1979).

*Notes*

65. Braud (1980).
66. Gruber (1980).
67. Braud (1980).
68. Winkelman (1983).
69. Ibid.
70. Loewe and Blacker (1981); Roney-Dougal (2006).
71. Loewe and Blacker (1981).
72. Winkelman (1983).
73. Bergson (1911/1921).
74. Haraldsson and Johnson (1979).
75. Palmer (1978).
76. Winkelman (1983).
77. Mishlove (1983).
78. Ibid.

## CHAPTER 3

1. Hubel *et al.* (1979).
2. Myers (1903/1915).
3. Kahn (1976).
4. Eisenbud (1970).
5. Zohar (1982). She has written several other books, all of which I recommend.(Zohar, 1990; Zohar & Marshall, 1994).
6. Zohar (1982, p. 139).
7. Costa de Beauregard (1975).
8. Pelletier, in Wilber (1982).
9. Chilton Pearce (1973).
10. Vaughan (1979).
11. Jung and Pauli (1955).
12. Adams (1979).
13. For what follows here I am deeply indebted to the makers of the film "What the bleep!?- Down the rabbit hole." (2006). *Revolver Entertainment*, London. www.the bleep.co.uk
14. Bohm (1982).
15. Zohar (1982, p. 126).
16. Margenau (1967).
17. Radin (2006).
18. LeGuin (1979/1985).
19. Freedman and Clauser (1972); Aspect and Grangier (1986).
20. This last quote was taken from a book on precognition written by Danah Zohar (1982), which has a really clear explanation of quantum phenomena and its associated world-view. I recommend it highly, as also I recommend Arthur Koestler's *The Roots of Coincidence* (1972).
21. Ghosh, Rosenbaum, Aeppli & Coppersmith (2003); Salart *et al* (2008); Rudolph (2008). For a new book on physics and consciousness see Rosenblum & Kuttner (2007).
22. Carr (2008).
23. Zohar (1982).
24. Bohm (1982).
25. Capra (1975).
26. Zukav (1979).
27. Bohm (1982).
28. Ibid.
29. Ibid.
30. Ibid.
31. Pribram (1969, 1973).
32. Penfield and Milner (1958).
33. Pribram, in Wilber (1982).
34. Ibid.
35. Wilber (1984).
36. Krippner, in Wilber (1982).
37. Wilber (1982).
38. Bohm (1982).
39. Ibid.
40. Ibid.
41. Lovelock (1979, 1988).

42. Piaget (1977).
43. Zukav (1979).
44. Bohm (1982).
Strangely enough a book published by Michael Talbot (1991) at the same time as the original publication of this book used the title 'The Holographic Universe,' identical to the title of this chapter – obviously an idea whose time had come – and now as I revise for this third edition a DVD has appeared called 'Down the rabbit hole,' echoing my first subtitle of 'Through the Looking Glass.' (see note 13 for ref.)
45. Davies (1983).
46. Ibid.

## Chapter 4

1. e.g. Honorton (1977); Braud and Braud (1974); Krippner, Honorton and Ullman (1972). Technical versions of this chapter were published in the *Journal of the Society for Psychical Research* (Roney-Dougal, 1989a) and *Caduceus* (Roney-Dougal 1989b).
2. Satyananda (1972a, pp. 8, 12).
3. Rivier and Lindgren (1971).
4. Deulofeu (1967).
5. Rivier and Lindgren (1971).
6. Hochstein and Paradies (1957).
7. There is an excellent review of this by Airaksinen and Kari (1981a and 1981b) in addition to the Rivier and Lindgren review (1971).
8. Naranjo (1973, 1978).
9. Naranjo (1967).
10. Harner (1973/1978, p. 158).
11. Ibid., p. 160.
12. Kensinger, in Harner (1973/1978, p. 12).
13. Ibid.
14. Wurtman (1979).
15. Wiener (1968); Quay (1974).
16. Electricity and the Nervous System: A nerve impulse is an electrical impulse travelling at 100 feet per second. It is formed by positive sodium ions and negative ions moving across the nerve membrane. This creates an electric potential across the membrane which gives the energy for the impulse to travel down the membrane which discharges the potential. There is then a lag while work energy is expended by the cell to restore the ion potential. Providing the stimulus has a certain threshold strength the nerve impulse will be triggered. At the end of the axon secretes chemicals called hormones or neurotransmitters like adrenalin or acetylcholine. These pass the message on to the next stage — e.g. muscle — or another nerve.
Then they are destroyed by enzymes. If they were not destroyed the neurotransmitter would continue to trigger nerve impulses and the nerve system would run wild. There are three types of nerve: sensory with dendrites connected to sense receptors, their axons connecting to other nerve cells; motor nerves, whose dendrites connect to axons of other nerves, but whose axons connect to muscles or glands; and association nerves which connect from one nerve to another.
In the brain these nerves connect together to perform various functions as follows: the medulla controls breathing, heart function, blood pressure and digestive system. The cerebellum coordinates muscle movement. The thalamus and hypothalamus control the passage of sensory information, and regulate body temperature, appetite, sleep and similar functions as well as being the seat of emotions. And the cerebrum is the seat of conscious sensation, voluntary movements, memory and intelligence, the right and the left halves of the cerebrum (or cerebral hemispheres) are concerned with slightly different though overlapping functions. Thus the left is concerned with language, writing, logical analytical thought, whilst the right processes music, art, poetry and global holistic type thought such as dreams. Epilepsy occurs when the electrical pulses in the nerve tissues of the cerebrum, quite often what is called the temporal lobe which is closely linked to the thalamus, discharge themselves excessively and in concerted bursts. Some of the symptoms of seizure are similar to those from an electric shock: muscles convulse, breathing can stop and unconsciousness results. In the Grand Mal, the sufferer can often predict the onset of a fit because moods can change abnormally or the person experiences strange emotions or an altered state of consciousness. In some tribal societies people who suffered from epilepsy have been considered to be particularly special people to the society, because in these states they are often highly psychic. During and after a seizure the brain function is lowered and it is not uncommon for sufferers to have no recollection of the fit at all. (Shallis, 1988.) This was up to date when I wrote it in 1991 – I am sure that it is basically still true, but so much research has occurred in this area in the past twenty years that much more could be said here!
17. Wiener (1968).

## Notes

18. Wurtman and Moskowitz (1977a and b).
19. Dixon (1979).
20. Most (1986).
21. Ibid.
22. Ott (1976).
23. Arendt (1985).
24. Ullman, Krippner and Vaughan (1973).
25. Buckholtz (1980); Rollag (1982); Naranjo (1967).
26. Pähkla, Zilmer, Kullisaar and Rägo (1998).
27. Callaway (2006).
28. Strassman (2001).
29. Jacob & Presti (2005); Shulgin & Shulgin (1997).
30. Quay (1974).
31. Mclsaac (1961).
32. Mclsaac, Khairallah and Page (1961).
33. Barker et al. (1981). These findings were confirmed independently by Honecker and Rommelspacher (1978).
34. Langer et al. (1984).
35. Prozialeck et al. (1978).
36. Barker et al. (1981).
37. Rimon et al. (1984).
38. Hider, Smart and Suleiman (1981).
39. Muller et al. (1981).
40. Roll and Montagno (1985); Neppe (1983).
41. Airaksinen and Kari (1981a and b).
42. Airaksinen and Kari (1981b).
43. Piñol-Ripoll et al. (2006).
44. Garciá et al. (1999).
45. Frederiksen et al. (1998).
46. Pähkla et al. (1996).
47. Injections of pinoline increase plasma aldosterone concentration and this is preceded by an increase of plasma renin activity. An increase iss also found in $\beta$-endorphin/$\beta$-lipotrophin concentration. Airaksinen et al. (1984).
48. Pähkla et al. (1998).
49. Naranjo (1967).
50. Middleton, Stone and Arendt (2002).
51. Rhine (1981).
52. Reppert et al. (1988).
53. Reiter (1982).
54. Golombek, Escolar and Cardinali, (1985).
55. Callaway (1988).
56. Whitehouse (1985).
57. e.g. Spinelli (1983).
58. Callaway (1988).
59. Pavel et al. (1981).
60. Arendt (1985).
61. Riemersma-van der Lek et al. (2008).
62. Shuttle and Redgrove (1978/1986); Lacey (1974/1976).
63. Reiter (1982).
64. Carman, Post, Buswell, and Goodwin (1976); Boyce (2007).
65. Smith (1978).
66. Rimon et al. (1984).
67. Airaksinen and Kari (1981b, p. 204).
68. Friedman, Becker and Bachman (1963).
69. Reiter (1982).
70. www.SpiritualCrisisNetwork.org.uk;
71. www.psi-researchcentre.co.uk/articles/Walking Between the Worlds.
72. Tart (1975).
73. Paranjpe (1984).
74. Davidson (1987 and 1988).
75. Satyananda (1972b).
76. Ostrander and Schroeder (1970).
77. Quay (1974).
78. Johnson (1982).

79. Reiter (1982).
80. Satyananda (1972b, p. 73).
81. Maestroni *et al.* (1989); Majewski *et al.* (2005).
82. Martikuinen *et al.* (1985).
83. Allain *et al.* (1986).
84. Reiter (1981).
85. Barkers *al.* (1981).
86. Airaksinen and Kari (1981b).
87. Leino *et al.* (1984).
88. Satyananda (1972b, p. 56).
89. Shuttle and Redgrove (1978/1986).
90. Reiter (1984).
91. Wurtman (1979).
92. Arendt (1978).
93. Matthews (1981).
94. Reppert *et al.* (1988).
95. Airaksinen and Kari (1981b).
96. Schmitt and Stanford (1978); Keane and Wells (1979).
97. Roll (1977); Gauld and Cornell (1979); Spinelli (1983).
98. Dodt and Heard (1962).
99. Gaer-Luce (1973).

## CHAPTER 5

1. Shallis (1988, p. 255).
2. Playfair and Hill (1978); Shallis (1988); Graves (1978 revised 1986); Lonegren (1986); McGillion (1980); Robins (1985 and 1988); Devereux (1982); Becker (1985).
3. Shallis (1988, p. 102).
4. Hasted (1981); Hasted and Robertson (1981).
5. A Fortean type of event is one that is disregarded by all scientific disciplines as being so zany and improbable that it just could not have happened - like falls of frogs or fishes from a clear sky. It is not so long ago that meteorites were 'Fortean' in that the scientific establishment denied firmly that anything solid could fall from the skies and the peasants who reported such things were gullible, credulous, and superstitious. As most people know, meteorites are now accepted by the scientific establishment, but no one apologised to the peasants. Instead when peasants tell scientists about seeing lights in the sky they are called gullible, etc. or are told that they are hallucinating, not in the positive sense in which I have just used the word as being our unique human faculty for seeing psychic phenomena or connecting with the spirit world, but in the modern derogatory sense of being nuts.
6. Shallis (1988).
7. Miller and Creim (1997).
8. Bender (1969;1974). The Rosenheim poltergeist.
9. In order to illustrate this let me give you some quotes from Michael Shallis's book: 'For the population at large, the electrical revolution has been totally transforming. Electricity has radically altered the way we work, the way we live, the way we communicate and even the way we think ... oscillating electric currents that run through our wired society reflect our own addiction for the new, the different. The sustaining power of constancy has given way to the perpetually changing ... our minds have become caught up in the electronic circuits around us ... The web of electrical circuitry is an extension of our own nervous systems, whose effects remain subliminal, whose message is pure information. Electricity has become an unseen insidious pollutant as damaging as any noxious chemical or radioactive discharge. We remain electrical beings in an unnatural electrical world ... our loss of subtlety is an indication of the polluting effect of the electrical environment, just as our apparent lack of wisdom is a sign we have mistaken information for knowledge. Our polluted electrical surroundings disturb the subtle forces that operate on and in us. We have removed the peace of mind that comes from a harmony with the rhythms of nature and plugged ourselves in to a weird and wired construction, whose effects we ignore at our peril' (Shallis, 1988, pp. 157—9).
10. Lacy-Hubert, Metcalfe and Hesketh (1998); Ahlbom and Feychting (2003); Feychting, Ahlbom and Kheifets (2005). Some of these studies have shown associations between exposure to power-frequency (50–60 Hz) magnetic fields and increased health risk (Valberg, Kavet & Rafferty (1997); Wertheimer, Savitz and Leeper (1995)); but other studies have not shown such a link (Preece, Hand, Clarke and Stewart (2000)).
11. Litovitz, Montrose and Wang (1992); Repacholi and Greenebaum (1999); Foster (2003); Juutilainen (2003); Sastre, Graham, and Cook (2000); Szmigielski *et al.* (1998); Graham, Sastre *et al.* (2000); Jauchem *et al.* (1999); Miller and Creim (1997).

## Notes

12. Luo *et al.* (2004); Bellossi *et al.* (1998); Kumosani and Qari (2003).
13. Di Carlo, White and Litovitz (2000); Hook *et al.* (2004); Dachà *et al.* (1993); Yokus *et al.* (2005);
14. Shallis, (1988, pp. 77-8).
15. Ibid., p. 243
16. Davenas *et al.* (1988).
17. Schwartz *et al.* (1986).
18. Dean (1983)
19. Playfair and Hill (1978). John Henry Nelson (1940s) discovered that 'certain behavioural characteristics of sunspots could be associated with *planetary positions*. It seemed that when certain planets were in certain positions relative to the Sun their combined influence was sufficient to cause effects' — and so affect Earth. 'There is a drop in the Earth's magnetism when the Moon is full... When the Earth shares certain positions relative to other planets - such as Venus, Mars and Jupiter — magnetic changes are noted on our planet' (Playfair and Hill, 1978, p. 90).
20. Gribbin (1978).
21. 'Burr's L-fields have the same shaping property as Sheldrake's morphogenetic fields. If the "form field" affects the material world, it would be surprising if it did not have a material correlation. We would expect to be able to find a measurable and detectable aspect of the form field, and I assert the L-field is just that ... If Sheldrake's form fields are non-material, then they coexist in the space around us but in an unseen dimension. Like the spirit world, however, the form fields operate in the material realm and must therefore (by definition) be detectable at a material level. What I am suggesting here is that the electrodynamic fields of life are those material aspects of the form field. They share the same properties, do the same job. The fields of life and the fields of form affect growth, maintain pattern and are in turn affected by the world in which they are manifest. The mechanism, if you like, is electromagnetic' (Shallis, 1988, p. 247). See also Burr (1972). 'The electromagnetic universe is the physical component of the unseen realms ... The form fields that govern ideas connect with our electric brains, but our electric brains are more than straightforward, mechanical computers. The L-fields and their link with the form fields show us that there is more to reality than the mechanical world. We cannot explain everything by electricity, except that electricity is a reflection of a more subtle reality' (Shallis, 1988, p. 247).
22. Sheldrake (1980; 1988). New editions of these are now available (2008, 2010).
23. Healers feel they are a channel for some other energy — they often feel a warmth and electric tingle in the right hands, e.g. Harry Oldfield. Aura massage stimulates and gives strength if done with the right hand and tranquillises and anaesthetises doing with the left hand. Healers often see aura colours and this can tell them what is wrong. Halos are an example of the head auras that shine when someone is in an 'illuminated' or 'enlightened' state. See how our language is telling us all the time: radiant people, their face lighting up, etc. Halos and auras are described in most of the world's cultures. For example, in the Hindu religion in India there are considered to be five auras; that of one's health, the vital or life force, the karmic aura of fate or destiny, the character and the spiritual aura. In the traditional Christian religion four auras are delineated: the halo, nimbus, aureola, and glory. There are strong suggestions that this aura effect is the electrical life field of the person.
24. Burr (1972); Shallis (1988, p. 164).
25. Shallis (1988, p. 165).
26. Ibid., pp. 244-5; cf. astrology.
27. Ibid., p. 203.
28. Robins (1985).
29. Roney-Dougal (1986a, 1988, 1989a and 1989c).
30. Persinger (1989); Roney-Dougal and Vögl (1993).
31. Roney-Dougal (1987).
32. Schmidt-Koenig and Keeton (eds) (1978).
33. Mather and Baker (1980).
34. Levine and Bluni (1994).
35. Wiltschko *et al.* (2007); Ritz, Adem, and Schulten (2000); Ritz (2001).
36. Baker (1981).
37. Murphy (1989).
38. Stutz (1971).
39. Rudolph *et al.* (1985).
40. Walker (1984).
41. Rajaram and Mitra (1981).
42. Subrahmanyam, Sanker Narayan and Srinivasan (1985).
43. Shallis (1988).
44. Playfair and Hill (1978).
45. Semm *et al.* (1981).
46. Adams (1986); Persinger (1986).
47. Persinger (1986); Persinger and Krippner (1989); Krippner and Persinger (1996); Spottiswoode

(1990). Another Californian (and it is remarkable how much of this research involves people or experiments in California, right where the San Andreas fault is which affects local geomagnetic fields very strongly) called Charles Tart (1988) did two psi studies which he analysed for possible correlation with geomagnetic effects. Again, lower values of GMF activity on days preceding more successful psi performance were found, though not at a statistically significant level. One of the studies showed more successful psi on days of quiet geomagnetic activity, but not the other. If you're interested in the Maimonedes dream research I suggest you read Ullman, Krippner and Vaughan's *Dream Telepathy* (1973).

48. Gearhart and Persinger (1986).
49. Persinger (1988, 1989).
50. Nichols and Roll (1998, 1999); Maher (2000); Maher and Hansen (1997); Braithwaite (2005, 2008).
51. Persinger and Schaut (1988).
52. Persinger (1989); Nelson and Dunne (1987).
53. Wilkinson (1989).
54. Radin, McAlpine and Cunningham (1994).
55. Haraldsson and Gissurarson (1987).
56. The effects of lunar gravity in terms of ocean tides are, of course, well known. But the Moon also pulls at the solid ground as well... This is bound to contribute to the pressures and strains at weak points on the Earth's surface such as geological fault lines' [The sun also seemed related to earthquake incidence (Playfair and Hill, 1978)]. Tidal effects are also exerted on all bodies containing water, not just the ocean - so however minutely our bodies respond, they do still respond! '... By intensively measuring screened and unscreened test tubes of distilled water Piccardi found that certain chemical reactions were significantly different between the two sets of water samples. He was able to discover that solar activity, lunar phase and cosmic radiation all affected processes in water. The medium through which these cosmic influences got through to the water molecules was by alterations in the Earth's own magnetic field. This discovery means that vegetable, animal and human systems — all heavily composed of water — must be susceptible to cosmic influences' (Devereux, 1982, pp. 88—9). Burr's L-fields respond to lunar phases. Puharich (1973) conducted a series of telepathy experiments for a full month, attempting to control all other variables! He found peak psi scores at full Moon and dark of the Moon. Also that a negative ion environment, such as one gets near waterfalls or near bonfires, is more conducive to psi.
57. Krippner, Radin and Rebman (1994).
58. Radin and Rebman (1998).
59. Etzold (2003).
60. Sturrock & Spottiswoode (2007).
61. Fraser-Smith (1982); Stenning, Carmody and Du (2002).
62. Randall and Randall (1991).
63. Fuller *et al.* (1995); Dobson *et al.* (2000).
64. Persinger, (1995); Persinger, Tiller and Koren (2000); Persinger, Koren and O'Connor (2001); Persinger & Koren (2001); Persinger and Healey (2002); Persinger (1995); Persinger, Richards and Koren (1997); Cook and Persinger (2001); Booth, Charette, and Persinger (2002); Persinger and Healey (2002).
65. Hubbard and May (1987).
66. Semm *et al.* (1980).
67. Functional magnetic resonance imagery (fMRI) has been used recently with meditators, who, with their ability to maintain concentration on one thing for lengthy periods of time, have shown those areas of the brain related to compassion, anger, etc. (Davidson, et al. 2003; Lazar *et al.* 2000; Lutz, Brefczynski-Lewis, *et al.* 2008; Lutz, Slagter, *et al* 2008)
68. Barr *et al.* (1983).
69. Welker *et al.* (1983).
70. Henshaw and Reiter (2005).
71. Burch *et al* (2000)
72. Weydahl *et al*
73. Cremer-Bartels *et al.* (1984).
74. Wever (1973).
75. Cremer-Bartels *et al.* (1983); Reuss and Olcese (1986).
76. Elsworthy (1986); Braud, Shafer and Andrews (1993).
77. Olcese *et al.* (1985).
78. Clark *et al.* (2007).
79. Reiter *et al.* (1988).
80. Graham *et al.* (2000). Gobba *et al.* (2006).
81. Williamson (1987).
82. Ibid. See also Hansen (1982) for a review of dowsing.
83. Betz (1995).

## Notes

84. Devereux (1982).
85. If you want to read about this research in detail, see Paul Devereux's *Earth Lights* (1982), *Earth Lights Revelation* (1989) and *Places of Power* (1990); and Don Robins's *Circles of Silence* (1985) and *The Secret Language of Stone* (1988).
86. Persinger and Lafreniere (1977).
87. Persinger and Lafreniere (1977); Persinger and Cameron (1986).
88. Devereux (1982, pp. 70-4).
89. Vallée (1970, 1988); Jung (1959/1977).
90. Devereux (1982, pp. 70-4).
91. This is discussed in greater depth in my book *The Faery Faith* (2002).
92. Devereux (1982, p. 216).
93. 'We must not confuse the exterior, objective nature of the UFO 'carrier' with the imagery it sometimes translates itself into ... The UFO 'hardens' into the form that is finally identified as a spaceship, robot or little green man' (ibid., p. 212).
94. Lovelock (1979, 1988).
95. Devereux (1982, p. 220).
96. Persinger and Cameron (1986).
97. Devereux (1982).
98. Interestingly there is a St Michael Church by Croft Hill. Many of the ancient sacred site hill tops had St Michael churches built on them by the Christians in the thirteenth century. Burrow Mump, Glastonbury Tor and St Michael's Mount at Penzance are just three more examples that spring readily to mind. And St Michael was a dragon slayer, so all these hills are connected with dragon energy, Feng Shui as the Chinese call it; all these hills are places of unusual Earth energy.
99. Graves (1986).
100. Devereux (1982, p. 158).
101. Robins (1985).
102. Devereux (1982); Robins (1985, 1988).
103. Devereux (1982, p. 152).
104. Ibid., p. 154; c.f. Fibonacci series of spirals; chaos theory (e.g. Stewart, 1989; May, 1989; Mullin, 1989; Murray, 1989).
105. Robin Heath's books show the geometry of Stonehenge (1993, 2008), linking it with Lundy Isle and the Prescelly Hills where the bluestones came from, and suggesting that maybe this macro geometry was the reason for bringing the bluestones.
106. Graves (1986).
107. Watkins (1925); Devereux and Thomson (1979).
108. Devereux (1982, p. 229).
109. Graves (1986).
110. David-Neel (1965/1971).
111. Lessing (1979).
112. Shallis (1988, p. 188).
113. Ibid., pp. 261-2..
114. Ibid., p. 263.
115. Ibid., p.185.
116. Devereux (1982, p. 226).

# CHAPTER 6

1. This chapter has been expanded into a whole book in it's own right: The Faery Faith, *Green Magic*, 2003.
2. Evans Wentz (1977, p. 171).
3. Yeats (1902/1977, p. 2).
4. The fairy books by Andrew Lang (e.g. 1950, 1970) are the best that there are for children at present, and I recommend you to get hold of them. For adults the collection of Irish legends by Lady Gregory (1902/1970; 1904/1979; 1920/1979) are well worth investigating. Also there is the Welsh Mabinogian (Jones and Jones, 1949/1982), which feels to me to be a much later post-Iron Age Celtic collection, and Campbell's (1890/1940) collection of Scottish tales.
5. The word 'faerie' or 'feerie' is derived from the Latin word fatare, past participle fae, meaning to enchant. The word Fay or fey is still used in English to mean someone who is quixotic. Elf is Anglo-Saxon. In old English fairy is used to mean illusion or enchantment. In Scotland Fairyland is called Elfame from Norse Elfheim.
6. Kirk (1691/1976); Stewart (1990).
7. Chambers (1841, p. 324).

8. Evans Wentz (1977, p. 275).
9. Spence (1981, p. 6).
10. Evans Wentz (1977, p. 490, p. 335).
11. Gregory (1979, p. 28).
12. Evans Wentz (1977, p. 343).
13. Ibid., p. 372. For example, in the tale of Conn, he entered a rath, beheld the sun-god Lugh seated beside a woman of great beauty described as the Sovereignty of Erinn till the day of doom. She married Lugh at the annual feast of Lughnasad, nowadays also called Lammas (1 August). 'The earthly monarch of Ireland typified this god Lugh; he was his worldly representative, and he was ritually married to this goddess, or to a woman symbolizing her, at the time of his Coronation.' (Spence, 1981, p. 35.) This seems very similar to wiccan, pagan ideas of the goddess and her consort, and tales about early kings of Britain.
14. Evans Wentz (1977, p. 243).
15. Ibid., p. 480.
16. Ashe (1990).
17. *Fortean Times* (1990).
18. Evans Wentz (1977, p. 404).
19. Ibid., p. 407.
20. More recent work on crop circles includes that by Haselhoff (2001); Glickman (1999) and Pringle (1999).
21. Evans Wentz (1977, p. 408).
22. Ibid., p. 276.
23. Lessing (1979).
24. Unpublished. If interested contact: bruce@glastonbury.co.uk
25. Spence (1981, p. 26).
26. Evans Wentz (1977, p. 278).
27. Ibid., p. 315.
28. Zimmer Bradley (1978).
29. Spence (1981, p. 34).
30. Elworthy (1986).
31. E.g. Starhawk (1979; 1982; 1987; 2004); Starhawk & Macha (1997).
32. Spence (1981, p. 16).
33. Evans Wentz (1977, p. 413).
34. Ibid., p. 414.
35. Ibid., p. 416; see also Olcott (1911 and 1914).
36. Ibid., p. 383.
37. Ibid., p. 388.
38. Williams (1978, p. 19).
39. Stevenson (1987, 1997).
40. Evans Wentz (1977, p. 365).
41. Ibid., p. 504.
Ibid., p. 171.

# CHAPTER 7

1. Shuttle and Redgrove (1978/1986).
2. Oatley (1989).
3. Merlin Stone has written a book called *The Paradise Papers* which investigates this in some depth (Stone, 1979).
4. Starhawk (1982); Sjöö and Mor (1989); Hole (1977/1986).
5. Glass (1971).
6. e.g. Roney-Dougal (1984); this chapter is based on a lecture first given in Lisbon, Portugal (Roney-Dougal, 1983).
7. Tart (1984).
8. Glass (1971, pp. 13-14).
9. Some people object very strongly to the use of the word black to denote bad, primarily on racist grounds in that negroes are often called black people. However, in general I am trying to avoid using words out of their specific meaning, not just black to mean bad and white to mean good. Other examples of this common vernacular usage of words which incorporate good—bad overtones include higher and lower, positive and negative, right and left, light and dark, rational and emotional, etc.
10. Murray (1962, p. 97).
11. Glass (1971).
12. Glass (1971, pp. 37-8).

## Notes

13. When did human sexuality shift to a potentially daily joy rather than the 'on heat' copulation of the higher mammals?
14. Marija Gimbutas (1974). There can be no reasonable doubt that paganism or witchcraft was flourishing when Christianity came to Europe, and continued to flourish at least in country districts until the 1600s for kings, popes, and Church councils issued edict after edict condemning it. King Canute and Charlemagne both conducted vigorous campaigns against all the pagan worship and eventually churches, statues of Virgin Mary or local saints were placed at the sacred sites (Glass, 1971; Hole, 1986).
15. Sjöö and Mor (1987) describe what follows in far greater detail.
16. Leland (1974).
17. Fyfe, (1968); Lievegoed (1951/1983).
18. Ostrander and Schroeder (1970).
19. Lacey (1974/1976).
20. Kingsley (1863).
21. Grahame (1908/1957).
22. Ashe (1976).
23. There are lots of good books available today about witchcraft, paganism and the Western mystery tradition. In this chapter I quote extensively from Justine Glass (1971), but for those who wish to read further I recommend Starhawk (1979, 1982, 1987, 2004); Marian Green (1980, 1987, 1988); Matthews (1987, 1988, 1989, 1990, 2005). For articles aimed at providing some attempt at a modern synthesis see the *Quest* journals (1970-2010). There are many references in other books to Lethbridge's *Witches* (1962); and a clear summary appears in Christina Hole's *Witchcraft in Britain* (1977/1986). There are now several small publishers who specialise in this genre, e.g., Green Magic publishers: www.greenmagicpublishing.com.
24. Muir (1981).
25. Kenton (1987).
26. Glass (1971, pp. 19-20).
27. Stone (1979, p. 82); Reader (1988).
28. McMoneagle (1997, 2002).
29. Yolen (1980).
30. Harner (1973/1978).
31. Harner (1978, p. 146).
32. Starhawk (1979).
33. Benor (2004); Astin, Harness and Ernst, (2000); Dossey (2001).
34. Glass (1971, pp. 57-8).
35. Batcheldor (1983).
36. Braud and Schlitz (1983, 1991); Braud (2003). A fascinating experiment was done replicating the remote staring research with a sceptical experimenter, in which Marilyn Schlitz still got the effect but the sceptic got nothing! (Wiseman and Schlitz, 1997).
37. Dean (1983); Braud (1990, 2003); LeShan (1990).
38. Stone (1979).
39. Lethbridge (1962).
40. Glass (1971, p. 126).
41. This is discussed more fully my book *The Faery Faith* (2002), and also in the article, "Celtic Lunar Calendar" on my web site: www.psi-researchcentre.co.uk/articles. I was introduced to the Celtic Lunar calendar by Keith Mylchreest;
42. Graves (1977).
43. Glass (1971, pp. 18-19).
44. Cf. Capra (1975); Zukav (1979). And in the past two decades by so many books I don't know where to start! Alan B. Wallace (2008) and Fred Alan Wolf (2000) are two physicists who have embraced a spiritual aspect to their work.
45. Cf. Fromm (1962); Robinson (1975); Gardner (1978).

## Chapter 8

1. Wilber (1984, 2006).
2. A fairly complete work on Eckhart is Pfeiffer (1947), though there are several more modern versions; Blake (1994).
3. Huxley (1974).
4. Capra (1975); Wallace (2008); Wolf (2000).
5. Bohm (1982).
6. Wilber (1982); and Talbot (1991).
7. Wilber (2002, 2006, 2007)

8. Heisenberg, in Wilber (1984, pp. 42-3).
9. Sir Arthur Eddington, in Wilber (1984, p. 169).
10. Sir James Jeans, in Wilber (1984, p. 144).
11. Grad (1965); Le Shan (1990); Braud (1990); Inglis (1979, 1980); Benor (2001, 2004).
12. Sheldrake (1999).
13. Roney-Dougal (2002).
14. White (1982a); Rogo (1982); Evans (1984).
15. Einstein, in Wilber (1984, pp. 103-111).
16. Evans Wentz (1911/1977).
17. Lovelock (1979).
18. As in the Holographic Paradigm, e.g. Wilber (1982).
19. Whilst I am about it I would like to say how much I abhor the present commercial crystal trade. I find it unbearable that Mother Earth should have her self blasted to pieces just so that trendy New Agers can have masses of crystals to play with. I consider that much of this crystal craze is based on ignorance, egotism, glamour and illusion; the negative aspect of magical workings. If you find a crystal in a special place at a special moment, then this is being given to you by the Mother, and will be a healing tool with which to work and play and enjoy with fond memories of the moment when you found it; but to buy a fancy crystal out of a shop which comes from a quarry or mine from fat commercialists who are only in it for the profit and who are destroying the Mother in the process is abhorrent.
20. Capra (1975).
21. Wilber (1982).
22. Bohm (1982).
23. Sheldrake (1981).
24. Jung (1977/1987).
25. Jung (1987).
26. May (1988).
27. Bohm (1982).
28. There has been three decades of incredible research at the Princeton Engineering Anomalies Research (PEAR) Laboratory by Robert Jahn, Brenda Dunne, Roger Nelson and assorted colleagues, which culminated in the Global Consciousness Project: www.noosphere.princeton.edu. Their work shows the extent to which mind directly affects our material world we inhabit and perhaps even begins to show the centrality of consciousness in changing subtly the energy around us. Worth checking their work out. (Jahn and Dunne, 1988; Dunne and Jahn, 1992; Radin and Nelson, 1989).
29. White (1982b).
30. Giesler (1985).
Inglis (1986).

# EPILOGUE

1. Satyananda (2000).
2. Thrangu Rinpoche (2001, p.34).
3. Wilber (2001).
4. Satyananda (2000, p.19).
5. Ibid., p.18.
6. Thrangu Rinpoche (2001).
7. Tai Situ-pa Rinpoche (1996).
8. Radin (1997, 2006); Jahn & Dunne (1988); Harris Walker (1977).
9. Satyananda (2000, p.19).
10. Wilber (1984).
11. Satyananda (2000).
12. Ibid., p.19.
13. Niranjanananda (2002, p.25).
14. Adapted from Yogakanti (1999).
15. Lakshman (1988).
16. Yogakanti (1999).
17. Ibid.
18. Braud (1978); Honorton (1981).
19. Roney-Dougal & Solfvin, (2008).
20. Carr (2008).

# References

Adams, D. (1979). *The Hitchhiker's Guide to the Galaxy*, London, Britain: Pan Books.
Adams, M. H. (1986). Variability in remote-viewing performance: possible relationship to the geo magnetic field. In Weiner, D. H. and Radin, D. I. (eds), *Research in Parapsychology 1985*, Metuchen, NJ: Scarecrow Press, p. 25.
Ahlbom, A. and Feychting, M. (2003). Electromagnetic radiation. *British Medical Bulletin*, 68, 157-165.
Airaksinen, M. M. and Kari, I. (1981a). Beta-carbolines, psychoactive compounds in the mammalian body; Part 1: Occurrence, origin and metabolism, *Medical Biology*, 59, 21-34. (1981b). Beta-carbolines, psychoactive compounds in the mammalian body; Part II: Effects, *Medical Biology*, 59, 190-211.
—— Sainio, E.-L., Leppäluoto, J. and Kari I. (1984). (6-methoxy-tetrahydro-β-carboline (pinoline): effects on plasma renin activity and aldosterone, TSH, LH and β-endorphin levels in rats, *Acta Endocrinologica*, 107(4), 525-530.
Allain, D. *et al.* (1986). Effects of pinealectomy on photoperiodic control of hair follicle activity in the limousine ram: Possible relationship with plasma prolactin levels, *Journal of Pineal Research*, 3, 25-32.
Arendt, J. (1978). Melatonin assays in body fluids, *Journal of Neural Trans.*, Suppl. 13, 265-78.
——(1985). The pineal: a gland that measures time?, *New Scientist*, 1466, 25 July, 36-9.
Ashe, G. (1976). *The Virgin*, London, Britain: Routledge and Kegan Paul.
Aspect, A. and Grangier, P. (1986). Experiments on Einstein-Podolsky-Rosen-type correlations with pairs of visible photons. In Penrose, R. and Isham, C. J. (eds.), *Quantum Concepts in Space and Time*, Oxford, Britain: Oxford University Press.
Astin, J., Harness, E. and Ernst, E. (2000). The efficacy of "distant healing": A systematic review of randomized trials. *Annals of Internal Medicine*, 132, 903-10.
Axelrod, J. *et al.* (1965). Control of HIOMT activity in the rat pineal gland by environmental lighting, *Journal of Biol. Chem.*, 240, 249.
Baker, R. R. (1981). *Human Navigation and the Sixth Sense*, London, Britain: Hodder & Stoughton.
Barker, S. *et al.* (1981). Identification and quantification of 1, 2, 3, 4-Tetrahydrobetacarboline, 2-Methyl-1, 2, 3, 4-Tetrahydrobetacarboline, and 6-Methoxy-1, 2, 3, 4-Tetrahydrobetacarboline as invivo constituents of rat brain and adrenal gland, *Biochemical Pharmacology*, 30, 9-17.
Barr, F. E. *et al.* (1983). Melanin: The organising molecule, *Medical Hypotheses*, 11, (1).
Batcheldor, K. J. (1968). *Micro-PK in Group Sittings: Theoretical and Practical Aspects*. Society for Psychical Research Library (unpub.).
——(1984). Contributions to the theory of PK induction from sitter-group work, *Journal of the American Society for Psychical Research*, 78 (2), 105-122.
Becker, R. O. and Seldon, G. (1985). *The Body Electric,* New York, USA: Morrow,
Beloff, J. (1979). Using the Scientific Method to Probe the Limits of Science. A symposium paper presented at the 22nd Parapsychological Association Convention, J.F.K. Univ. California.
Bellossi, A., Pouvreau-Quillien, V., Rocher, C. and Ruelloux, M. (1998). Effect of pulsed magnetic fields on triglyceride and cholesterol levels in plasma of rats. *Panminerva Med.*, 40. 276-279.

# References

Bem, D. Palmer, J. and Broughton, R. (2001). Updating the Ganzfeld Database: A victim of its own success, *Journal of Parapsychology*, 65(3), 207-218.
Bender, H. (1969). New Developments in Poltergeist Research. In *Proceedings of Parapsychological Association Convention*, 6, 81.
——(1974). Modern Poltergeist Research – A plea for an unprejudiced approach. In Beloff, J. (ed.), *New Directions in Parapsychology*, London, Britain: Elek Science.
Benor, D.J. (2001). *Healing Research: Vol.1, Spiritual Healing: Scientific Validation of a Healing Revolution*, Southfield, MI, USA: Vision Pubs.
——(2004). *Healing Research: Vol. II., Consciousness, Bioenergy and Healing: Self-Healing and Energy Medicine for the 21st Century*, Bellmawr, NJ, USA: Wholistic Healing Publications.
Bentov, I. (1979). *Stalking the Wild Pendulum*, London, Britain: Fontana.
Bergson, H. (1911/1921). *Matter and Memory*, London, Britain: George Allen & Unwin.
Betz, H-D. (1995a). Unconventional Water Detection: Field Test of the Dowsing Technique in Dry Zones: Part 1. *Journal of Scientific Exploration*, 9(1), 1- 44.
——(1995b). Unconventional Water Detection: Field Test of the Dowsing Technique in Dry Zones: Part 2. *Journal of Scientific Exploration*, 9(2), pp.159-189.
Binkley S. (1979). A time keeping enzyme in the pineal gland, *Scientific American*, 240 (4), 50-5.
Blake, W. (1994). *The Works of William Blake*, Ware, Britain: The Wordsworth Poetry Library.
Bloom, W. (1986). *Devas, Fairies and Angels*, Glastonbury, Britain: Gothic Image.
Bohm, D. (1982). *Wholeness and the Implicate Order*, London, Britain: Routledge and Kegan Paul.
Booth, J. N., Charette, J. C., and Persinger, M. A. (2002). Rankings of stimuli that evoked memories in significant others after exposure to circumcerebral magnetic fields: correlations with ambient magnetic activity. *Perceptual and Motor Skills*, 95, 555-558.
Boyce, P.M. (2007). 6-Sulphatoxy melatonin in melancholia, *Lipids Health Dis.*, 6, 31.
Braithwaite, J.J. (2005). Using digital magnetometry to quantify anomalous magnetic fields associated with spontaneous strange experiences: The magnetic anomaly detection system (MADS), *Journal of Parapsychology*, 69 (1), 151-171.
——(2008). Putting magnetism in its place: A critical examination of the weak-intensity magnetic field account for anomalous haunt-type experiences, *Journal of the Society for Psychical Research*, 72(1), 34 -50.
Braud, L. W. and Braud, W. G. (1974). Further studies of relaxation as a psi-conducive state, *Journal of the American Society for Psychical Research*, 68, 229-45.
Braud, W. G. (1975). Psi-conducive States, *Journal of Communication*, 25, 142-52.
——(1978). Psi-conducive conditions: Explorations and interpretations. In Shapin, B. and Coly, L. (eds.), *Psi and States of Awareness*, New York, USA: Parapsychology Foundation, pp.1 – 41.
——(1980). Lability and inertia in conformance behaviour, *Journal of the American Society for Psychical Research*,74, 297-318.
——(1990). Distant mental influence of rate on hemolysis of human red blood cells, *Journal of the American Society for Psychical Research*,84, 1-24.
——(2003). *Distant Mental Influence*, Charlottesville, VA, USA: Hampton Roads. Publs.
—— and Dennis, S. P. (1989). Geophysical variables and behaviour: LVIII. Autonomic activity,hemolysis, and biological psychokinesis: possible relationships with geomagnetic field activity, *Perceptual and Motor Skills*, 68, 1243-1254.
—— Shafer, D. and Andrews, S. (1993). Further studies of autonomic detection of remote staring; Replication, new control procedures and personality correlates, *Journal of Parapsychology*, 57(4), 391- 409.
—— and Schlitz, M. (1983). Psychokinetic influence on electrodermal activity, *Journal of Parapsychology*, 47, 95-119.
——(1989). Possible role of intuitive data sorting in electrodermal biological psychokinesis (Bio-PK), *Journal of the American Society for Psychical Research*, 83, 289-302.
Braud, W. G. and Schlitz, M. (1991). Consciousness interactions with remote biological systems: anomalous intentionality effects. *Subtle Energies*, 2, 1-46.
Broughton, R. S. (1992). *Parapsychology: The Controversial Science*. New York, USA: Ballantine.
Buckholtz, N. S. (1980). Mini-review, neurobiology of beta-carbolines, *Life Sciences*, 27, 893-903.
Burch, J.B., Reif, J.S., Noonan, C.W. and Yost, M.G. (2000). Melatonin metabolite levels in workers exposed to 60-Hz magnetic fields: work in substations and with 3-phase conductors, *Journal of Occupational and Environmental Medicine*, 42 (2), 136-142.
Burr, H. S. (1972). *Blueprint for Immortality: The Electric Pattern of Life*, London, Britain: Neville Spearman.
Callaway J. C. (1988). A proposed mechanism for the visions of dream sleep, *Medical Hypotheses*, 26, 119-24.
——(2006). Phytochemistry and neuropharmacology of ayahuasca, in Metzner, R. (ed.), *Sacred vine of the spirits: Ayahuasca* (pp. 94-116). Rochester, VT, USA: Park Street.
Campbell J. F. (1890/1940). *Popular Tales of the West Highlands*, Britain: Anthropological and Folklore

Society of Scotland.
Capra, F. (1975). *The Tao of Physics,* London, Britain: Wildwood House.
Carman, J.S., Post, R.M., Buswell, R. and Goodwin, F.K. (1976). Negative effects of melatonin on depression, *American Journal of Psychiatry,* 133, 1181-118.
Carr, B. (2008). Worlds Apart? Can psychical research bridge the gulf between matter and mind?, *Proceedings of the Society for Psychical Research,* vol. 59, part 221.
Cavanna, R. and Servadio, E. (1964). ESP experiments with LSD 25 and Psilocybin: a methodological approach, *Parapsychology Monographs, no.* 5, New York, USA: Parapsychology Foundation Inc.
Chambers, R. (1841) *Popular Rhymes of Scotland,* Edinburgh, Britain: Chambers.
Child, I. L. and Levi, A. (1979). Psi-missing in free-response settings, *Journal of the American Society for Psychical Research,* 73, 273-90.
Clark, M.L., Burch, M.S., James, B., Yost, M.G., Zhai, Y., Bachand, A.M., Fitzpatrick, C.T., Ramaprasad, J., Cragin, L.A. and Reif, J.S. (2007). Biomonitoring of Estrogen and melatonin metabolites among women residing near radio and television broadcasting transmitters. *Journal of Occupational and Environmental Medicine,* 49 (10), 1149–1156.
Cook, C.M. and Persinger, M.A. (2001). Geophysical variables and behavior: XCII. Experimental elicitation of the experience of a sentient being by right hemispheric, weak magnetic fields: interaction with temporal lobe sensitivity. *Perceptual and Motor Skills,* 2, 447-448.
Costa de Beauregard, O. (1975). Quantum paradoxes and Aristotle's twofold information concept. In Oteri, L. (ed.), *Quantum Physics and Parapsychology,* New York, USA: Parapsychology Foundation Inc.
Cremer-Bartels, G. *et al.* (1983) EMF and quails, *Graefe's Arch. Clinical Experimental Opthalmology,* 22, 248.
——(1984). Magnetic field of the Earth as additional zeitgeber for endogenous rhythms, *Naturwissenshaften,* 71, 567-74.
Dale, L. A., White, R. and Murphy, G. (1962). A selection of cases from a recent survey of spontaneous ESP phenomena, *Journal of the American Society for Psychical Research,* 56, 3-47.
Dachà, M., Accorsi, A., Pierotti, C., Vetrano, F., Mantovani, R., Guidi, G., Conti, R. and Nicolini, P. (1993). Studies on the possible biological effects of 50 Hz electric and/or magnetic fields: evaluation of some glycolitic enzymes, glycolitic flux, energy and oxido-reductive potentials in human erythrocytes exposed in vitro to power frequency fields. *Bioelectromagnetics,* 14, 383-391.
Davenas, E. *et al.* (1988). Human basophil degranulation triggered by very minute antiserum against IgE, *Nature,* 333, 816-18.
David-Neel, A. (1965/1971). *Magic and Mystery in Tibet,* Corgi Books.
Davidson, J. (1987). *Subtle Energy,* Saffron Walden, Britain: C.W. Daniel.
——(1988) *The Web of Life,* Saffron Walden, Britain: C.W. Daniel.
Davidson, R.J *et al.* (2003). Alterations in brain and immune function produced by mindfulness meditation, *Psychosomatic Medicine,* 65 (4), 564-570.
Davies, P. (1983). *God and the New Physics,* New York, USA: Touchstone, Simon and Schuster.
Davis, G. and Braud, W. G. (1980). Autonomic recognition of ESP targets. In Roll, W. G. (ed.), *Research in Parapsychology 1979,* Metuchen, NJ, USA: Scarecrow Press.
Dean, D. (1966). Plethysmograph recordings as ESP responses, *International Journal of Neuropsychiatry,* 2, 439-47.
——(1983). An examination of infra-red and ultra-violet techniques to test for changes in water following the laying-on-of-hands, Doctoral dissertation, Saybrook Institute, San Francisco.
Deulofeu, V. (1967). Chemical compounds from Banisteriopsis and related species, in Effron *et al.* (eds), *Ethnopharmalogic Search for Psychoactive Drugs,* NIMH, US Dept. of Health, Education and Welfare, pp. 393-401.
Devereux, P. (1982). *Earth Lights,* Winnipeg, Canada: Turnstone Press.
——(1989). *Earth Lights Revelation,* Poole, Britain: Blandford Press.
——(1990). *Places of Power,* Poole, Britain: Blandford Press.
—— and Thomson, I. (1979). *The Ley Hunter's Companion,* London, Britain: Thames & Hudson.
Di Carlo, A.L., White, N.C. and Litovitz, T.A. (2000). Mechanical and electromagnetic induction of protection against oxidative stress. *Bioelectrochemistry,* 53, 87–95.
Dixon, N. F. (1971). *Subliminal Perception: The Nature of a Controversy,* New York, USA: McGraw Hill.
——(1979). Subliminal perception and parapsychology. In Coly, L. and Shapin, B. (eds), *Brain/Mind and Parapsychology,* New York, USA: Parapsychology Foundation Inc., pp. 206-20.
——(1981). *Preconscious Processing,* Hoboken, USA: Wiley.
Dobson, J., St.Pierre, T., Wieser, H.G. & Fuller, M. (2000). Changes in paroxysmal brainwave patterns of epileptics by weak-field magnetic stimulation. *Bioelectromagnetics,* 21, 94-99.
Dodt, E. and Heard, E. (1962). Mode of action of pineal nerve fibres in frogs, *Journal of Neuropsychiatry,* 25 (3), 405-29.
Don, N.S. *et al.* (1996). Psi, Brain Function and "Ayahuasca" (Telepathine). Paper presented at 39th Parapsychological Association Convention, San Diego, pp 315-334.
—— and Moura, G. (1997). Topographic brain mapping of UFO experiencers, *Journal of Scien-*

## References

*tific Exploration*, 11 (4), 435-453.
Dossey, L. (2001). *Healing Beyond the Body*, Boston, USA: Shambhala.
Drury, N. (1982). *The Shaman and the Magician: Journeys between the Worlds*, London, Britain: Routledge & Kegan Paul.
Dunne, B.J. and Jahn, R.G. (1992). Experiments in Remote Human /Machine Interaction, *Journal of Scientific Exploration*, 6(4), 311-332.
Edge, H. L. et al. (1986). *Foundations of Parapsychology*, London, Britain: Routledge and Kegan Paul.
Eisenbud, J. (1970). *Psi and Psychoanalysis*, New York, USA: Grune and Stratton Inc.
Elkin, A. (1977). *Aboriginal Men of High Degree*, New York, USA: St Martin's Press.
Elworthy, F. T. (1986). *The Evil Eye: An Account of This Ancient and Widespread Superstition*, New York, USA: Julian Press.
Emmons, C. F. (1982). *Chinese Ghosts & ESP: A Study of Paranormal Beliefs and Experiences*, Metuchen, USA: Scarecrow Press.
Etzold, E. (2005). Solar-periodic full moon effect in the Fourmilab retropsychokinesis project experiment data: An exploratory study, *Journal of Parapsychology*, 69(2), 233-262.
Evans, C. (1973). Parapsychology — what the questionnaire revealed, *New Scientist*, 25 Jan., 209.
Evans, H. (1984). *Visions, Apparitions, Alien Visitors: A Comparative Study of the Entity Enigma*, Wellingborough, Britain: Aquarian Press.
Evans-Pritchard, E. E. (1937/1958). *Witchcraft, Oracles and Magic among the Azande*, Oxford, Britain: Clarendon Press.
Evans Wentz, W. A. (1911/1977). *The Fairy Faith in Celtic Countries*, Gerrards Cross, Britain: Colin Smythe.
Farrell, G. and McIsaacs, W. M. (1961). Adrenoglomerulotropin, *Arch. Biochemistry and Biophysics*, 94, 543.
Feychting, M., Ahlbom, A. and Kheifets, L. (2005). EMF and health. *Annual Revue of Public Health*, 26, 165-189.
*Fortean Times* (1980-92). Box 2409, London NW5 4NP.
Foster, K.R. (2003). Mechanisms of interaction of extremely low frequency electric fields and biological systems. *Radiation Prot. Dosimetry*, 106, 301-310.
Fraser-Smith, A.C. (1982). Is there an increase of geomagnetic activity preceding total lunar eclipses? *Journal of Geophysical Research*, 87, 895.
Frazer, J. (1929). *The Golden Bough, Vol. I*, New York, USA: Book League of America.
Frederiksen, T. J. P., Pless, G., Garcia, J. J. and Reiter, R. J. (1998). Pinoline and melatonin protect against $H_2O_2$-induced lipid peroxidation in rat brain homogenates, *Neuroendocrinology Letters*, 19 (3), 117-123.
Freedman, S. and Clauser, J. (1972). Experimental test of local hidden variables theories, *Physical Review Letters*, 28, p. 938-41.
Friedman, H., Becker, R.O. and Bachman, C.H. (1963). Geomagnetic parameters and psychiatric hospital admission, *Nature*, 200, 626-628.
Fromm, Erich (1962). *The Art of Loving*, London, Britain: Allen and Unwin.
Fuller, M., Dobson, J., Wieser, H.G. and Moser, S. (1995). On the sensitivity of the human brain to magnetic fields: Evocation of epileptiform activity. *Brain Research Bulletin*, 36, 155-159.
Fyfe, A. (1968). *Moon and Plant*, Arlesheim Switzerland: Soc. for Cancer Research.
Gaer-Luce, G. (1973). *Body Time*, Boulder, USA: Paladin Press.
García J.J., Reiter R.J., Pié J., Ortiz G.G., Cabrera J., Sáinz R.M. and Acuña-Castroviejo D. (1999). Role of pinoline and melatonin in stabilizing hepatic microsomal membranes against oxidative stress, *Journal of Bioenergetics and Biomembranes*, 31 (6), 609-616.
Gardner, R. (1978). *Evolution through the Tarot*, Newburyport, USA: Weiser.
Gauld, A. and Cornell, A.R. (1979) *Poltergeists*, London, Britain: Routledge and Kegan Paul.
Gauquelin, M. (1974). *Cosmic Influences on Human Behaviour*, London, Britain: Gemstone Press.
Gearhart, L. and Persinger, M.A. (1986). Geophysical variables and behaviour: XXXIII. Onsets of historical and contemporary poltergeist episodes occurred with sudden increases in geomagnetic activity, *Perceptual and Motor Skills*, 62, 463-466.
Ghosh, S., Rosenbaum, T.F., Aeppli, G. and Coppersmith, S.N. (2003). Entangled quantum state of magnetic dipoles, *Nature*, 425, 48.
Giesler, P. (1985). Parapsychological Anthropology: II. A Multi-Method Study of Psi and Psi-Related Processes in the Umbanda Ritual Trance Consultation, *Journal of the American Society for Psychical Research*, 79, 113-166.
Gimbutas, M. (1974). *The Gods and Goddesses of Old Europe: 7000-3,000 BC*, London, Britain: Thames & Hudson.
Glass, J. (1971) *Witchcraft: The Sixth Sense*, Santa Monica, USA: Wilshire.
Glickman, M. (1999). Crop Circles, Glastonbury, Britain: Wooden Books.
Gobba, F., Bravo, G., Scaringi, M. and Roccatto, L. (2006). No association between occupational exposure to ELF magnetic field and urinary 6-sulfatoximelatonin in workers, *Bioelectromagnetics*,

27 (8), 667-673.
Golombek, D. A., Escolar E. and Cardinali, D. P. (1985). Melatonin-induced depression of locomotor activity in hamsters: Time-dependency and inhibition by the central-type benzodiazepine antagonist, *American Journal of Psychiatry*, 142, 125-127.
Grad, B. (1965). Telekinetic Effect on Yeast Activity, *Journal of Parapsychology*, 29, 285-6.
Graham, C., Cook, M.R., Sastre, A., Riffle, D.W. and Gerkovich, M.M. (2000). Multi-night exposure to 60 Hz magnetic fields: Effects on melatonin and its enzymatic metabolite, *Journal of Pineal Research*, 28 (1), 1-8.
—— Sastre, A., Cook, M.R. and Kavet, R. (2000). Heart rate variability and physiological arousal in men exposed to 60 Hz magnetic fields. *Bioelectromagnetics*, 21, 480-482.
Grahame, K. (1908/1957). *The Wind in the Willows*, London, Britain: Methuen.
Graves, R. (1977). *The White Goddess*, London, Britain: Faber & Faber.
Graves, T. (1978) *Needles of Stone*, Winnipeg, Canada: Turnstone Press.
——(1986). *Needles of Stone Revisited*, Glastonbury, Britain: Gothic Image.
Green, E. and Green, A. (1977). *Beyond Biofeedback*, New York, USA: Delacorte Press.
Green, M. (1980). *A Harvest of Festivals*, Harlow, Britain: Longman.
——(1987). *The Gentle Arts of Aquarian Magic*, Dartford, Britain: Aquarian Press.
——(1988). *The Path through the Labyrinth*, Shaftesbury, Britain: Element Books.
Gregory, Lady (1902/1970) *Cuchulain of Muirthemne*, Gerrards Cross, Britain: Colin Smythe.
——(1904/1979). *Gods and Fighting Men*, Gerrards Cross, Britain: Colin Smythe.
——(1920/1979). *Visions and Beliefs in the West of Ireland*, Gerrards Cross, Britain: Colin Smythe.
Gribbin, J. (1978). *The Climatic Threat*, London, Britain: Fontana.
Gruber, E. (1980). Conforming of pre-recorded group behaviour with disposed observers, in Roll, W.G(ed.), *Research in Parapsychology 1979*, Metuchen, USA: Scarecrow Press.
Halaris, A. (ed.) (1987). *Chronobiology and Psychiatric Disorders*, New York, USA: Elsevier.
Hansen, G.P. (1982). Dowsing: A Review of Experimental Research, *Journal of the Society for Psychical Research*, 51, 343-67.
Hansen, G.P. (2000). *The Trickster and the Paranormal*, Philadelphia, USA: XLibris Corp.
Haraldsson, E. (1987). *Miracles Are My Visiting Cards*, London, Britain: Century Hutchinson.
——and Gissurarson, L.R. (1987). Does geomagnetic activity affect extrasensory perception? *Journal of Personality and Individual Differences*, 8, 745-747.
——and Johnson, M. (1979). ESP and the defense mechanism test (DMT). Icelandic study no. III. A case of the experimenter effect? *European Journal of Parapsychology*, 3, 11-20.
Harner, M. J. (ed.) (1973/1978). *Hallucinogens and Shamanism*, Oxford, Britain: Oxford Univ. Press.
——(1980). *The Way of the Shaman: A Guide to Power and Healing*, New York, USA: Harper & Row.
Harris Walker, E. (1977). Comparison of Some Theoretical Predictions of Schmidt's Mathematical Theory and Walker's Quantum Mechanical Theory of Psi, *The Journal of Research in Psi Phenomena*, 2 (1), 54-70.
Haselhoff, E.H. (2001). *The Deepening Complexity of Crop Circles*, Berkeley, USA: Frog Ltd.
Hasted, J.B. (1981). *The Metal Benders*, London, Britain: Routledge and Kegan Paul.
——and Robinson, D. (1981). Paranormal electrical effects, *Journal of the Society for Psychical Research*, 51, 75-87.
Heath, R. (1993). *Stonehenge: the Marriage of the Sun and the Moon*, Britain: Bluestone Press.
——(2008). *A Key to Stonehenge*, Britain: Bluestone Press.
Henshaw, D.L. & Reiter, R.J. (2005). Do magnetic fields cause increased risk of childhood leukaemia via melatonin disruption? *Bioelectromagnetics*, 26, S86-87.
Heywood, R. (1964). *The Infinite Hive*, London, Britain: Chatto and Windus.
——(1959/1971). *The Sixth Sense*, London, Britain: Chatto and Windus.
Hider, R.C., Smart, L. and Suleiman, M.S. (1981). The effect of harmaline and related beta-carbolines on the acetylcholine-stimulated contractions of guinea-pig ileum, *European Journal of Pharmacology*, 70, 429-36.
Hochstein, F.A and Paradies, A.M. (1957). Alkaloids of Banisteria caapi and Prestonia amazonicum, Journal of the *American Chemical Society*, 79, 5735-6.
Hole, C. (1977/1986). *Witchcraft in Britain*, Boulder, USA: Paladin.
Hook, G.J., Spitz, D.R., Sim, J.E., Higashikubo, R., Baty, J.D., Moros, E.G. and Roti, J.L. (2004). Evaluation of parameters of oxidative stress after *in vitro* exposure to FMCW and CDMA modulated radio frequency radiation fields. *Radiation Research*, 162, 497-504.
Honecker, H. and Rommelspacher, H. (1978). Tetrahydro-harmane (Tetrahydrobetacarboline), a physiologically occurring compound of indole metabolism, *Naunyn-Schmiedeberg's Archs. Pharmacology*, 305, 135.
Honorton, C. (1977). Psi and internal attention states. In Wolman, B.B. (ed.), *Handbook of Parapsychology*, New York, USA: Van Nostrand Rheinhold, pp. 435-72.
——(1981). Psi, Internal Attention States and the Yoga Sutras of Patanjali. In Shapin, B. & Coly, L. (eds.), *Concepts and Theories of Parapsychology*, New York, USA: Parapsychology Foundation, pp.55

# References

———- 68.
———(1985). Meta-analysis of psi Ganzfeld research, *Journal of Parapsychology*, 49, 51-92.
———and Barksdale, W. (1972). PK Performance with waking suggestions for muscle tension versus relaxation, *Journal of the American Society for Psychical Research*, 66, 208-14.
Hsu, L.L. (1984). Pineal aryl acylamidase: effects of melatonin, serotonin-related compounds, beta-carbolines, R04 — 4602 and antidepressants, *Res. Commun. Chem. Pathol. Pharmacol.*, 43 (2), 223-34.
Hubbard, G.S. and May, E.C. (1987). Aspects of the measurement and application of geomagnetic indices and extremely low frequency electromagnetic radiation for use in parapsychology. In Weiner, D.H. and Nelson, R.D. (eds.), *Research in Parapsychology 1986*, Metuchen, USA: Scarecrow Press.
Hubel, D.H. et al. (1979). The brain, *Scientific American*, Sept., 241, 38-188.
Hurley, F. (1985). *Sorcery*, London, Britain: Routledge and Kegan Paul.
Huxley, A. (1946/1974). *The Perennial Philosophy*, London, Britain: Chatto and Windus.
Hyman, R. (1985). The Ganzfeld psi experiment: a critical appraisal, *Journal of Parapsychology*, 49, 3-50. Inglis, B. (1978). *Natural and Supernatural*, Abingdon, Britain: Hodder and Stoughton.
———(1979/80). *Natural Medicine*, London, Britain: Fontana.
———(1980). The controversial and the problematical, *The Times*, 20 Dec, p. 12.
———(1984). *Science and Parascience*, Abingdon, Britain: Hodder & Stoughton.
———(1986). *The Hidden Power*, London, Britain: Jonathan Cape.
Irwin, H.J. (1979). *Psi and Mind: An Information Processing Approach*, Metuchen, USA: Scarecrow Press.
———(2004). *An Introduction to Parapsychology: 4$^{th}$ Edition*, Jefferson, USA: McFarland.
Isaacs, J. (1984). The Batcheldor approach: some strengths and weaknesses, *Journal of the American Society for Psychical Research*, 78 (2) 123-32.
Jacob, M. S. and Presti, D. E. (2005). Endogenous psychoactive tryptamines reconsidered: An anxiolytic role for dimethyltryptamine. *Medical Hypotheses*, 64, 930-937.
Jahn, R.G. and Dunne, B.J. (1988). *Margins of Reality: The Role of Consciousness in the Physical World*, Orlando, USA: Harcourt Brace.
James, W. (1902/1958). *The Varieties of Religious Experience*, New York, USA: Mentor.
Jauchem, J.R., Frei, M.R., Ryan, K.L., Merritt, J.H. and Murphy, M.R. (1999). Lack of effects on heart rate and blood pressure in ketamine-anesthetized rats briefly exposed to ultra-wideband electromagnetic pulses. *IEEE Trans Biomedical Engineering*, 46:117–120.
Johnson, L.Y. (1982). The pineal as a modulator of the adrenal and thyroid axes. In Reiter, R.J. (ed.), *The Pineal Gland, vol. III, Extra-reproductive Effects*, Florida, USA: CRC Press Inc.
Johnson, M. and Haraldsson, E. (1984). The Defense Mechanism Test as a predictor of ESP scores, *Journal of Parapsychology*, 48, 185-200.
Jones, G. and Jones, T. (1949/1982) *The Mabinogion*, Holland: Dragon's Dream.
Jouvet, M. (1974). Monoaminergic regulation of the sleep-waking cycle in the cat. In Schmitt, F.O. and Worden, F.G. (eds), *The Neurosciences*, Cambridge, USA: MIT Press.
Juutilainen, J. (2003). Developmental effects of extremely low frequency electric and magnetic fields. *Radiation Prot Dosimetry*, 106, 385–390
Jung, C.G. (ed.) (1964/1990). *Man and His Symbols*, London, Britain: Arkana.
———(1977/1987). *Psychology and the Occult*, London, Britain: Ark Paperbacks.
———(1959/1977). *Flying Saucers: A Modern Myth of Things Seen in the Sky*, London, Britain: Routledge and Kegan Paul.
———and Pauli, W. (1955). *The Interpretation of Nature and the Psyche*, New York, USA: Bollingen Series, Pantheon Books.
Kahn, S.D. (1976). Myers' problem revisited. In Schmeidler, G.R. (ed.), *Parapsychology: Its Relationship to Physics, Biology, Psychology and Psychiatry*, Metuchen, USA: Scarecrow Press.
Keane, P. and Wells, R. (1979). An examination of the menstrual cycle as a hormone related physiological concomitant of psi performance. In Roll, W.G. (ed.), *Research in Parapsychology 1978*, Metuchen, USA:Scarecrow Press.
Kelly, E.F. and Locke, R.G. (1981). Altered states of consciousness and psi: an historical survey and research prospectus, *Parapsychological Monographs, no. 18,* New York, USA: Parapsychological Foundation Inc.
Kennedy, J. (1979). Redundancy in psi information: implications for the goal-oriented hypothesis and for the application of psi, *Journal of Parapsychology*, 43, 290-314.
———and Taddonio, J. (1976). Experimenter effects in parapsychological research, *Journal of Parapsychology*, 40, 1-33.
Kensinger, K.M. (1978). Banisteriopsis usage among the Peruvian Cashinahua. In Harner, M.J. (ed.), *Hallucinogens and Shamanism*, Oxford, Britain: Oxford Univ. Press, pp. 9-14.
Kenton, W. (1987). *The Anointed One*, London, Britain: Arkana.
Kingsley, C. (1863). *The Water Babies*, Glasgow, Britain: Blackie and Son.
Kirk, R. (1691/1976). *The Secret Commonwealth of Elves, Fauns and Fairies; and a short treatise of

*charms and spells,* Cambridge, Britain: D.S. Brewer for the Folklore Society.
Koestler, A. (1972). *Roots of Coincidence,* London, Britain: Hutchinson.
——(1978). *Janus: A Summing Up,* London, Britain: Hutchinson.
Kooy, J.M.J. (1984). On the mental background of the human personality, *Research Letter of the Utrecht Parapsychology Lab., No. 12.*
Krippner, S. (1980). Folk healing and Parapsychological investigation, *American Society for Psychical Research Newsletter,* 6 (1), 3—4.
——(ed.) (1977-97). *Advances in Parapsychological Research,* vols. 1-3, New York, USA: Plenum Publishing; vols, 4 – 8, Jefferson, USA: McFarland.
——(1982). In Wilber, K. (ed.), *The Holographic Paradigm and Other Paradoxes,* Boulder, USA: Shambala.
——Honorton, C. and Ullman, M. (1972). A second precognitive dream study with Malcolm Bessent, *Journal of the American Society for Psychical Research,* 66, 269-79.
——and Persinger, M. (1996). Evidence for enhanced congruence between dreams and distant target material during periods of decreased geomagnetic activity. *Journal of Scientific Exploration,* 10 (4), 487-493.
——Radin, D.I. and Rebman, J.M. (1994). Lunar correlates of normal, abnormal, and anomalous human behavior, *Subtle Energies and Energy Medicine,* 5, 209.
Kumosani, T.A. and Qari, M.H. (2003). The effect of magnetic field on the biochemical parameters of mice blood. *Pakistan Journal of Medical Science,* 19, 36–40.
Lacy-Hubert, A., Metcalfe, J.C. and Hesketh, R. (1998). Biological responses to electromagnetic fields. *FASEB J.,*12, 395–420.
Lacey, L. (1974/1976). *Lunaception,* New York, USA: Warner Books.
Lakshman Jee, Sw. (1988). *Kashmir Shaivism: The Secret Supreme,* Delhi, India: Sri Satguru Publications, pp. 71-85.
Lang, A. (1950/1970). *The Orange Fairy Book,* London, Britain: Longman.
Langer, S.Z. et al. (1984). Possible endocrine role of the pineal gland for 6-Methoxy-tetrahydro-betacarboline, a putative endogenous neuromodulator of the (3H) Imipramine recognition site, *European Journal of Pharmacoogy,* 102, 379-80.
Lansky, P. (1979). Neurochemistry and the awakening of Kundalini. In White, J. (ed.), *Kundalini, Evolution and Environment,* New York, USA: Omega Books.
Lazar, S.W., Bush, G., Gollub, R.L., Fricchione, G.L., Khalsa, G and Benson, H. (2000). Functional brain mapping of the relaxation response and meditation, *NeuroReport,* 11(7), 1581-1585.
Le Guin, U. (1979/1985). *The Earthsea Trilogy,* London, Britain: Penguin.
Leino, M. et al. (1984). Effects of melatonin and 6-MeOTHBC in light induced retinal damage: a computerized morphometric method, *Life Sciences,* 35, 1997-2001.
Leland, G. (1974). *Aradia: The Gospel of the Witches,* London, Britain: C.W. Daniels.
LeShan, L. (1974). *The Medium, The Mystic and the Physicist,* Northampton, Britain: Viking Press.
——(1990). Explanations of psychic healing, *American Society for Psychical Research Newsletter,* 16, 1-3.
Lessing, D. (1979/1981). *Shikasta,* Chelmsford, Britain: Grafton Books.
Lethbridge, T.E. (1962). *Witches: Investigating an Ancient Religion,* London, Britain: Routledge and Kegan Paul.
——(1963). *Ghosts and Divining Rod,* London, Britain: Routledge and Kegan Paul.
Levine R. L. and Bluni, T. D. (1994). Magnetic field effects on spatial discrimination learning in mice, *Physiology and Behaviour,* 55 (3), 465-467.
Levi-Strauss, C. (1963). *Structural Anthroplogy,* New York, USA: Basic Books.
Lievegoed, C.B.J. (1951/1983). *The Working of the Planets and the Life Processes in Man and Earth,* Stourbridge, Britain: Bio-Dynamic Agricultural Association.
Litovitz, T.A., Montrose, C.J. and Wang, W. (1992). Dose-response implications of the transient nature of electromagnetic-field-induced bioeffects: theoretical hypotheses and predictions. *Bioelectromagnetics,* 12, 237–246.
Llinás, R. and Paré, D. (1998). Coherent oscillations in specific and non-specific thalamocortical networks and their role in cognition, *Thalamus,* 501-516.
Loewe, M. and Blacker, C. (eds) (1981). *Divination and Oracles,* Sydney, Australia: Allen & Unwin.
Lonegren, S. (1986) *Spiritual Dowsing,* Gothic Image.
Lovelock, J.E. (1979). *Gaia - A New Look at Life on Earth,* Oxford, Britain: Oxford Univ. Press.
——(1988). *The Ages of Gaia,* Oxford, Britain: Oxford Univ. Press.
Luke, D.P., Delanoy, D. and Sherwood, S. (2008). Psi may look like luck: Perceived luckiness and beliefs about luck in relation to precognition, *Journal of the Society for Psychical Research,* 72 (4), 193-207.
Luo, E.P., Jiao, L.C., Shen, G.H., Wu, X.M. and Cao, Y.X. (2004). Effects of exposing rabbits to low-intensity pulsed electromagnetic fields on levels of blood lipid and properties of haemorheology. *Chinese Journal of Clinical Rehabilitation,* 8, 3670–3671.
Lutz, A., Brefczynski-Lewis, J. Johnstone, T. and Davidson, R.J. (2008). Regulation of the neural

# References

circuitry of emotion by compassion meditation: Effects of Meditative Expertise, *PLoS ONE*, 3(3), e1897.

Lutz, A., Slagter, H.A., Dunne, J.D. and Davidson, R.J. (2008). Attention regulation and monitoring in meditation, *Trends in Cognitive Sciences*, 12 (4), 163 – 169.

Maestroni, G.J.M. et al (1989). Pineal Melatonin, its fundamental immunoregulatory role in aging and cancer. *Annals of the New York Academy of Sciences*, 140 - 148.

Maher, M.C. (2000). Quantitative investigation of the General Wayne Inn. *Journal of Parapsychology*, 64 (4), 365-390.

——and Hansen, G.P. (1997). Quantitative investigation of a legally disputed "haunted house." *Proceedings of Presented Papers: The Parapsychological Association 40$^{th}$ Annual Convention held in conjunction with the Society for Psychical Research*, 184-201.

Majewski, P., Adamska, I., Pawlak, J., Baranska,A. and Skwarlo-Sonta, K. (2005). Seasonality of pineal gland activity and immune functions in chickens, *Journal of Pineal Research*, 39(1), 66 -72.

Malinowski, B. (1925/1954). *Magic, Science and Religion*, New York, USA: Anchor Press.

Margenau, H. (1967). ESP in the framework of modern science. In Smythies, J.R. (ed.), *Science and ESP*, London, Britain: Routledge and Kegan Paul.

Martikuinen, H. *et al.* (1985). Circannual concentrations of melatonin, gonadotrophins, prolactin and gonadal steroids in males in a geographical area with a large annual variation in daylight. *Acta Endocrinology (Copenh)*, 109, 446-450.

Mather, J.G. and Baker, R.R. (1980). A demonstration of navigation by rodents using an orientation cage, *Nature*, 284, 259-62.

Matthews, J. and Matthews, C. (1987). *The Western Way: A practical Guide to the Western Mystery Tradition, Vol. 1: The Native Tradition*, London, Britain: Arkana.

——(1988). *The Western Way: A practical guide to the Western Mystery Tradition, Vol. 2: The Hermetic Tradition*, London, Britain: Arkana.

Matthews, C. (1989). *The Elements of the Celtic Tradition*, Shaftesbury, Britain: Element Books.

——(1990). *The Elements of the Goddess*, Shaftesbury, Britain: Element Books.

——(2005). *The Psychic Protection Handbook*, Llandeilo, Wales: Cygnus Books.

Matthews, C.D. *et al.* (1981). Melatonin in humans. In Biran, N. and Schloot, W. (eds), *Melatonin: Its Current Status and Perspectives. Advances in the Biosciences*, vol. 29, Oxford, Britain: Pergamon Press.

Mauss, M. (1906/1972). *A General Theory of Magic*, New York, USA: Norton.

Mavromatis, A. (1987). *Hypnagogia*, London, Britain: Routledge and Kegan Paul.

May, J. (1988). *Intervention*, London, Britain: Pan.

May, R. (1989). The Chaotic Rhythms of Life, *New Scientist*, 124 (1691), 37-41.

——and Mead, R.A. (1986). Evidence for pineal involvement in timing implantation in the Western Spotted Skunk, *Journal of Pineal. Research*, 3, 1-8.

McGillion, F. (1980). *The Opening Eye*, Fort Collins, USA: Coventure Press.

McIsaac, W.M. (1961). Formation of l-Methyl-6-Methoxy-l,2,3,4-retrahydrobeta-carboline under physiological conditions, *Biochim. Biophys. Acta*, 54, 607-9.

——Khairallah, P.A. and Page, I.H. (1961). 10-methoxy harmalan, a potent serotonin antagonist which affects conditioned behaviour, *Science*, 134, 674—5.

McKee, H.L. (1982). Psi-related beliefs among the rural people of Swaziland, *Parapsychology Review*, 13 (3), 11-16.

McMoneagle, J. (1997). *Mind trek: Exploring consciousness, time and space through remote viewing*, Charlotttesville, USA: Hampton Roads Publishing Co.

——(2002). *Memoirs of a psychic spy: The remarkable life of a US government remote viewer 001*, Charlotttesville, USA: Hampton Roads Publishing Co.

Middleton, B., Stone, B. M. and Arendt J. (2002). Human circadian phase, *Neuroscience Letters*, 329 (1), 41-44.

Michell, J. (1967). *The Flying Saucer Vision*, London, Britain: Sidgwick and Jackson.

——(1969/1975). *The View over Atlantis*, Auckland, New Zealand: Abacus.

Miller, D.L. and Creim, J.A. (1997). Comparison of cardiac and 60 Hz magnetically induced electric fields measured in anesthetized rats. *Bioelectromagnetics*, 18, 317–323.

Milton, J. and Wiseman, R. (1999). Does Psi Exist? Lack of Replication of an anomalous process of information transfer, *Psychological Bulletin*, 125(4), 387-391.

Mishlove, J. (1983). *Psi Development Systems*, Jefferson, USA: McFarland and Co.

Mori, Y. and Okamura, H. (1986). Plasma prolactin levels in ruminant seasonal variations, *Journal of Pineal Research*, 3, 77-86.

Morris, R. L.(1980). New directions in parapsychological research: the investigation of psychic development procedures, *Parapsychology Review*, 11 (2), 1-4.

——and Harnaday, J. (1981). An attempt to employ mental practice to facilitate PK, in Roll, W.G. and Beloff, J. (eds), *Research in Parapsychology 1980*, Metuchen, USA: Scarecrow Press.

———Nanko, M. and Phillips, D. (1979). Intentional observer influences upon measurement of a quantum mechanical system: A comparison of two imagery strategies. In Roll, W.G. (ed.), *Research in Parapsychology 1978*, Metuchen, USA: Scarecrow Press.
Most, A. (1985). *Peganum Harmala: The Hallucinogenic Herb of the American Southwest*, Denton, USA: Venom Press,
———(1986). *Eros and the Pineal: The Layman's Guide to Cerebral Solitaire*, Denton, USA: Venom Press.
Muir, R. (1981). *Riddles in the British Landscape*, London, Britain: Thames & Hudson.
Muller, W.E., Fehske, K.J. *et al.* (1981). On the neuropharmacology of harmane and other beta-carbolines, *Pharmacology, Biochemistry. and Behaviour*, 14, 693-9.
Mullin, T. (1989). Turbulent times for fluids, *New Scientist*, 124 (1690), 52-55.
Murphy, G. (1989). The emergence of a magnetic personality, *New Scientist*, 124 (1684), 40.
Murray, C. (1989). Is the solar system stable, *New Scientist*, 124 (1692), 60-64.
Murray, M. (1962). *The God of the Witches*, London, Britain: Background Books, The Daimon Press.
Myers, F.W.H. (1903/1915). *Human Personality and its Survival of Bodily Death*, 2 Vols. London, Britain: Longmans, Green.
Naranjo, C. (1967). Psychotropic properties of the harmala alkaloids, in Effron *et al.* (eds), *Ethnopharmacological Search for Psychoactive Drugs*, NIMH, US Dept. for Health, Education and Welfare, pp. 385-91.
———(1973). *The Healing Journey: New Approaches to Consciousness*, New York, USA: Ballantine Books.
———(1978). Psychological aspects of the yagé experience in an experimental setting. In Harner, M.J.(ed.), *Hallucinogens and Shamanism*, Oxford, Britain: Oxford Univ. Press.
Nelson, R. D., and Dunne, B. (1987). Attempted correlation of engineering anomalies with global geomagnetic activity. In Weiner, D.H. and Nelson, R. D., (eds.), *Research in Parapsychology 1986*, Metuchen, USA: Scarecrow Press, p. 82-85.
Neppe, V.M. (1980). Subjective paranormal experience and temporal lobe symptomatology, *Parapsychology Journal of South Africa*, 1 (2), 78-98.
———(1983). Temporal lobe symptomatology in subjective paranormal experiments, *Journal of the American Society for Psychical Research*, 77, 1-29.
Nichols, A. and Roll, W.G. (1998). The Jacksonville water poltergeist: Electromagnetic and neuropsychological aspects. *Proceedings of Presented Papers: The Parapsychological Association 41$^{st}$ Annual Convention*,
———and Roll, W.G. (1999). Discovery of electromagnetic anomalies in two reputedly haunted castles in Scandinavia. *Proceedings of Presented Papers: The Parapsychological Association 42$^{nd}$ Annual Convention*,
Niranjanananda Saraswati, Sw. (1993/2002). *Yoga Darshan: Vision of the Yoga Upanishads*, Munger, India: Yoga Publications Trust.
Ng, T.B. and Wong, CM. (1986). Pineal and lipid metabolism, *Journal of Pineal Research*, 3, 55-66.
Oatley, K. (1989). The importance of being emotional, *New Scientist*, 123 (1678), 33-6.
Olcese, J., Reuss, S. and Vollrath, L. (1985). Evidence for the involvement of the visual system in mediating magnetic field effects on pineal melatonin synthesis in the rat, *Brain Research*, 333, 382-384.
———Reuss, S. and Semm, P. (1988). Minireview: geomagnetic field detection in rodents, *Life Sciences*, 42, 605-13.
Olcott, W.T. (1911). *Star Lore of all Ages*, New York, USA: G.P. Putnam's Sons.
———(1914). *Sun Lore of all Ages*, New York, USA: G. P. Putnam's Sons.
Ostrander, S. and Schroeder, L. (1970). *PSI: Psychic Discoveries behind the Iron Curtain*, Aukland, New Zealand, Abacus.
Ott, J. (1976). *Hallucinogenic Plants of North America*, Berkeley, USA: Wingbow Press.
Pähkla, R., Masso, R., Zilmer, M., Rägo, L. and Airaksinen, M. M. (1996). Autoradiographic localization of [$^3$H]-pinoline binding sites in mouse tissues, *Methods and findings in experimental and clinical pharmacology*, 18, (6), 359-366.
———Zilmer, M., Kullisaar, T. and Rägo, L. (1998). Comparison of the antioxidant activity of melatonin and pinoline in vitro. *Journal of Pineal Research*, 24, 96–101.
Palmer, J. (1978). Extrasensory perception: research findings, in Krippner, S. (ed.), *Advances in Parapsychology*, vol. 1, New York, USA: Plenum Press.
———(1980). Parapsychology as a probabilistic science: facing the implications. In Roll, W.G. (ed.), *Research in Parapsychology 1979*, Metuchen, USA: Scarecrow Press.
———(2003). ESP in the Ganzfeld, *J. of Consciousness Studies*, 10.
———and Dennis, M. (1975). A community mail survey of psychic experiences. In Morris, J.D. Roll, W.G.and Morris, R.D. (eds), *Research in Parapsychology 1974*, Metuchen, USA: Scarecrow Press, pp.130-3.
Paranjpe, A.C. (1984). *Theoretical Psychology: The Meeting of East and West*, New York, USA: Plenum Press.
Parker, A. (1975). *States of Mind: ESP and Altered States of Consciousness*, London, Britain: Malaby

# References

Press.
Pavel, S. *et al.* (1981). Melatonin, vasotocin and REM sleep in prepubertal boys, in *Melatonin: Current Status and Perspectives; Advances in Biochemistry*, vol. 29, Oxford, Britain: Pergamon Press, pp. 343-7.
Pearce, J. C. (1973). *The Crack in the Cosmic Egg: Challenging Constructs of Mind and Reality*, Melbourne, Australia: Lyrebird Press Ltd.
Pedler, K. (1981). *Mind over Matter*, London, Britain: Thames, Methuen.
Pelletier, K.R. (1982). Uncertainty Principle factors in holographic models of neurophysiology. In Wilber, K. (ed.) *The Holographic Paradigm and Other Paradoxes*, Boston, USA: Shambhala.
Penfield, W. and Milner, B. (1958). Memory deficit produced by bilateral lesions of the hippocampal zone, *Arch. Neurology and Psychiatry*, 74.
Persinger, M.A. (1986). Intensive subjective telepathic experiences occur during days of quiet global geomagnetic activity. In Weiner, D.H. and Radin, D.I. (eds), *Research in Parapsychology 1985*, Metuchen, USA: Scarecrow Press, p. 32.
——(1987). Spontaneous telepathic experiences from *Phantasms of the Living* and low global geomagnetic activity, *Journal of the American Society for Psychical Research*, 81, 23-36.
——(1988). Increased geomagnetic activity and the occurrence of bereavement hallucinations: evidence for melatonin-mediated microseizuring in the temporal lobe?, *Neuroscience Letters*, 88, 271-4.
——(1989). Psi phenomena and temporal lobe activity: the geomagnetic factor. In Henkel, L.A. and Berger, R.E. (eds), *Research in Parapsychology 1988*, Metuchen, USA: Scarecrow Press.
——(1995). Out-of-body-like experiences are more probable in people with elevated complex partial epileptic-like signs during periods of enhanced geomagnetic activity: a non-linear effect. *Perceptual and Motor Skills*, 80, 563-569.
——and Cameron, R.A. (1986). Are earth faults at fault in some poltergeist-like episodes?, *Journal of the American Society for Psychical Research*, 80, 49-74.
——and Healey, F. (2002). Experimental facilitation of the sensed presence: possible intercalation between the hemispheres induced by complex magnetic fields. *The Journal of Nervous and Mental Disease*, 190 (8), 533-541.
——and Koren, S.A. (2001). Predicting the characteristics of haunt phenomena from geomagnetic factors and brain sensitivity: evidence from field and experimental studies. In Houran, J. and Lange, R. (eds.) *Hauntings and Poltergeists: Multidisciplinary Perspectives*, 179-194. Jefferson, USA: McFarland.
——Koren, S.A. and O'Connor, R.P. (2001). Geophysical variables and behaviour: CIV. Power-frequency magnetic field transients (5 microtesla) and reports of haunt experiences within an electronically dense house. *Perceptual and Motor Skills*, 3, 673-674.
——and Krippner, S. (1989). Dream ESP experiments and geomagnetic activity. *Journal of the American Society for Psychical Research*, 83, 101-116.
——and Lafreniere, G.F. (1977). *Space-Time Transients and Unusual Events*, Chicago, USA: Nelson-Hall.
——Richards, P.M. and Koren, S.A. (1997). Differential entrainment of electroencephalographic activity by weak complex electromagnetic fields. *Perceptual and Motor Skills*, 84, 527-536.
——and Schaut, G.B. (1988). Geomagnetic factors in subjective telepathic, precognitive, and postmortem experiences, *Journal of the American Society for Psychical Research*, 82, 217-36.
——Tiller, S.G. and Koren, S.A. (2000). Experimental stimulation of a haunt experience and elicitation of paroxysmal electroencephalographic activity by transcerebral complex magnetic fields: Induction of a synthetic "ghost"? *Perceptual and Motor Skills*, 90, 659-674.
Pfeiffer, F. (trans. C. de B. Evans) (1947). *Meister Eckhart*, London, Britain: Watkins.
Piaget, J. (1977). *The Essential Piaget*, London, Britain: Routledge and Kegan Paul.
Piñol-Ripoll, G., Fuentes-Broto, L., Millán-Plano, S., Reyes-Gonzáles, M., Mauri, J.A., Martínez-Ballarín, E., Reiter, R.J. and García, J.J. (2006). Protective effect of melatonin and pinoline on nitric oxide-induced lipid and protein peroxidation in rat brain homogenates, *Neuroscience Letters*, 405(1-2), 89-93.
Playfair, G. L. and Hill, S. (1978). *The Cycles of Heaven*, London, Britain: Souvenir Press.
Pratt, G., Rhine, J.B. *et al.* (1940). *Extrasensory Perception after 60 years*, Boston, USA: Bruce Humphries.
Preece, A.W., Hand, J.W., Clarke, R.N. and Stewart, A. (2000). Power frequency electromagnetic fields and health. Where's the evidence? *Physics in Medicine and Biology*, 45, 139–154.
Pribram, K.H. (1969). The neurophysiology of remembering, *Scientific American*, 220, 73-86.
——(1973). *Psychology of the Frontal Lobes*, New York, USA: Academic Press,.
Pringle, L. (1999). *Crop Circles: The greatest mystery of modern times*, London, Britain: Thorsons.
Prozialeck, W.C. *et al.* (1978). The fluorometric determination of 5-Methoxytryptamine in mammalian tissues and fluids, *Journal of Neurochemistry*, 30, 1471.
Puharich, A. (1973). *Beyond Telepathy*, Garden City, USA: Anchor Books.
Quay, W.B. (1974). *Pineal Chemistry*, Springfield, USA: C.C. Thomas.

Quest Journal (1970-92). London, Britain: BCM-SCL Quest.
Quest Magazine (1980). *Witchcraft Anthology*, London, Britain: BCM-SCL Quest.
——(1982). *Introduction to Witchcraft*, London, Britain: BCM-SCL Quest.
Radin, D. (1997). *The Conscious Universe: The Scientific Truth of Psychic Phenomena*. New York, USA: Harper Edge.
——(2005). The sense of being stared at: A preliminary meta-analysis, *Journal of Consciousness Studies*, 12(6), 95-100.
——(2006). *Entangled Minds: Extrasensory Experiences in a Quantum Reality*. New York, USA: Simon & Schuster.
——McAlpine, S. and Cunningham, S. (1994). Geomagnetism and psi in the ganzfeld. *Journal of the Society for Psychical Research*, 59, 352-363.
——and Rebman, J.M. (1998). Seeking psi in the casino, *Journal of the Society for Psychical Research*, 62, 193-219.
——and Nelson, R.D. (1989). Evidence for consciousness-related anomalies in random physical systems, *Foundations of Physics*, 19, 1499-1514.
Rajaram, M. and Mitra, S. (1981). Correlations between convulsive seizures and geomagnetic activity, *Neuroscience Letters*, 24, 187-91.
Randall, W. and Randall, S. (1991). The solar wind and hallucinations – a possible relation due to magnetic disturbances. *Bioelectromagnetics*, 12, 67.
Rao, K.R. (1969). Yoga and psi. In Cavanna, R. and Ullman, M. (eds), *Psi and Altered States of Consciousness*, New York, USA: Parapsychology Foundation Inc.
——(1978). Theories of Psi. In Krippner, S. (ed.), *Advances in Parapsychological Research, vol. II. ESP*, New York, USA: Plenum Press.
Reader, J. (1988). Human ecology: how land shapes society, *New Scientist*, 119 (1629), 50-7.
Reichbart, R. (1978). Magic and psi: Some speculations on their relationship, *Journal of the American Society for Psychical Research*, 72, 153-75.
Reiter, R.J. (1981). *The Pineal, vol. 6. Annual Research Reviews*, Fountain Valley, USA: Eden Press.
——(1982). *The Pineal Gland, vol. III: Extra-reproductive Effects*, Florida, USA: CRC Press Inc.
——(1984). *Pineal Research Reviews, vol. 2*, New York, USA: Alan R. Liss,.
——*et al.* (1988). Reduction of the nocturnal rise in pineal melatonin levels in rats exposed to 60-Hz electric fields in utero and for 23 days after birth, *Life Sciences*, 42, 2203-6.
Repacholi, M. and Greenebaum, B. (1999). Interaction of static and extremely low frequency electric and magnetic fields with living systems: health effects and research needs. *Bioelectromagnetics*, 20, 133-160.
Reppert, S. *et al.* (1988). Human clock pinpointed, *New Scientist*, 119 (1638), 33.
Reuss, S. (1986). Ganglionectomy effects on pineal electrical activity, *Journal of Pineal Research*, 3, 87-94.
——and Olcese, J. (1986). Magnetic field effects on the rat pineal gland: role of retinal activation by light, *Neuroscience Letters*, 64, 97-101.
Rhine, L.E. (1977). Research methods with spontaneous cases. In Wolman, B.B. (ed.), *Handbook of Parapsychology*, New York, USA: Van Nostrand Rheinhold,.
——(1981). *The Invisible Picture*, Jefferson, USA: McFarland.
Riemersma-van der Lek, R. F., Swaab, D. F., Twisk, J., Hol, E. M., Hoogendijk, W. J. G. and Van Someren, E. J. W. (2008). Effect of Bright Light and Melatonin on Cognitive and Noncognitive Function in Elderly Residents of Group Care Facilities: A Randomized Controlled Trial, *Journal of the American Medical Association*, 299(22), 2642 - 2655.
Rimon, R. *et al.* (1984). Pinoline, a beta-carboline derivative in the serum and cerebrospinal fluid of patients with schizophrenia, *Annals of Clinical Research*, 16, 171—5.
Ritz, T. (2001). Disrupting Magnetic Compass Orientation with Radio Frequency Oscillating Fields. In *Proceedings of RIN01: Orientation & Navigation - Birds, Humans & other Animals*. Oxford, Britain: Royal Institute of Navigation.
Ritz, T., Adem, S. and Schulten K. (2000). A model for photoreceptor-based magnetoreception in birds. *Biophysical Journal*, 78, 707-718.
Rivier, L. and Lindgren, J.E. (1972). "Ayahuasca", the South American hallucinogenic drink: an ethnobotanical and chemical investigation, *Journal of Economic Botany*, 29, 101-129.
Robins, D. (1985). *Circles of Silence*, London, Britain: Souvenir Press.
——(1988). *The Secret Language of Stone*, London, Britain: Rider/Century Hutchinson.
Robinson, D. (1985). Stress and psi: a preliminary model involving epilepsy, sexuality and shamanism, in White, R.A. and Solfvin, J. (eds), *Research in Parapsychology 1984*, Metuchen, USA: Scarecrow Press.
Robinson, O. (1975). *The Call of Isis*, Enniscorthy Ireland: Cesara Publications,
Rogo, D. S. (1982). *Miracles: A Parascientific Enquiry into Wondrous Phenomena*, New York, USA: Dial Press.
——(1983). *Leaving the Body: A Practical Guide to Astral Projection*, Upper Saddle River, USA:

## References

Prentice Hall.
——(1984). *Our Psychic Potentials,* Upper Saddle River, USA: Prentice Hall.
——(1985). *The Search for Yesterday: A Critical Examination of the Evidence for Reincarnation,* Upper Saddle River, USA: Prentice Hall.
——(1986a). *On the Track of the Poltergeist,* Upper Saddle River, USA: Prentice Hall.
——(1986b). *Life after Death: The Case for Survival of Bodily Death,* Wellingborough, Britain: Aquarian Press.
Roll, W.G. (1977). Poltergeists. In Wolman, B.B. (ed.), *Handbook of Parapsychology,* New York, USA: Van Nostrand Rheinhold,.
——and Montagno, E. de A. (1985). Neurophysical aspects of psi. In White, R.A. and Solfvin, J. (eds), *Research in Parapsychology 1984,* Metuchen, USA: Scarecrow Press.
Rollag, M.D. (1982). Ability of tryptophan derivatives to mimic melatonin's action upon the Syrian hamster reproductive system, *Life Sciences,* 31, 2699-2707.
Rommelspacher, H. and Susilo, R. (1985). Tetrahydroisoquinolines and beta-carbolines: putative natural substances in plants and mammals, *Progress in Drug Research,* 29, 415—59.
Roney-Dougal, S.M. (1979a). *The key to the subconscious.* Lecture for the Society of Psychical Research, London. (1979b). *Psi research (as I see it).* Lecture for the Society of Light Conference, London.
——(1979c). Quest conference questionnaire, *Quest,* 38.
——(1980). Psi research in the 80's, *Quest,* 42, 7-10.
——(1981). The Interface between Psi and Subliminal Perception, *Parapsychology Review,* 12, 4, 12 - 18.
——(1983). Witchcraft: a parapsychological perspective. Paper presented at the 1st Int. Spiritualistic Congress, Lisbon, Portugal.
——(1984). Occult conference questionnaire, *Journal of the Society for Psychical Research,* 52, 379-82.
——(1985). Relevance of the implicate/explicate view of physics to psi research Paper presented at the 12th International Parascience Conf., Oxford, Britain.
——(1986a). Some speculations on a possible psychic effect of harmaline. In Weiner, D.H. and Radin, D.I. (eds), *Research in Parapsychology 1985,* Metuchen, USA: Scarecrow Press, pp. 120—3.
——(1986b). Subliminal and psi perception: a review of the literature, *Journal of the Society for Psychical Research,* 53, 405-34.
——(1987). A comparison of subliminal and psi perception: exploratory and follow-up studies, *Journal of the American Society for Psychical Research,* 81, 141-82.
——(1988). The pineal gland's possible role as a psi-conducive neuromodulator. In *Proceedings of International Conference. on Paranormal Research,* Colorado State Univ., Colorado, USA.
——(1989a). Recent findings relating to the possible role of the pineal gland in affecting psychic ability, *Journal of the Society for Psychical Research,* 56, 313-28.
——(1989b). A psychophysiology of the yogic chakra system, *Caduceus,* 8, 8-12.
——(1989c). Geomagnetism and the pineal gland: some speculations. Paper presented at the 32nd Parapsychological Association Conference, San Diego, USA.
——(1991). *Where Science and Magic Meet,* Shaftesbury, Britain: Element Books.
——(1999). A Possible Psychophysiology of the Yogic Chakra System, *Journal of Indian Psychology,* 17(2), 18 - 40.
——(2003). *The Faery Faith,* Glastonbury, Britain: Green Magic.
——(2006). Taboo and Belief in Tibetan Psychic Tradition, *Journal of the Society of Psychical Research,* 70 (4), 193-210.
——and Solfvin, J. (2006). Yogic Attainment in Relation to Awareness of Precognitive Targets, *Journal of Parapsychology,* 70(1), 91 -120.
——Solfvin, J. and Fox, J. (2008). An exploration of degree of meditation attainment in relation to psychic awareness with Tibetan Buddhists, *Journal of the Society for Scientific Exploration,* 22 (2), 161-178.
——and Solfvin, J. (2008). Exploring the relationship between two Tibetan meditation techniques, the Stroop Effect and precognition, *Proceedings of the 51st Annual Convention of the Parapsychological Association,* Winchester, Britain, pp.187-203.
——and Vögl, G. (1993). Some speculations on the effect of geomagnetism on the pineal gland, *Journal of the Society for Psychical Research,* 59, 1-15.
Rosenblum, B and Kuttner, F (2007). *Quantum Enigma: Physics Encounters Consciousness,* London, Britain: Duckworth.
Rudolph, K. *et al.* (1985). Effect of 60 Hz electric fields on rodent behaviour, *Physiology and Behaviour,* 35, 505-8.
Rudolph. T.G. (2008). Quantum Mechanics: The speed of instantly, *Nature,* 454, 831-832.
Salart, D. et al. (2008). Testing the speed of 'spooky action at a distance,' *Nature,* 454, 861-864.
Sastre, A., Graham, C. and Cook, M.R. (2000). Brain frequency magnetic fields alter cardiac autonomic control mechanisms. *Clinical Neurophysiology,* 111, 1942-1948.
Satyananda Saraswati, Swami (1972a). *The Pineal Gland (Ajna chakra),* Bihar, India: Bihar School of

Yoga.
——(1972b). *Kundalini Yoga*, Bihar, India: Bihar School of Yoga.
——(1972). *Ajna Chakra: The Pineal Gland*, Munger, India: Bihar School of Yoga.
——(1976/2000). *Four Chapters on Freedom*, Munger, India: Yoga Publications Trust.
Schlitz, M. and Braud, W. (1997). Distant intentionality and healing: Assessing the evidence. *Alternative Therapies*, 3 (6), 62-73.
Schlitz, M., Morris, R.L. and White, R.A. (1982). Psi induction rituals: their role in experimental Parapsychology. In Roll, W.G. (eds.), *Research in Parapsychology 1981*, Metuchen, USA: Scarecrow Press, pp. 39-40.
——and Gruber, E. (1981). Transcontinental remote viewing. In Roll, W.G. and Beloff, J. (eds), *Research in Parapsychology 1980*, Metuchen, USA: Scarecrow Press.
Schmeidler, G.R. (ed.) (1976). *Parapsychology: Its Relation to Physics, Biology, Psychology and Psychiatry*, Metuchen, USA: Scarecrow Press.
——(1986). Subliminal Perception and ESP: Order in diversity? *Journal of the American Society for Psychical Research*, 80, 214-264.
——and Murphy, G. (1946). The influence of belief and disbelief in ESP upon individual scoring levels, *Journal of Experimental Psychology*, 36, 271-6.
——and McConnell, R.A. (1958). *ESP and Personality Patterns*, New Haven, USA: Yale Univ. Press.
Schmidt, S., Schneider, R., Utts, J., and Walach, H. (2004). Distant intentionality and the feeling of being stared at: Two meta-analyses, *British Journal of Psychology*, 95, 235-247
Schmidt-Koenig, K. and Keeton, W. T. (eds.) (1978). *Animal Migration, Navigation and Homing*, Heidelberg, Germany: Springer.
Schmitt, M. and Stanford, R.G. (1978). Free-response ESP during ganzfeld stimulation: the possible influence of menstrual cycle phase, *Journal of the American Society for Psychical Research* 72, 177-82.
Schwartz, S. *et al.* (1986). Infrared spectra alteration in water proximate to the palms of therapeutic practitioners, Los Angeles, USA: *The Mobius Society*.
Semm, P. *et al.* (1981). Electrical response of pineal cells to thyroid hormones and parathormone, *Neuroendocrinology*, 33, 212-17.
Shallis, M. (1988). *The Electric Shock Book*, London, Britain: Souvenir Press.
Sheldrake, R. (1981). *A New Science of Life: The Hypothesis of Morphic Resonance*, Vermont, USA: Park St. Press.
——(1988). *The Presence of the Past: Morphic resonance and the Habits of Nature*, New York, USA: Times Books.
——(1999). *Dogs that know when their owners are coming home and other unexplained powers of animals*, New York. USA: Crown Pubs.
Shevrin, H., Bond, J.A., Brakel, L.A.W., Hertel, R.K. and Williams, W.J. (1996). *Conscious & Unconscious Processes: Psychodynamic, Cognitive & Neurophysiological Convergences*, NewYork, USA: Guilford Press.
Shulgin, A. T. and Shulgin, A. (1997). *TIHKAL: The Continuation*. Berkeley, USA: Transform Press.
Shuttle, P. and Redgrove, P. (1978/1986). *The Wise Wound: Menstruation and Everywoman*, Boulder, USA: Paladin.
Sidgwick, H. and Committee (1894). Report on the census of hallucinations, *Proceedings of the Society for Psychical Research*, 10, 25-422.Sjöö, M. and Mor, B. (1989). *The Great Cosmic Mother of All*, San Francisco, USA: Harper and Row.
Smith, J.A. (1978). The pineal gland: its possible significance in schizophrenia. In Hemmings, G. (ed.), *The Biological Basis of Schizophrenia*, Cambridge, USA: MIT Press Ltd.
Snodgrass, M., Bernat, E. and Shevrin, H. (2004). Unconscious perception: a model –based approach to method and evidence, *Perception & Psychophysiology*, 66(5), 846-867.
Spence, L. (1946/1981) *British Fairy Origins: The Genesis and Development of Fairy Legends in British Tradition*, Wellingborough, Britain: Aquarian Press.
Spinelli, E. (1983). Paranormal cognition: its summary and implications, *Parapsychology Review*, 14 (5), 5-8.
Spottiswoode, S. J. P. (1997). Geomagnetic fluctuations and free-response anomalous cognition: a new understanding. *Journal of Parapsychology*, 61 (1), 3-12.
Stanford, R.G. (1974a). An experimentally testable model for spontaneous psi events. I. ESP, *Journal of the American Society for Psychical Research*, 68, 34-57,
——(1974b). An experimentally testable model for spontaneous psi events.II. PK, *Journal of the American Society for Psychical Research*, 68, 321-56.
——(1977a). Experimental psychokinesis: A review from diverse perspectives. In Wolman, B. (ed.), *Handbook of Parapsychology*, New York, USA: Van Nostrand Rheinhold, pp. 324-381.
——(1977b). Conceptual frameworks of contemporary psi research. In Wolman B. (ed.), *Handbook of Parapsychology*, New York, USA: Van Nostrand Rheinhold, pp. 823-58.
——(1977c). Are parapsychologists paradigmless in psiland? In Shapin, B. and Coly, L. (eds), *The*

## References

*Philosophy of Parapsychology*, New York, USA: Parapsychology Foundation.
——(1978). Toward reinterpreting psi events, *Journal of the American Society for Psychical Research*, 72, 197-214.
——(1979). Are we shamans or scientists? In Roll, W.G. (ed.) *Research in Parapsychology 1979*, Metuchen, USA: Scarecrow Press.
Starhawk (1979). *The Spiral Dance: A Rebirth of the Ancient Religion of the Great Goddess*, San Francisco, USA: Harper & Row.
——(1982). *Dreaming the Dark: Magic, Sex & Politics*, Boston, USA: Beacon Press.
——(1987). *Truth or Dare: Encounters with Power, Authority and Mystery*, San Francisco, USA: Harper & Row
——(2004). *The Earth Path: Grounding your Spirit in the Rhythms of Nature*, Harper, USA: San Francisco.
——and Macha, M. (1997). *The Pagan Book of Living and Dying*, San Francisco, USA: Harper.
Stenning, R. J., Carmody, C. and Du, J. (2002). Simulating the lunar geomagnetic variations. *Journal of Geophysical Research 107 No. A7*, SIA 12-1 – SIA 12-11.
Stevenson, I. (1987). *Children who Remember Previous Lives*, Charlottesville, USA: Univ. Press of Virginia.
Stevenson, I. (1997). *Where Reincarnation and Biology Intersect*, Westport, USA: Praeger.
Stewart, I. (1989). Portraits of chaos, *New Scientist*, 124 (1689), 42-7.
Stewart, R.J. (1990). *Robert Kirk: Walker between Worlds, Shaftesbury, Britain*: Element Books.
Stockmeier, C. and Blask, D. (1986). Catecholamines and convulsions from pinealectomy, *Journal of Pineal Research*, 3, 67-76.
Stone, M. (1979). *The Paradise Papers: The Suppression of Women's Rites*, London, Britain: Virago Press.
Sturrock, P. A. and Spottiswoode, S. J. P. (2007). Time-series spectrum analysis of performance in free response anomalous cognition experiments. *Journal of Scientific Exploration*, 21, 47-66.
Strassman, R.J. (1990). The Pineal Gland: Current Evidence for its Role in Consciousness. In Lyttle, T. (ed.), *Psychedelic Monographs and Essays. Vol. 5*, Boynton Beach, USA: PM&E Pub.
Strassman, R. (2001). *DMT: The Spirit Molecule*, South Paris, USA: Park St. Press.
Stutz, A.M. (1971). Circadian rhythm in gerbils relative to geomagnetic variation, *Ann. New York Academy of Science*, 188, 312-23.
Subrahmanyam, S., Sanker Narayan, P.V. and Srinivasan, T.M. (1985). Effects of magnetic micropulsations on the biological systems - a bioenvironmental study, *International Journal of Biometeorology*, 29, 283-305.
Szmigielski, S., Bortkiewicz, A., Gadzicka, E., Zmyslony, M. and Kubacki, R. (1998). Alteration of diurnal rhythms of blood pressure and heart rate to workers exposed to radiofrequency electromagnetic fields. *Blood Pressure Monitor*, 3, 323-330.
Tai Situ-pa Rinpoche (1996). *Prajnaparamita: Tape 1*, Kagyu Samye Ling, Britain: Rokpa Trust.
Talbot, M (1991). *The Holographic Universe*, New York, USA: HarperCollins,
Targ, R. and Harary, K. (1984). *The Mind Race: Understanding and Using Psychic Abilities*, New York, USA: Villard Books.
——and Puthoff, H. (1977). *Mind Reach*, New York, USA: Delacorte Press.
Tart, C.T. (1975). *States of Consciousness*, Boston: USA: E.P. Dutton.
——(ed.) (1975). *Transpersonal Psychologies*, London, Britain: Routledge and Kegan Paul.
——(1984). Acknowledging and dealing with the fear of psi, *Journal of the American Society for Psychical Research*, 78, 133-43.
——(1988). Geomagnetic effects on GESP: two studies, *Journal of the American Society for Psychical Research*, 82, 193-216.
Thrangu Rinpoche, Khenchen (trans Roberts, P.) (2001). *Transcending Ego: Distinguishing Consciousness from Wisdom. A treatise of the third Karmapa*, Boulder, USA: Namo Buddha Pub
Tylor, E. (1924/1971). *Primitive Culture*, New York, USA: Brentano.
Ullman, M., Krippner, S. and Vaughan, A. (1973). *Dream Telepathy*, New York, USA: Macmillan.
Valberg, P.A., Kavet, R. and Rafferty, C.N. (1997). Can low-level 50/60 Hz electric and magnetic fields cause biological effects? *Radiation Research*, 148, 2-21.
Vallée, J. (1970/1975). *Passport to Magonia: From Folklore to Flying Saucer*, London, Britain: Tandem.
——(1988/1996). *Dimensions: A Casebook of Alien Contacts*, London, Britain: Souvenir Press.
Vaughan, A. (1979). *Incredible Coincidences*, New York, USA: J.P. Lippincott,.
Vaughan, G.M. and Reiter, R.J. (1986). Pineal dependence of the Syrian hamster's nocturnal serum melatonin surge, *Journal of Pineal Research*, 3, 9-14.
Walker, M.W. (1984). Learned magnetic field discrimination in yellowfin tuna, Thunnus Abacares, *Journal of Comparative Physiology A*, 155, 673-9.
Walker, R.F. *et al.* (1986). Temporal effects of norepinephrine on pineal serotonin in vitro, *Journal of Pineal Research*, 3, 33-40.
Wallace, A. B. (2008). *Hidden Dimensions: The Unification of Physics and Consciousness*, New York, USA: Columbia University Press.

Watson, L. (1974). *The Romeo Error*, London, Britain: Hodder & Stoughton.
—— (1979). *Lifetides*, London, Britain: Hodder and Stoughton.
Weber, R. (1982). The enfolding-unfolding universe: a conversation with David Bohm. In Wilber, K., (ed.), *The Holographic Paradigm and other Paradoxes*, Boston, USA: Shambala.
Welker, H.A. *et al.* (1983). Effects of an artificial magnetic field on serotonin N-acetyltransferase activity and melatonin content of the rat pineal gland, *Experimental Brain Research*, 50, 426-32.
Wertheimer, N., Savitz, D.A. and Leeper, E. (1995). Childhood cancer in relation to indicators of magnetic fields from ground current sources. *Bioelectromagnetics*, 16, 86-96.
West, D.J. (1946-9). The investigation of spontaneous cases, *Proceedings of the Society for Psychical Research*, 48, 264-300.
Wever, R. (1973). Human circadian rhythms under the influence of weak electric fields and the different aspects of these studies, *International Journal of Biometeorology*, 17, 227-32.
Weydahl, A., Sothern, R.B., Cornélissen, G. and Wetterberg, L. (2000). Geomagnetic activity influences the melatonin secretion at latitude 70° N, *Biomedicine and Pharmacotherapy*, 55, S1, s57 - s62.
White, R.A. (1964). A comparison of old and new methods of response to targets in ESP experiments, *Journal of the American Society for Psychical Research*, 58, 21-56.
—— (1977). The influence of the experimenter's motivation, attitudes and methods of handling subjects on psi test results. In Wolman, B. (ed.), *Handbook of Parapsychology*, New York, USA: Van Nostrand Rheinhold, pp. 273-304.
—— (1980). On the genesis of research hypotheses in parapsychology. In Roll, W.G. (ed.), *Research in Parapsychology 1979*, Metuchen, USA: Scarecrow Press.
—— (1982a). Saintly psi: a study of spontaneous ESP in saints. In Roll, W.G., Morris, R.L. and White, R.A. (eds), *Research in Parapsychology 1981*, Metuchen, USA: Scarecrow Press, pp. 124-7.
—— (1982b). Parapsychology and the mystic path, in Roll, W.G., Morris, R.L. and White, R.A., *Research in Parapsychology 1981*, Metuchen, USA: Scarecrow Press, pp. 45-6.
Whitehouse, E. (1985). personal communication.
Wiener, H. (1968). External chemical messengers: IV. Pineal gland, *New York State Journal of Medicine*, 68 (7), 912-938.
Wilber, K. (ed.) (1982). *The Holographic Paradigm and Other Paradoxes*, Boston, USA: Shambhala.
—— (ed.) (1984). *Quantum Questions*, Boston, USA: Shambhala.
—— (2001). *A Theory of Everything*, Bath, Britain: Gateway Books.
—— (2002). *The Spectrum of Consciousness*, Delhi, India: Motilal Banarsidass Pub.
—— (2006). *Integral Spirituality: A starting new role for religion in the modern and post-modern world*, Kerala India: Integral Books.
—— (2007). *The Integral Vision: A very short Introduction to the Revolutionary Integral Approach to Life, God, the Universe and Everything*, Boston, USA: Shambahala.
Wilkinson, H. (1989). Geomagnetic variations and spontaneous psi phenomena, Paper presented at the 13th Society for Psychical Research Conf., Bournemouth, Britain.
Williams, H. (1978). *The Immortalist*, London, Britain: John Calder
Williamson, T. (1987). A sense of direction for dowsers?, *New Scientist*, 118 (1552), 40-3.
Wilson, C. (1973). *Strange Powers*, London, Britain: Latimer New Dimensions.
Wiltschko, W. et al. (2007). The magnetic compass of domestic chickens, Gallus gallus. *Journal of Experimental Biology*, 210, 2300-2310.
Winkelman, M. (1982a). Is it advantageous for parapsychologists to investigate popular occult practices? In Roll, W.G., Morris, R.L. and White, R.A. (eds), *Research in Parapsychology 1981*, Metuchen, USA: Scarecrow Press.
—— (1982b). Parapsychology and the social sciences, *American Society for Psychical Research Newsletter*, 8 (4), 25-6.
—— (1983). The anthropology of magic and parapsychological research, *Parapsychological Review*, 14 (2), 13-19.
—— (1992). Shamans, priests and witches: A cross-cultural study of magico-religious practitioners, *Anthropological Research Papers, no.44*, Tempe, Arizona, USA: Arizona State University.
—— (2000). *Shamanism: The neural ecology of consciousness and healing*, Westport, USA: Bergin and Garvey.
Wiseman, R. and Schlitz, M. (1997). Experimenter effects and the remote detection of staring, *Journal of Parapsychology*, 61, 197 – 207.
Wolf, F.A. (2000). *Mind into Matter: A new Alchemy of Science and Spirit*, Needham, USA: Moment Point Press.
Wolman, B.B. (ed.) (1977). *Handbook of Parapsychology*, New York, USA: Van Nostrand Rheinhold.
Wurtman, R.J. (1979). Rhythms in melatonin secretion: their possible role in reproductive function, in Zichella, L. and Pancheri, P. (eds), *Psychoneuroendocrinology in Reproduction*, Amsterdam, Holland: Elsevier Biomedical Press.

## References

———and Moskowitz, M.A. (1977a). The pineal organ. Part 1, *New England Journal of Medicine,* 296 (23), 1329-33.
———(1977b). The pineal organ. Part 2, *New England Journal of Medicine,* 296 (24), 1383-6.
Yeats, W.B. (1902/1977). *Irish Fairy and Folk Tales,* Gerrards Cross, Britain: Colin Smythe.
Yogakanti Saraswati, Sw. (1999). *The Advayatarakopanishad,* Unpub. MA Dissertation thesis in Yoga Philosophy, Bhagalpur Uni., Bihar, India, pp.38-50.
Yokus, B., Cakir, D.U., Akdag, M.Z., Sert, C. and Mete, N. (2005). Oxidative DNA damage in rats exposed to extremely low frequency electromagnetic fields. *Free Radical Research*, 39, 317-323.
Yolen, J. (1980). *The Moon Ribbon and Other Tales,* London, Britain: Dent.
Zimmer, Bradley, M. (1978). *Stormqueen,* New York, USA: Daw,.
Zohar, D. (1982). *Through the Time Barrier: A Study of Precognition and Modern Physics,* London, Britain: Heinemann.
———(1990). *The Quantum Self: Human Nature and consciousness defined by the new physics,* London, Britain: Bloomsbury and New York, USA: Morrow.
———and Marshall, I. (1994). *The Quantum Society: Mind, Physics and a new social vision,* London, Britain: Bloomsbury and New York, USA: Morrow.
Zukav, G. (1979). *The Dancing Wu Li Masters,* New York, USA: Rider.

# Index

6-MeOTHBC (6-methoxytetrahydro-betacarboline) – 103, 106
   see also: Beta-carbolines
acupuncture, acupuncturists – 164
Adams, D. – 70
Adams, M. – 144
adrenal glands – 106, 115, 120
adrenalin (epinephrine) – 99, 120
agriculture – 185, 186, 212
ajna chakra – 93
altered States of Consciousness (ASCs) - 27, 59
amanita muscaria – 27
Amazonian Indians – 101
amber – 129, 130
American Indians – 167
Annwn – 175, 177
anxiety – 53
ancestors – 173, 178, 191
animism – 170, 172
anthropological parapsychology – 33, 25, 37, 94, 96, 108, 219
apple – 177
*Aradia: the Gospel of the Witches* – 211
archaeological – 210
archetypes – 20, 21, 36, 79, 86, 153, 157, 175, 210, 244
Arendt J. – 109
ASCs: see altered states of consciousness – 35, 42, 60, 90, 97, 108,
Ashe, G. – 181
astral travelling: see out-of-body experience – 219
astrology – 70, 148
attitude – 10, 38, 40, 44, 74, 220, 247
aura – 137, 139,
Aurora Borealis – 137, 143, 158
autogenic technique – 26
automatic pilot – 19
autonomic nervous system (ANS) – 22, 99, 101, 107, 113, 147, 222
Awareness – 255, 256, 257, 259, 269
Avebury – 163
ayahuasca – 27, 93, 94, 95, 96, 98, 103

Baker, R. – 142, 143
Balanowski, E. – 160
ball lightning – 152
Banisteriopsis (Banisteria) caapi – 94, 96, 101, 105
banshee – 172
Barker *et al* – 37
Barron, Z. – 94
barrows – 132, 165, 166, 173, 175, 176, 177, 184, 185, 213
   West Kennet long barrow – 166
Batcheldor, K. – 46, 47, 51, 52, 54, 222
belief – 44, 61, 232, 247
Bell's Theorem – 76
belladonna – 27
Belloff, J. – 37
Beltane – 163, 227
Bergson H. – 61
beta-carbolines – 102, 104, 105, 110, 118, 120, 125
biofeedback – 25, 41
Black Death – 203, 204
Blake, William – 234
Bodisattva – 262
Bohm, D. – 72, 77, 78, 79, 80, 83, 85, 86, 88, 89, 199, 234, 238, 246, 256
bone – 124, 144
brain – 22, 14, 65, 76, 78, 83, 84, 97, 99, 153
   amygdala – 99
   cerebral cortex– 22, 105, 112, 120,
   corpus callosum – 99 forebrain – 100
   fornix – 99
   hemispheres – 24
   hippocampus – 99
   hypothalamus – 24, 97, 99, 100, 107, 120
   limbic system – 22, 24, 97, 99, 100
   midbrain – 22, 97, 100
   neurones – 65, 66, 74, 76, 97, 102
   old brain – 22
   opiate receptors – 105
   pons – 100
   reticular activating system (RAS) – 100, 108, 110

stem – 108
subcortical structures – 24
synapses – 65, 66, 67, 76, 83, 97, 98, 104
ventricles – 99
bride – 176
Brigid – 226
Braud W. – 42, 55, 223, 237
Braud and Schlitz - 222
Bronze Age peoples – 165, 166, 185, 211, 213
Brownies – 171, 180, 186
Buddhism – 69, 198, 234, 235, 249, 253, 254, 255, 256, 257, 259, 260, 261
Burr, H.S. – 138, 139, 166, 167
Butler, W. – 32

Callaway, J.C. – 102, 107, 108
Campbell – 22
Capra, F. – 78
causality – 71
Celtic – 176, 177, 200, 226
 calendar – 185, 186
Celts – 163, 178, 187
central nervous system (CNS) – 22, 101, 113
Cernunnos – 213
chakras: chakra system – 30, 92, 112, 113, 114, 115, 116, 120, 122, 123, 144, 261
Chambers, Robert – 172
Changeling – 190, 191
channellers; channelling – 27
chant; chanting – 182, 193
chaos; chaos theorists – 69
chi – 138, 164, 167, 168
Child and Levi – 48
Chilton Pearce, J. – 44, 67
China – 60
circle dancing – 181, 183
Claims of the Paranormal (CSICOP) – 18
clairvoyance – 6, 15, 85, 89, 95, 235, 257
cognitive dissonance – 18
coincidences: – 30, 70
Collective Unconscious – 21, 36, 82, 86, 153, 155, 210, 221, 144, 145
Committee for the Scientific Investigation for compassion – 262
concentration – 13, 40
cone of power – 224
Conformance Theory – 45, 55, 155
consciousness – 253, 254, 255, 256, 257, 258
 dream states – 13
crop circles – 181, 182
corticosterone – 120
cortisol – 120
Costa de Beauregard – 67
covens – 225
*Crack in the Cosmic Egg, The* – 44, 67
creativity – 14, 19
Cremer-Bartels *et al.* – 148
cross quarter days – 185, 225
Crowley, A. – 35, 49, 150
crows – 176
crystals – 12
*Cycles of Heaven, The* – 128, 136

daemonic – 179

dancing – 189
*Dancing Wu Li Masters, The* – 78
Daime – 27
Darwinism – 88,
David-Neel, A. – 49, 164
Davidson, J. – 112
Davies, P. – 90
Dean, D. – 135, 223
death – 172, 177, 178, 181, 185, 213
Defence Mechanism Test (DMT) – 16, 17, 19, 61
Delphi – 59, 126
Descartes – 93
devas – 157, 179
Devereux, P. – 128, 152, 153, 156, 157, 159
divination – 58, 70, 94, 95, 144, 177, 178, 179, 233
Dixon, Prof. N. – 100, 108
DMT (dimethoxytryptamine) – 94, 95, 96, 100, 103, 104, 107, 108, 110, 155
dolmens – 183, 184
dowsing – 59, 142, 143, 150, 151, 159, 160
dragon lines – 163, 166
Dragon Project – 152, 160
*Dreaming the Dark* – 203
dreams – 14, 18, 20, 21, 90, 102, 107, 108, 153, 177, 180, 258, 259
druids – 131, 168, 177, 187, 214, 228
Drury, N. – 36, 39

*Earth Lights* – 128, 152
Earth's Magnetic Field (GMF) – 124, 128, 146, 140, 142, 143, 144, 145, 146, 147, 149, 150, 151, 152, 166, 167
earthquakes – 152, 153, 154, 159
ebony – 130,
Eckhart, Meister – 234
ectoplasm – 157
EDA (electrodermal activity) – 23, 55, 139
ego – 53
Egyptian ritual magic – 36
Einstein – 7, 174, 239, 243, 245
Einstein-Podolsky-Rosen (EPR) paradox – 76
*Electric Shock Book, The* – 128
electromagnetism – 129, 131, 133, 138, 179, 184
electrostatics – 129, 130, 131, 137, 158, 160
elementals – 36, 157, 179, 180
Eleusinian mysteries – 27, 194
Elkin – 48
elves – 174, 184, 186
*Emerald Forest, The* – 94
EMF – 132, 133, 134,
emotion – 22, 41, 211, 232, 246, 247
emptiness – 256
endocrine system – 114, 116, 118, 120, 133, 147
energy – 261
energy-matter – 245
Enlightenment – 259
EPR paradox: see Einstein-Podolsky-Rosen
 paradox – 13, Equinoxes – 225, 227
ergot – 27
ESP: see also: psi – 61, 84
ethical codes – 35, 186, 207
Evans, H. – 238
evil eye – 49, 190

## Index

*Evolution through the Tarot* – 15
experimenter effect – 51,
fairy – 157, 158, 171, 172, 173, 187, 198
   faith – 36, 170
   folk – 153, 176, 178, 228
   lore – 132, 175, 200
fairy food – 173
fairy kingdoms – 175
fairy land – 177
fairy mound – 184, 190
fairy ointment – 171
fairy paths – 181, 185
faith – 247
Faraday chamber – 132, 166, 167
*Fauns and Fairies* – 171
fear – 187, 202, 204, 205, 206
feathers – 144,
Feng Shui – 60
festivals – 225
filter theory – 16, 18, 61
flying ointment – 219
folklore – 170
Fortune, D. – 32, 49
fourth state of consciousness – 258
Fox sisters, the – 205
Freedman and Clauser – 76, Freud S. – 8. 14, 166

Gaia – 242
Gaia Hypothesis – 86, 156,
Ganzfeld – 11, 25, 42, 102, 126, 145
Gardner, Dr. G.
Gardner, R. – 15, 26
Gauquelin M. – 37
Geisler – 249
geological faulting – 150, 152, 158, 159, 160, 161, 162
ghosts – 158, 159
*Ghosts and Divining Rods* – 224
glamour – 218, 235
glamourie – 171
goal oriented strategies – 53
God – 232
Goddess – 178, 182, 201, 211, 212, 213, 233
Gorsedd – 183
Grad, B. – 237
Grahame, K. – 150,213
Graves, R. – 228
Great Chain of Being – 235, 237, 238
Great Mother – 210, 212, 214
green – 188
Green Man – 153, 214
Green, E. and A. – 25,
green spirituality – 187
Gruber, E. – 50, 58
Gwynn ap Nudd – 175

Hallowe'en: see Samhain – 225
hallucinations – 96, 144, 146, 156, 159, 180
hallucinogens – 100, 101, 103, 104, 107, 108, 110, 182, 219
hallucinogens and shamanism – 219
Hansen, G. – 47
harmala alkaloids – 94, 96, 106
harmaline – 96, 101, 103

Harner, M. – 219
Hasted, J. – 130
hauntings – 140, 144, 155, 158, 159, 163, 179, 186
   see also: ghosts
healing – 84, 99, 133, 134, 135, 140, 174, 188, 221, 222, 233
Heisenberg – 71, 72, 73, 75
henbane – 27
Hinduism – 42, 79
Hochstein and Paradies – 94
Hole, C. – 203
*Holographic Paradigm, The* – 14, 84
holographic paradigm – 82, 238, 243
holographic philosophy – 243
holographic principles – 81, 82, 83, 84, 85
holographic universe – 256, 257, 262
holy wells – 135, 223
Honorton, C. – 42, 43, 145
Honorton and Barksdale – 53,
hormonal system – 22
horn – 144
Horned God – 211, 213,214
Hubbard and May – 146,
hunting – 178, 213
Hurley, F. – 27, 29, 31
Huxley, A. – 234
Hyman, R. – 43
hypnagogic state, 24, 25, 42, 102, 106, 126
hypnosis – 11, 26, 220

I Ching – 59, 60
illusion – 171, 221
imagination – 26, 41, 222
Imbolc (Candlemas) – 163, 226
*Immortalist, The* – 196
immune response systems – 118
Implicate Order – 72, 78, 79, 80, 81, 83, 86, 238, 256
Inglis – 250
initiation ceremonies – 188, 195
inquisition – 203, 204
invisibility – 171
Irish – 22
iron – 184, 185
Irwin, A. – 48

Jagrat – 258, 259
Jahn, R. and Dunne, B. – 145
Jee, Lakshman – 259
Jung, C. G. – 8, 21, 36, 45, 153, 210, 244, 245

karma – 262
Kennedy, J. – 55
Kennedy and Taddonio – 48
Kensinger, K.M. – 96,
Kirk, R. – 171
Kirlian photography – 224
Koestler, A. – 15,
Krippner, S. – 37, 102, 145
kundalini – 112, 122

L-fields – 138, 139, 158, 166, 167
Lacey, L. – 212

Lammas – 163, 195, 213, 227
Lang, A. – 22
Langer *et al.* – 104
LeShan, L. – 15, 223, 237
Leland, G. – 211
Lessing, D. – 165
Lethbridge, T.C. – 224
Lewis, B. – 160
ley lines – 152, 158, 159, 162, 163, 164, 173, 179, 181, 185
light – 174, 246l
lightning – 131, 158
Lonegren, S. – 128
love – 209, 210, 228, 248, 262
Lovelock, J. – 156, 242
LSD – 99, 106
*Lunaception* – 212

*Mabinogion, The* – 22
magic – 79, 90, 129, 135, 142, 156, 170, 171, 182, 194, 200, 215, 228
  ceremonial ritual – 36, 102, 172
    Egyptian – 42, 45
    Jewish – 42,
*Magic and Mystery in Tibet* – 49, 164
magnetism – 129, 142
magnetosphere – 136, 145
Mala – 60
Malinowski – 46, 48
mana – 38, 187
mandrake – 27
Manannan – 175, 177
Manning, M. – 12, 56
MAO: see monoamine oxidase – 100, 102, 104, 106
marriage – 189, 190
Mather and Baker – 142, 143
matriarch – 178, 187, 190
matriarchical – 212
Mauss, M. – 46
Mavromatis, A. – 24
May, Julian – 245
maya – 79, 85
May, Ed – 43
McIsaac W. – 103
meditation – 26, 73, 102, 108, 115, 238, 247, 260, 261
  chambers – 167
mediumistic Seances – 180
mediums – 8, 9, 11, 12, 101, 150, 157, 180, 200, 206
megaliths see also: barrows, stone circles – 184
Melatonin – 99, 101, 102, 107, 108, 109, 110, 114, 118, 120, 122, 123, 124, 126, 146, 147, 149
memory – 82
menstrual cycle – 118, 122, 124, 126, 134, 139, 212
metal bending – 12
mind – 254, 255, 256, 258
  subliminal – 10, 14, 15, 16, 17, 18, 19, 21, 22, 23, 25, 28, 29, 31, 44, 65, 78, 86, 97, 142, 155, 200
miracle – 223, 234
Mishlove, J. – 35, 36, 50, 61, 62

monoamine oxidase (MAO) – 100
moon cycle – 118, 119, 124, 145, 147, 168, 175, 183, 212, 226, 227
Mor, Barbara – 203
morality – 186, 188, 207, 248
morphogenetic field – 136, 166
Murphy, G. – 143,
mushroom – 182
music – 177, 189, 193
motivation – 53
Myers, W. – 65
mystics – 28, 34, 90, 110, 170, 198, 232, 243
*Mythology of the British Isles* – 181
myths – 22, 86, 170, 175, 184, 211

Nadis – 261
Naranjo, C. – 93, 94
natural magic – 56
*Nature* – 194, 216, 228
*Needles of Stone* – 128
*Needles of Stone Revisited* –164
neuroscientists – 20
neocortex – 24
neolithic peoples – 162, 163, 164, 167, 175, 176, 177, 185, 186, 189, 190, 213
*New Science of Life, A* – 138, 243
*New Scientist* – 151, 211
Newton, Sir Isaac – 88
noradrenaline (norepinephrine) – 99

OBE: see out-of-body experience –146
occult – 15, 26, 28, 33, 34
oestrogen –118,120, 123, 125, 149
Omens – 60,
*Opening Eye, The* –128
orgone chambers –132, 166, 167
out-of-body experience (OOBE, OBE) – 95, 110, 219
ownership resistance – 52

paganism – 46, 27, 38, 228
  animistic – 188, 198, 210
pagans – 175, 182, 186, 191, 209, 234
Palaeolithic – 212, 213
Palmer, J. – 37, 43
Pan – 213
Paramitas – 41
Paranjpe A. – 112
*Parapsychology Review* – 33, 37, 38
parasympathetic nervous system – 22, 113
parathyroids – 114
Patanjali – 92, 260
patriarchical – 201, 230
*Pattern of the Past, The*
Pavel *et al.* – 108
Pelletier, K. – 67,83
Penfield, W. – 83
perceptual defence – 11, 16, 17
perennial philosophy – 187, 196, 234, 239, 243
Persinger, M. – 144, 145, 146, 150
Persinger and Cameron – 158
Persinger and Lafreniere – 152
personality change and development – 10, 35
personality traits, as influences on psychic

297

## Index

experiences – 11
peyote – 27
philosophy of magic – 234
Piaget, J. – 86
Piezo-electric effect – 12, 124, 130, 144, 152, 160
pineal – 30, 92, 96, 98, 99, 101, 102, 104, 107, 110, 111, 112, 113, 115, 118, 120, 123, 133, 141, 143, 144, 146, 147, 150, 152, 155
pinoline: see 6-MeOTHBC
pituitary – 98, 99, 114, 118, 120, 125
PK: see psychokinesis – 48, 49, 52, 52, 53, 54, 55, 58, 83, 84, 99
placebo – 44
*Places of Power* – 152
plague – 204
Playfair, G.L. and Hill, S. – 128
poetry – 222
poltergeists – 12, 126, 132, 144, 152, 158, 180, 186, 247
possession – 27
Prana – 137, 261
precognition – 6, 15, 66, 70, 77, 89, 171, 209, 257
    see also: ESP; psi
primary process – 14, 18, 20, 24, 156, 157
Princeton engineering research – 249
Pritchard, E. – 55
probability – 70, 73, 75
progesterone – 122, 123
prolactin – 118, 123
prophets – 171, 174
Prozialeck *et al* – 104
psilocybin – 27, 220
psi: see also: – 6, 37, 85, 90, 92, 95, 102, 103, 109, 126, 134, 145, 147, 156, 233, 257, 260
    active – 10, 11
    development methods – 35, 36
    receptive – 146
psi-conducive – 43, 59 s
Psi-Mediated Instrumental Response (PMIR) – 28
psi-missing – 18
psychedelics – 100, 108
psychic – 105, 114, 140, 142, 144, 157, 158, 159, 160, 167, 173, 180, 185, 200, 216, 233
    experiences and personality traits – 156
    healing – 23
    surgeons – 47
    psychological – 10,
psychology of occultism – 40
psychokinesis (PK) – 6, 84, 130, 155
psychosis – 100, 108, 109, 110
psychotropic plants – 27, 42
puberty – 110, 114, 118, 122, 123, 126, 143, 180
Puck – 171, 175, 179, 190

quantum – 71, 72, 74, 76, 78, 100, 103, 257, 261
    indeterminacy in neuron firing – 67
    mechanics – 46, 67, 70, 72, 73, 78
    physics – 78
quartz – 129, 130, 160
Quay, W.B. – 103
*Quest* – 33

Radin, D – 42
radionics – 140
random-event generators – 12
rapid eye movement (REM) sleep – 107, 108, 110
Reich, W. – 166, 167
Reichbart, R. – 47
reincarnation – 191, 196, 197
Reiter – 109, 147
Reiter *et al.* – 149
relaxation – 40
release-of-effort effect – 13, 54
remote viewing – 11, 43, 53
Reppert *et al.* – 107, 124
retina – 101, 114
Rhine, J.B. – 9, 12, 28, 48
Rhys, Sir John – 182
Rinpoche, Tai Situ-pa – 256
Rivier and Lindgren – 94
ritual – 50, 52, 53
Robertson, D. – 130
Robins, D. –128, 140, 141, 160, 162
Rocard, Y. – 151
Rogo, Scott – 238
Rollright Stones – 160
Rudolph *et al.* – 143
runes – 59

sacred sites – 158, 159, 163, 175, 179
SAD: see Seasonal Affective Disorder – 101
Sahasrara chakra – 112
Samhain – 163, 225, 226
Saraswati, Swami Satyananda – 93, 112, 113, 114, 116, 120, 121, 123, 253, 254, 257, 258
*Sceptical Inquirer, The* – 18
schizophrenia – 110
Schlitz, M. – 50, 53, 55
Schmeidler, G. – 44
Schrodinger – 19
Scottish – 22
scrying – 59
Seasonal Affective Disorder (SAD) – 101, 109
second sight – 29, 93, 158, 162, 171, 189, 200
*Secret Commonwealth of Elves* – 171
*Secret Language of Stone, The* – 128
seers – 180
self development – 217, 218
self-discipline – 193, 194, 221
self-transcendence – 14, 19
Semm *et al.* – 144
serotonin – 99, 101, 102, 107, 108, 110
Shallis, M. – 128, 132, 133, 134, 135, 140
*Shaman and the Magician, The* – 39
shamanism – 27, 39, 46, 200
    shamans – 11, 47, 94, 95, 108, 110, 219
shape shifting – 180
Sheldrake, R. – 138, 166, 238, 243
Sheep-Goat Effect – 44
*Shikasta* – 165, 184
Shuttle, P. and Redgrave, P. - 122
Sidhe – 172, 174, 175, 184
siddhis – 28, 93, 235, 257
Signal Detection Theory (SDT) – 66
silicon – 129. 130, 133
Simmons – 151

298

sixth sense – 122
Sjöö, M. and Mor, B. – 203
Smith, C. – 110
Smith, J. – 133
social influences – 61
Society for Psychical Research(SPR) – 7, 32, 200
Society of Light (SOL) – 32
solstices – 168, 186, 225, 226, 227
song in shamanistic trance – 27
sorcery – 27, 29, 219
soul – 255, 257
spells – 30, 152, 171, 187, 222
*Spiritual Dowsing* – 128
SPR: see Society for Psychical Research – 65
Spring Equinox – 226
Stanford, R.G. – 29, 34, 41, 45, 53, 54, 55, 155
states of consciousness – 11
stargate – 21
Starhawk – 203, 220, 224
Stevenson, I. – 197
stigmata – 45, 233
strain guage – 12
stress – 22,
stone circles – 141, 152, 158, 159, 160, 161, 162, 183, 184, 185, 213, 256
Stonehenge – 163, 165, 182
Spence – 191
state of consciousness – 177, 261
Stewart, Bob – 171
Stutz, A. – 143
subconscious – 8, 13, 14, 15, 19, 22, 26, 28, 40, 64, 86, 176, 222
subliminal perception – 108, 139, 143
  advertising – 10
  mind – 6, 8, 14
Sufism – 42, 234
sun – 175, 183, 186, 227
sunspot cycle – 136, 139
Sushupna – 258, 259
Swapna – 258, 259
sweat lodges – 165, 167
symbolism – 20, 21, 59, 221
sympathetic magic – 212
sympathetic nervous system – 22, 99, 107, 113
synchronicity – 31, 45, 58, 67, 68, 71, 79, 84, 89
Syrian rue (Peganum harmala) – 94, 106

*Tao of Physics, The* – 15, 78, 243
taboo – 187, 189
Targ and Puthoff – 11, 43
tarot – 59
Tart, C. – 112, 205
tectonic activity – 153, 160
telepathine: see harmaline – 94
telepathy see also: ESP; psitestosterone – 6, 15, 77, 84, 89, 93, 134, 144, 145, 180, 257
third eye  92, 93, 102
Thorn Apple – 27
Thom, Prof. A. – 162
thymus – 118
thyroid – 99, 114, 115, 120
Tibetan culture – 37, 40, 42, 44, 59, 69, 164, 252, 254
time – 173, 174

Tir na Nog – 175, 177
tobacco – 27
Toon, M. – 212
trance – 11, 27, 109, 177, 180, 219
*Transpersonal Psychologies* - 112
trickster, the – 46
Triple Goddess – 36
Tromp, S. – 151
Tuatha De Naan – 22,
Tuatha de Danaan – 175, 176, 177, 178, 186
tumuli see: barrows  – 132, 165, 166, 167
Turiya – 258, 259, 261
Tylor – 38
Tylwyth Teg – 172

UFOs – 132, 146, 152, 153, 154, 155, 156, 157, 159, 162, 163, 165, 179, 191
Ullman, M. – 102
unconscious – 16, 18, 19, 28, 259

Vallée, J. – 153
Vaughan, A. – 68, 102
Vedic – 253, 254, 255, 257, 261
visualisation – 26, 41, 48, 49, 53, 155

Walker, M. – 143
water – 223
*Water Babies, The* – 212
Watson, L. – 15, 27, 29
wave-particle – 246
Welsh – 22
Wentz, E. – 183, 193, 195, 242
White, R. – 26, 42, 48-238, 249
*White Goddess, The* – 228
Whitehouse, E. – 107, 108
*Wholeness and the Implicate Order* – 78, 243
Wigner – 19
Wilber, K. – 84, 232, 234, 235, 236, 254, 257
Wilkinson, H. – 145
will-o'-the-wisps – 152
Williams, H. – 196
Williamson, T. – 151
Wilson, Colin – 15
*Wind in the Willows, The* - 213
Winkelman, M. – 37, 58, 61
*Wise Wound, The* – 122
witch trials – 182, 189, 192, 202, 203, 204
witchcraft – 171, 175, 189, 206, 210, 213, 221
*Witchcraft in Britain* – 203
witches – 144, 188, 200, 203, 212
witness inhibition – 52
*Wizard of Earthsea, The* –75, 76
work in a group – 13

yage: see ayahuasca – 27
yoga – 25, 44, 77, 92, 93, 112, 113, 209, 235, 256, 257, 258, 260, 261
  yogis – 25, 44, 90, 116, 252, 253, 254, 255

Zen action – 41, 53
Zener cards – 9
Zohar, D. – 66, 100
Zukav, G. – 87

www.ingramcontent.com/pod-product-compliance
Ingram Content Group UK Ltd.
Pitfield, Milton Keynes, MK11 3LW, UK
UKHW020927040725
460395UK00002B/5